T0223681

Lecture Notes in Computer Science

Vol. 49: Interactive Systems. Proceedings 1976. Edited by A. Blaser and C. Hackl. VI, 380 pages. 1976.

Vol. 50: A. C. Hartmann, A Concurrent Pascal Compiler for Minicomputers. VI, 119 pages. 1977.

Vol. 51: B. S. Garbow, Matrix Eigensystem Routines – Eispack Guide Extension. VIII, 343 pages. 1977.

Vol. 52: Automata, Languages and Programming. Fourth Colloquium, University of Turku, July 1977. Edited by A. Salomaa and M. Steinby. X, 569 pages. 1977.

Vol. 53: Mathematical Foundations of Computer Science. Proceedings 1977. Edited by J. Gruska. XII, 608 pages. 1977.

Vol. 54: Design and Implementation of Programming Languages. Proceedings 1976. Edited by J. H. Williams and D. A. Fisher. X, 496 pages. 1977.

Vol. 55: A. Gerbier, Mes premières constructions de programmes. XII, 256 pages. 1977.

Vol. 56: Fundamentals of Computation Theory. Proceedings 1977. Edited by M. Karpiński. XII, 542 pages. 1977.

Vol. 57: Portability of Numerical Software. Proceedings 1976. Edited by W. Cowell. VIII, 539 pages. 1977.

Vol. 58: M. J. O'Donnell, Computing in Systems Described by Equations. XIV, 111 pages. 1977.

Vol. 59: E. Hill, Jr., A Comparative Study of Very Large Data Bases. X, 140 pages. 1978.

Vol. 60: Operating Systems, An Advanced Course. Edited by R. Bayer, R. M. Graham, and G. Seegmüller. X, 593 pages. 1978.

Vol. 61: The Vienna Development Method: The Meta-Language. Edited by D. Bjørner and C. B. Jones. XVIII, 382 pages. 1978.

Vol. 62: Automata, Languages and Programming. Proceedings 1978. Edited by G. Ausiello and C. Böhm. VIII, 508 pages. 1978.

Vol. 63: Natural Language Communication with Computers. Edited by Leonard Bolc. VI, 292 pages. 1978.

Vol. 64: Mathematical Foundations of Computer Science. Proceedings 1978. Edited by J. Winkowski. X, 551 pages. 1978.

Vol. 65: Information Systems Methodology, Proceedings, 1978. Edited by G. Bracchi and P. C. Lockemann. XII, 696 pages. 1978.

Vol. 66: N. D. Jones and S. S. Muchnick, TEMPO: A Unified Treatment of Binding Time and Parameter Passing Concepts in Programming Languages. IX, 118 pages. 1978.

Vol. 67: Theoretical Computer Science, 4th GI Conference, Aachen, March 1979. Edited by K. Weihrauch. VII, 324 pages. 1979.

Vol. 68: D. Harel, First-Order Dynamic Logic. X, 133 pages. 1979.

Vol. 69: Program Construction. International Summer School. Edited by F. L. Bauer and M. Broy. VII, 651 pages. 1979.

Vol. 70: Semantics of Concurrent Computation. Proceedings 1979. Edited by G. Kahn. VI, 368 pages. 1979.

Vol. 71: Automata, Languages and Programming. Proceedings 1979. Edited by H. A. Maurer. IX, 684 pages. 1979.

Vol. 72: Symbolic and Algebraic Computation. Proceedings 1979. Edited by E. W. Ng. XV, 557 pages. 1979.

Vol. 73: Graph-Grammars and Their Application to Computer Science and Biology. Proceedings 1978. Edited by V. Claus, H. Ehrig and G. Rozenberg. VII, 477 pages. 1979.

Vol. 74: Mathematical Foundations of Computer Science. Proceedings 1979. Edited by J. Bečvář. IX, 580 pages. 1979.

Vol. 75: Mathematical Studies of Information Processing. Proceedings 1978. Edited by E. K. Blum, M. Paul and S. Takasu. VIII, 629 pages. 1979.

Vol. 76: Codes for Boundary-Value Problems in Ordinary Differential Equations. Proceedings 1978. Edited by B. Childs et al. VIII, 388 pages. 1979.

Vol. 77: G. V. Bochmann, Architecture of Distributed Computer Systems. VIII, 238 pages. 1979.

Vol. 78: M. Gordon, R. Milner and C. Wadsworth, Edinburgh LCF. VIII, 159 pages. 1979.

Vol. 79: Language Design and Programming Methodology. Proceedings, 1979. Edited by J. Tobias. IX, 255 pages. 1980.

Vol. 80: Pictorial Information Systems. Edited by S. K. Chang and K. S. Fu. IX, 445 pages. 1980.

Vol. 81: Data Base Techniques for Pictorial Applications. Proceedings, 1979. Edited by A. Blaser. XI, 599 pages. 1980.

Vol. 82: J. G. Sanderson, A Relational Theory of Computing. VI, 147 pages. 1980.

Vol. 83: International Symposium Programming. Proceedings, 1980. Edited by B. Robinet. VII, 341 pages. 1980.

Vol. 84: Net Theory and Applications. Proceedings, 1979. Edited by W. Brauer. XIII, 537 Seiten. 1980.

Vol. 85: Automata, Languages and Programming. Proceedings, 1980. Edited by J. de Bakker and J. van Leeuwen. VIII, 671 pages. 1980.

Vol. 86: Abstract Software Specifications. Proceedings, 1979. Edited by D. Bjørner. XIII, 567 pages. 1980

Vol. 87: 5th Conference on Automated Deduction. Proceedings, 1980. Edited by W. Bibel and R. Kowalski. VII, 385 pages. 1980.

Vol. 88: Mathematical Foundations of Computer Science 1980. Proceedings, 1980. Edited by P. Dembiński. VIII, 723 pages. 1980.

Vol. 89: Computer Aided Design - Modelling, Systems Engineering, CAD-Systems. Proceedings, 1980. Edited by J. Encarnacao. XIV, 461 pages. 1980.

Vol. 90: D. M. Sandford, Using Sophisticated Models in Resolution Theorem Proving.
XI, 239 pages. 1980

Vol. 91: D. Wood, Grammar and L Forms: An Introduction. IX, 314 pages. 1980.

Vol. 92: R. Milner, A Calculus of Communication Systems. VI, 171 pages. 1980.

Vol. 93: A. Nijholt, Context-Free Grammars: Covers, Normal Forms, and Parsing. VII, 253 pages. 1980.

Vol. 94: Semantics-Directed Compiler Generation. Proceedings, 1980. Edited by N. D. Jones. V, 489 pages. 1980.

Vol. 95: Ch. D. Marlin, Coroutines. XII, 246 pages. 1980.

Vol. 96: J. L. Peterson, Computer Programs for Spelling Correction: VI, 213 pages. 1980.

Vol. 97: S. Osaki and T. Nishio, Reliability Evaluation of Some Fault-Tolerant Computer Architectures. VI, 129 pages. 1980.

Vol. 98: Towards a Formal Description of Ada. Edited by D. Bjørner and O. N. Oest. XIV, 630 pages. 1980.

Vol. 99: I. Guessarian, Algebraic Semantics. XI, 158 pages. 1981.

Vol. 100: Graphtheoretic Concepts in Computer Science. Edited by H. Noltemeier. X, 403 pages. 1981.

Vol. 101: A. Thayse, Boolean Calculus of Differences. VII, 144 pages. 1981.

Vol. 102: J. H. Davenport, On the Integration of Algebraic Functions. 1–197 pages. 1981.

Vol. 103: H. Ledgard, A. Singer, J. Whiteside, Directions in Human Factors of Interactive Systems. VI, 190 pages. 1981.

Vol. 104: Theoretical Computer Science. Ed. by P. Deussen. VII, 261 pages. 1981.

Vol. 105: B. W. Lampson, M. Paul, H. J. Siegert, Distributed Systems – Architecture and Implementation. XIII, 510 pages. 1981.

Vol. 106: The Programming Language Ada. Reference Manual. X, 243 pages. 1981.

Lecture Notes in Computer Science

Edited by G. Goos and J. Hartmanis

147

RIMS Symposia on Software Science and Engineering

Kyoto, 1982
Proceedings

Edited by Eiichi Goto, Koichi Furukawa, Reiji Nakajima, Ikuo Nakata, and Akinori Yonezawa

Springer-Verlag
Berlin Heidelberg New York 1983

Editorial Board

D. Barstow W. Brauer P. Brinch Hansen D. Gries D. Luckham
C. Moler A. Pnueli G. Seegmüller J. Stoer N. Wirth

Editors

Eiichi Goto
Dept. of Information Science, Faculty of Science
University of Tokyo
7-3-1 Hongo, Bunkyo-ku, Tokyo 113, Japan

Koichi Furukawa
Inst. for New Generation Computer Technology
Mita Kokusai Building, 1–4–28 Mita, Minato-ku, Tokyo 108, Japan

Reiji Nakajima
Research Institute for Mathematical Sciences
Kyoto University, Kyoto 606, Japan

Ikuo Nakata
Inst. of Information Sciences and Electronics
University of Tsukuba, Sakuramura, Niihari-gun, Ibaraki 305, Japan

Akinori Yonezawa
Dept. of Information Science, Tokyo Institut of Technology,
Ookayama, Meguro-ku, Tokyo 152, Japan

CR Subject Classifications (1982): D.2

ISBN 3-540-11980-9 Springer-Verlag Berlin Heidelberg New York
ISBN 0-387-11980-9 Springer-Verlag New York Heidelberg Berlin

Library of Congress Cataloging in Publication Data. Main entry under title: RIMS symposia on software and engineering. (Lecture notes in computer science; 147) 1. Electronic digital computers–Programming–Congresses. 2. Computer architecture–Congresses. I. Goto, Eiichi, 1931-. II. Kyoto Daigaku. Súri Kaiseki Kenkyujo. III. Title: R.I.M.S. symposia on software and engineering. IV. Series. QA76.6.R55 1983 001.64'2 82-19669
ISBN 0-387-11980-9 (U.S.)

This work is subject to copyright. All rights are reserved, whether the whole or part of the material is concerned, specifically those of translation, reprinting, re-use of illustrations, broadcasting, reproduction by photocopying machine or similar means, and storage in data banks. Under § 54 of the German Copyright Law where copies are made for other than private use, a fee is payable to "Verwertungsgesellschaft Wort", Munich.

© by Springer-Verlag Berlin Heidelberg 1983

Printed in Germany

Printing and binding: Beltz Offsetdruck, Hemsbach/Bergstr.
2145/3140-543210

PREFACE

This volume contains selected papers from those presented at a series of symposia held at Kyoto University during the years of 1980 through 1982 under the title of "Software Science and Engineering". The symposia have been held once each year since 1979, sponsored by Research Institute for Mathematical Sciences (RIMS) of Kyoto University. The fields intended to be covered by the symposia include theoretical and practical aspects of programming languages and systems, programming styles and methodology, design and analysis of algorithms, database systems and machine architectures. The goal of the symposia is to promote research activities in software, to encourage publication of recent works by Japanese researchers and to circulate these results to the world-wide academic community.

The editors thank all those who contributed either as referees or in organizing the conferences. Without their efforts, publication of this volume would not have been possible.

Kyoto, October 1982

Editors

Editorial Board

Eiichi Goto (chief), University of Tokyo

Koichi Furukawa, ICOT

Reiji Nakajima, Kyoto University

Ikuo Nakata, University of Tsukuba

Akinori Yonezawa, Tokyo Institute of Technology

Referees

M. Amamiya, S. Arikawa, S. Goto, T. Katayama, S. Kawai

T. Mizoguchi, T. Sasaki, M. Sassa, N. Takagi, Y. Tanaka

M. Tokoro, S. Yajima, H. Yasuura, T. Yuasa

TABLE OF CONTENTS

Yoshihiko FUTAMURA 1
Partial Computation of Programs

Takuya KATAYAMA 35
Treatment of Big Values in an Applicative Language HFP

Mario TOKORO 49
Towards the Design and Implementation of Object Oriented Architecture

Shigeki GOTO 73
DURAL: an Extended Prolog Language

Taisuke SATO 88
An Algorithm for Intelligent Backtracking

Keiji KOJIMA 99
A Pattern Matching Algorithm in Binary Trees

Takeshi SHINOHARA 115
Polynomial Time Inference of Extended Regular Pattern Languages

Kenichi HAGIHARA, Kouichi WADA and Nobuki TOKURA 128
Effects of Practical Assumption in Area Complexity of VLSI Computation

Shuzo YAJIMA and Hiroto YASUURA 147
Hardware Algorithms and Logic Design Automation
- An Overview and Progress Report -

Makoto AMAMIYA, Ryuzo HASEGAWA and Hirohide MIKAMI 165
List Processing with a Data Flow Machine

Masaru KITSUREGAWA, Hidehiko TANAKA, and Tohru MOTO-OKA 191
Relational Algebra Machine GRACE

Yuzuru TANAKA 215
Vocabulary Building for Database Queries

Partial Computation of Programs

Yoshihiko Futamura

Central Research Laboratory, HITACHI,LTD.
Kokubunji, Tokyo, Japan

Abstract

This paper attempts to clarify the difference between partial and ordinary computation. Partial computation of a computer program is by definition "specializing a general program based upon its operating environment into a more efficient program". It also shows the usefulness of partial computation. Finally, the formal theory of partial computation, technical problems in making it practical, and its future research problems are discussed.

The main purpose of this paper is to make partial computation effectiveness widely known. However, two new results are also reported:

(1) a partial computation compiler, and

(2) a tabulation technique to terminate partial computation.

1. Introduction

Research on automating partial computation has been conducted in Sweden, Russia, the United States, Italy, Japan etc. for some time. Rapid progress has been made in this field in the last decade. Recent research achievements have indicated the practical utility of partial computation.

Partial computation of a computer program is by definition "specializing a general program based upon its operating environment into a more efficient program" (This is also called mixed computation[14] or partial evaluation [6, 16, 20, 23 etc.]). For example, let us assume that a programmer has written program f having two input data k and u. Then let f(k,u) be the computation performed by f. Let us further assume that f is often used with k=5. In this case, it is more efficient to produce a new program, say f_5, by substituting 5 for k in f and doing all possible computation based upon value 5, as well as iteratively compute $f_5(u)$, than to compute f(5,u) iteratively changing values of u.

Partial computation is an operation automatically producing f_5 when f(5,u) is given. Contrary to partial computation, ordinary computation described, for example, in FORTRAN or COBOL does not produce f_5 but produces an error message "u is undefined". Therefore, generating f_5 from f(5,u) must be done by a FORTRAN or COBOL programmer by himself. Thus, present computer's lack of partial computation ability forces programmers to do partial computation by themselves to write efficient programs. Since general or standard program specialization is often performed in everyday program production activity, automating the specialization is expected to greately improve program productivity.

This paper attempts to clarify the difference between partial and ordinary computation. It also shows the usefulness of partial computation. Finally, the formal theory of partial computation, technical problems in making it practical, and its future research problems are discussed.

2. Partial Computation Algorithm

Usually, a computer program is designed to produce desired result based upon given data. To make discussion easier, let us consider program f processing only two input data k (for known data) and u (for unknown data). Ordinary computation can start only when both k and u are given. This is called total computation (see Fig 1).

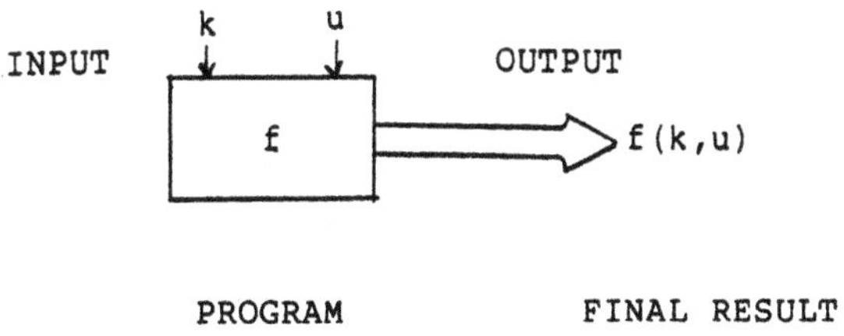

Fig. 1: Total computation produces a final result.

On the contrary, partial computation can start when one of the input data, say k, is given. Content of the computation is as follows:

All computation in f that can be performed based upon k are finished. All the computation that cannot be performed without u are left intact. Thus, a new program is produced (see Fig 2).

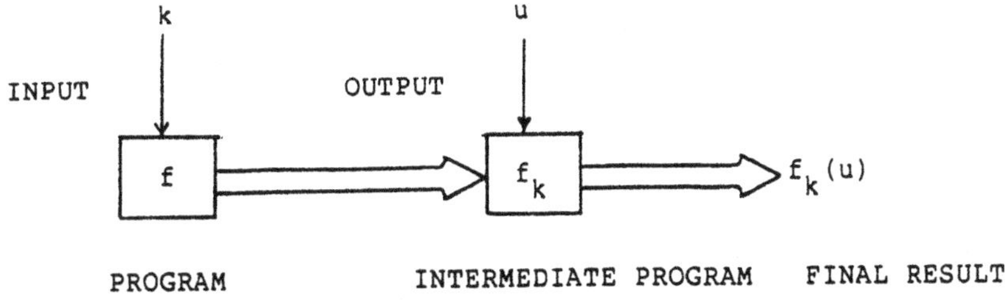

Fig. 2: Partial computation produces intermediate program f_k based upon f and partial data k.

For example, let

$$f(k,u)=k*(k*(k+1)+u+1)+u*u$$

and k=2 be given. Thus, an intermediate program (or projection of f at k=2) is

$$f_2(u)=2*(7+u)+u*u \ .$$

When data u is given to f_2, the $f_2(u)$ result is equal to that of f(2,u). Thus,the following equation holds for f_k and f:

$$f_k(u)=f(k,u) \quad \text{-----E1}$$

Both total computation and computation via partial computation give identical results. This f_k is called a projection of f at k.

Now, let us consider the reason partial computation is useful. Actually, there is a case in which partial computation is not useful.

When f is computed just once with k=2 and u=1 in the above example, computation via f_2 is less efficient than direct computation of f(2,3).

However, when f(k,u) is iteratively computed with k fixed to 2, and u varied from 1 to 1000 by 1, the situation may be reversed.

The reason is obvious:

Assume that each addition and multiplication takes time a, and producing an intermediate program f_2 takes time p (see Fig 3).

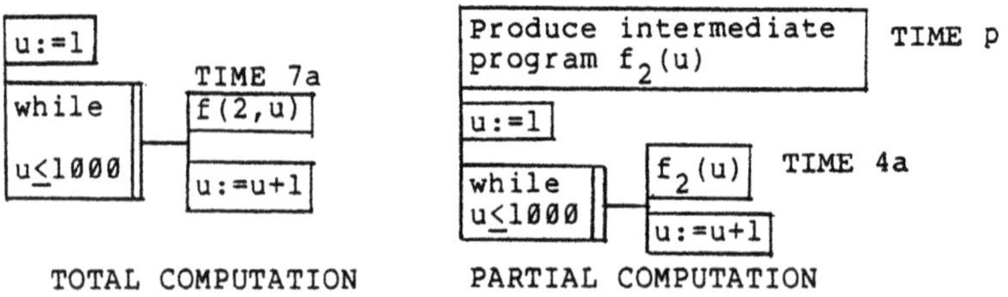

Fig. 3: Partial computation is often useful when f is computed repeatedly (Programs in this figure are described in PAD. See Appendix 1).

time for computing f(2,u) = 7a

time for computing $f_2(u)$ = 4a

Therefore, if computation is repeated 1000 times it takes 7000a for total computation, and p + 4000a for partial computation. Thus, if p is less than 3000a, partial computation is more efficient.

As described above, partial computation may be more efficient when a program is computed iteratively with a part of the input data fixd and the remaining data changed.

A little more complex example[8] of partial computation is given below to clarify its difference from algebraic manipulation and symbolic execution.

A program that computes u^5 and stores its value to z is shown in Fig 4. The partial computation result of f at k=5, i.e. an intermediate program, is shown in Fig 5.

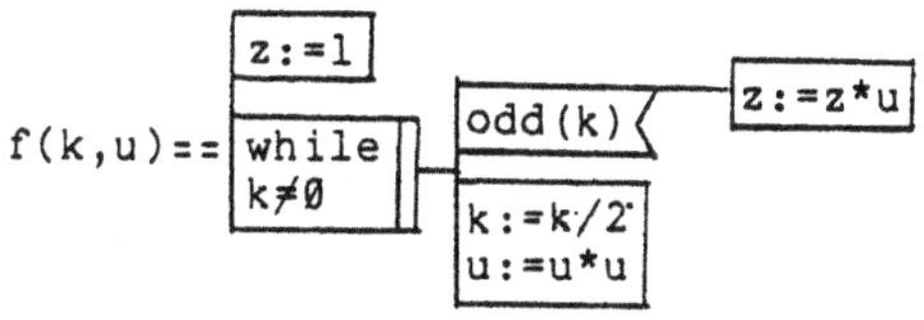

Fig. 4: f computes $z:=u^k$

$$f_5(u) = \begin{array}{|l|} \hline z:=1*u \\ u:=u*u \\ u:=u*u \\ z:=z*u \\ \hline \end{array}$$

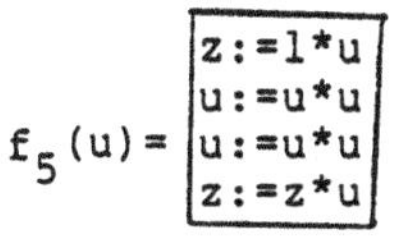

Fig. 5: $f_5(u)=f(5,u)$

Note that the result of partial computation (i.e. projection) is a program while that of algebraic manipulation or symbolic execution is an algebraic expression.

A very important point is that partial computation procedure itself can be a computer program. This partial computation program is called projection machine, partial computer or partial evaluator. By letting it be α , we get the below equation:

$\alpha(f,k) = f_k$ -----E2

Now, consider α_f, which is the result of partial computation of α at f.

From E1 we get $\alpha_f(k) = \alpha(f,k)$

From E2 we get $\alpha(f,k) = f_k$

Therefore

$\alpha_f(k) = f_k$ -----E3

This E3 is called the basic equation of partial computation.

The author hopes that readers have grasped the rough idea of partial computation thus far. To clarify the usefulness of partial computation, examples of its applications are described in the next chapter.

3. Applications of Partial Computation: Examples

Since a computer program usually contains repetitions, computation efficiency of a program can often be improved through partial computation. Therefore, the author believes that the basic comuptation mechanism of a computer should be established based on partial computation. He also believes that present computer's lack of partial computation ability is one of the biggest obstacle to program development. It forces programmers to do partial computation by themselves to write efficient programs. This is basically inefficient.

As Kahn pointed out in [23], program production using automatic partial computation mechanism, i.e. projection machine, could be made much easier than it is now. For readers to understand the power of partial computation, four examples are shown below. They are selected due to their clearness in application areas and effectiveness (See [3] and [12] for other examples).

3.1 Automatic theorem proving

Theorem prover is a program automaticlly proving a given theorem based upon a given set of axioms.

Let theorem prover be P, axioms be A, and theorems be T. Generally, a large number of theorems are proved based upon a fixed set of axioms, e.g., axioms of Euclidian geometry or the group theory. That is, compute P(A,T) with A fixed and T changed. In this case, generating P_A by partially computing P at A and iteratively computing $P_A(T)$ is more efficient than computing P(A,T) iteratively.

For example, consider theorem proving in Euclidian geometry. Program P_A is then a specialized theorem prover for Euclidian geometry ($P_A=\alpha(P,A)$). Theorems are proved by using this P_A.

Projection machine α for specializing P automatically was first developed by Dixon in 1971 to the best of the author's knowledge [5]. His α can partially compute programs written in pure LISP [38].

3.2 Pattern matching

Text editors like word processors have simple pattern matching functions. For example, consider the case when a text string like

THIS IS A BOOK.

is searched in order to substitute PEN in the pattern and to get the string

THIS IS A PEN.

This sort of operation for locating where a pattern is in a text string is pattern matching.

Let M be a pattern matching program, P be a pattern and T be a text string. Pattern matching is thus represented as M(P,T).

When a pattern, say BOOK, repeatedly appears in a text and when all the instances of the pattern are replaced by the other string ,say PEN, or when a pattern is used frequently, it is useful to partially compute M at the pattern, say P, to get M_P, and to perform $M_P(T)$ repeatedly.

Emanuelson[6] implemented such a projection machine α to produce M_P by performing $\alpha(M,P)$ for a complicated pattern matcher M written in LISP.

3.3 Syntax analyzer

In language processing, a program called syntax analyzer is necessary for parsing a given sentence based upon a given grammar. Let S be a syntax analyzer, G be a language grammar and T be a given text. Syntax analysis then can be represented as S(G,T).

When grammar G is fixed and T is varied, it is not efficient to compute S(G,T) repeatedly. Specialized analyzer S_G can be produced by partially computing S at G. The author discussed the partial computation method for BTN(Basic Transition Network) analyzer in [8]. Ershov's group also partilly computed LR(1) analyzer S at PASCAL grammar G[28].

Let G be FORTRAN grammar and T be a FORTRAN program. Syntax analysis in FORTRAN compiler can then perform the same function as S(G,T). Note that G is fixed and only T is varied in this case. Thus, $S_G(T)$ can be used instead of S(G,T).

Generally speaking, projection machine α produces less efficient syntax analyzer S_G than a good human partial computer nowadays. Therefore, very efficient S_G for popular programming languages like FORTRAN and COBOL are implemented by human programmers at the cost of a lot of man-hours. However, in the future it may be possible that S_G produced by $\alpha(S,G)$ will become more efficient than that made by human programmers.

3.4 Compiler generation

The three examples described above are generation of intermediate programs from data like axioms, patterns, or grammars. Therefore, they may be called compilation of axioms, patterns, or grammars instead of partial computation.

In this section, problems in automatically generating compilers are discussed. A compiler is a program to transform a program in some language (this is called source language) into a program in another language (this is called object language).

We often hear "A BASIC program runs slowly because it is executed by an interpreter. If we have a good BASIC compiler as FORTRAN, it will run much faster". What is the interpreter in this case?

Interpreter, say I, is a program to perform specified computations that analyze the meanings of a given program, say P, based upon given data, say D. Thus, running a program by interpreter means performing I(P,D). While a compiler, say C, translates a source program, say P, to an object program C(P). When this C(P) runs with data D, the result is the same as that of I(P,D). This produces the following equation:

$$C(P)(D)=I(P,D)$$

While $I_P=\alpha(I,P)$. Thus $C(P)(D)=I_P(D)$ and we get the below equation.

$$I_P=C(P)\text{-----}E4$$

(First equation of partial computation)

Therefore I_P can be regarded as an object program of P.

From the above discussion, it is clear that the repetitive computation of I(P,D) with P fixed and D varied is less efficient than that of $I_P(D)$ after producing I_P by compiler C.

Now if we compute I(P,D) once with both P and D fixed, is a compiler still useful ? Generally, partial computation is not useful when a computation is performed just once. However, it is also true that generally a program runs faster after it is compiled even if it is executed only once. Why is this ?

The reason is that a program P itself contains repetitive computation (i.e. loops). Usually, a program contains a loop for performing repetitive, say 10 or 10000 times, computation. When this program is executed by an interpreter, the interpreter repeats analysis of a loop. However, partial computation of interpreter I at program P (i.e. $\alpha(I,P)$) finishes the part of meaning analysis which can be performed without knowing data D, e.g. analysis of a program structure. So, when a loop in an object program I_P is performed repetitively, there remains just a small part of computation performed by an interpreter I. Thus, even when I(P,D) is computed just once for given data P and D, compiling P is often useful. If P contains no loops, compiling P is useless.

From the discussion above, readers may see the significance of a compiler. The remaining part of this section deals with a compiler generation method based upon partial computation.

If P is an object program, then can α which produces I_P from I and P be regarded as a compiler ? The answer is no. Generally, α is a complicated program and computation of $\alpha(I,P)$ takes a long time. Therefore, regarding α as a compiler is not practical.

Now, noting that α itself is a program with two data, let us look at α_I which is produced by partially computing α at I. From equation E3, equation E5 is derived by substituting each f and I for k and P, respectively.

$$\alpha_I(P)=I_P \text{-----E5}$$

(Second equation of partial computation)

Since I_P is an object program and α_I transforms a source program P to I_P, α_I has the same function as that of a compiler.

Furthermore, in α_I, computation of α concerning I has been finished. Thus, it is not wrong to regard α_I as a compiler. Generally, implementing an interpreter is much easier than that of a compiler [16]. Therefore, if we have a good partial computer α, a compiler can be made automatically through $\alpha(\alpha,I)$ after easily implementing I [16].

Now, what is the result of partial computation of α at α itself (i.e. α_α) ?

Equation E6 below is derived from equation E3 by substituting each α and I for f and k, respectively.

$\alpha_\alpha(I)=\alpha_I$-----E6

(Third equation of partial computation)

Since α_I is a compiler, from E6 above, α_α is regarded as a compiler-compiler which generates a compiler α_I from an interpreter I.

From the above discussion, the relationship between language processors like interpreter, compiler and compiler-compiler is clarified (see [15]).

Let I-language be a language defined by an interpreter I. From equation E6, I-language compiler can be produced automaticlly through $\alpha_\alpha(I)$. Now, let us substitute α for I in E6 or consider α as an interpreter. Then what is α-language ?

Let f be a program in α-language and k be data. From E1, the result of performing f with data k is:

$\alpha(f,k)=f_k$.

Substituting α for I in E6, we get:

$\alpha_\alpha(\alpha)=\alpha_\alpha$.

This means α_α is an α-language compiler. Therefore, $\alpha_\alpha(f)$ is an object program of f and we get equation E7 below.

$\alpha_\alpha(f)(k)=f_k$-----E7

(Forth equation of partial computation).

From the discussion above, we see that, in general, partial computation of f at k can be done more efficiently through compiling f by α_α than by directly computing α(f,k) (However, when f=α, computing α_α(k) is sufficient). Therefore α_α is the partial computation compiler desired. However, the author has not heard of a report concerning the execution of α (α,α) to produce α_α for practical α.

The author implemented an α and an interpreter I in LISP for ALGOL-like language, and tried to generate ALGOL compiler α_I in 1970 [16]. The generated compiler was not efficient enough to be used practically. Similar, but more advanced, experiment was conducted by Haraldsson [20]. He implemented partial computation program REDFUN in LISP [3,20]. REDFUN was used by Emanuelson to implement a pattern compiler [6].

Based upon partial computation and Prolog interpreter on LISP machine (LM-Prolog [37]), Kahn [23] tried to automatically generate Prolog compiler. The compiler translates a Prolog program into a LISP program.

To do so, from the discussion in this section, it is sufficient to have a LISP program α which can partially compute a LISP program. However, good α is very hard to write (Haraldsson told in [20] his partial evaluator REDFUN-2 was 120 pages in prettyprinted format). Therefore, Kahn tried to write his α in Prolog [35,37,39]. Since Prolog has a powerful pattern matching ability and theorem proving mechanism, it seems easier to write complicated α in Prolog than in LISP. The outline of Kahn's method is described below.

First, Kahn implemented the following two programs:

(1) L : Prolog interpreter written in LISP.

Let D be database and S be a statement in Prolog, then L performs L(D,S). Interpreter L is similar to a theorem prover described in

Section 2.1. Database D and S correspond to axioms and a theorem, respectively.

(2) P : LISP partial computation program (projection machine) written in Prolog.

Note that P is database for L. Let f be a LISP program, k and u be its input data, and [f,k] be a Prolog statement describing f and k. Then the equation below holds.

f_k=L(P,[f,k])-----E8

He then gets L_P by performing L(P,[L,P]) from E8 which is again a LISP program.

From E1:

L_P([f,k])=L(P,[f,k])-----E9

From E8 and E9, L_P([f,k])=f_k is derived. Therefore, L_P is a LISP program producing f_k from f and k. This means that L_P is a LISP partial computer described in LISP.

Let L_P be α. Then from equation E5, by performing $\alpha(\alpha,L)$, we can get a Prolog compiler α_L translating a Prolog program into a LISP program.

Kahn's method of producing a Prolog compiler seems very promising because:

(1) A partial computation method of a LISP program is becoming clearer through its long time research.

(2) A very convenient language like Prolog for describing partial computer is becoming available.

4. Theory of Partial Computation

It was in the 1930's that Turing, Church, Kleene, etc. proposed several computation models and clarified the mathematical meanings of mechanical procedure. A computation model, in plain language, is a sort of programming language. Typical computation models include the

Turing machine, lambda expression and the partial recursive function.

At that time, the main research interest concerned computability, i.e. computational power of the models, not computational complexty or efficiency. Since partial computation was the same as ordinary computation, as far as computability was concerned, it did not attract research attention.

However, equation E1 in Chapter 2 appeared in Kleene's S-m-n theorem (It is also called the parametarization theorem or iteration theorem) of the 1930's[41,45]. A procedure to produce f_k by fixing k in f was also described in the proof of the theorem. That procedure was just like putting assignment statement k:=(value of k), e.g. k:=5, in front of f(k,u):

```
          ┌────────┐
          │ k:=5   │
f (u)==   ├────────┤
 5        │ f(k,u) │
          └────────┘
```

Furthermore, the partial computation of α itself, which appears in the third equation of partial computation, was also used in the proof of Kleene's Recursion Theorem[41](1938). However, he only used partial computation in its simplest form to fix the value of a variable.

While Turing machines and partial recursive functions were formulated to describe total computation, Church's lambda expression was based upon partial computation. Inputs to a lambda expression are also lambda expressions, and the result is, again, a lambda expression. This computation is called lambda conversion. Furthermore, when a lambda expression corresponding to f(5,u),for example, is computed with u undefined, the result is a lambda expression corresponding to $f_5(u)$ in which computation concerning k=5 is finished. Lambda conversion of $f_5(u)$ with, again for example, u=6 produces the same result as f(5,6) (by the Church-Rosser Theorem). That is to say, $f_k(u)=f(k,u)$. Thus lambda conversion is partial computation.

As described above, the concept of partial computation already existed in the 1930's. However, implementation of a projection machine and its application to real world problems started in the 1960's after the programming language LISP began to be widely used (see Appendix 2).

Problems in making a practical projection machine will be discussed in the next chapter. The rest of this chapter deals with the theory behind the projection machine.

The relationships between a projection machine and a language processor, i.e. equations E4, E5 and E6, have been discussed by Futamura[16,17], Beckman[3], Ershov[8,12,14] and others.

Formal treatment of a projection machine and its proof of correctness were presented by Ershov in [9,13,14]. He dealt with ALGOL-like programs that included assignment, conditional and GOTO statements in [9] and a recursive program schema similar to pure LISP in [13]. The author believes [9] and [13] are landmark papers that establish the theoretical foundations for a projection machine.

It is important for projection machine α to satisfy the following three conditions:

(1) Correctness: Program α should satisfy the equation $\alpha(f,k)(u)=f(k,u)$.

(2) Efficiency improvement: Program α should perform as much computation as possible based upon given data k.

(3) Termination: Program α should terminate on partial computation of as many programs as possible. Termination at $\alpha(\alpha,\alpha)$ is most desirable.

In condition (2), the meaning of "as much computation as possible" is not mathematically clear. This can be rephrased as "all possible computation" for a simple language as a recursive program schema. However, for a language with assignment statements or side effects, it is not easy to make computations concerning k only without spoiling condition (1). Therefor, the author believes that it is of primary importance to establish a projection machine satisfying conditions (1), (2) and (3) for a recursive program schema that makes theoretical discussion easiest to do.

Ershov[13] has established α satisfying (1) and (2) for a recursive program schema. This paper attempts to describe a new α that satisfies all three conditions based upon Ershov[13]. The description is rather informal in order for readers to understand intuitively.

Roughly speaking, a recursive program schema is a program consisting of only three different components:

(1) Condition:

(2) Expression:

(3) function definition:

==

Six examples of recursive program schema are described below:

Example 1:

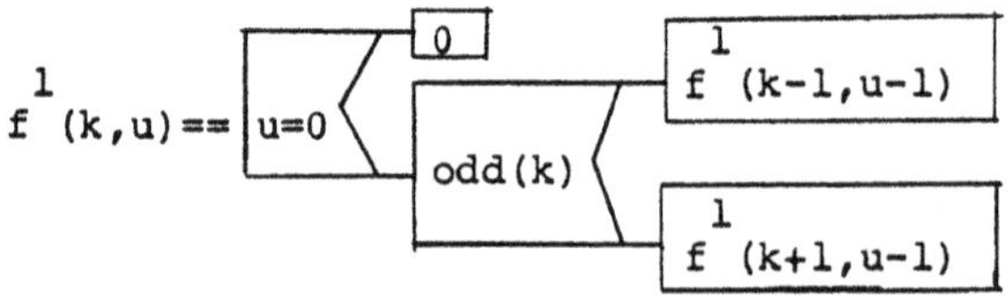

This means a program that takes value 0 if u>=0 or loop indefinitely if u<0; i.e.

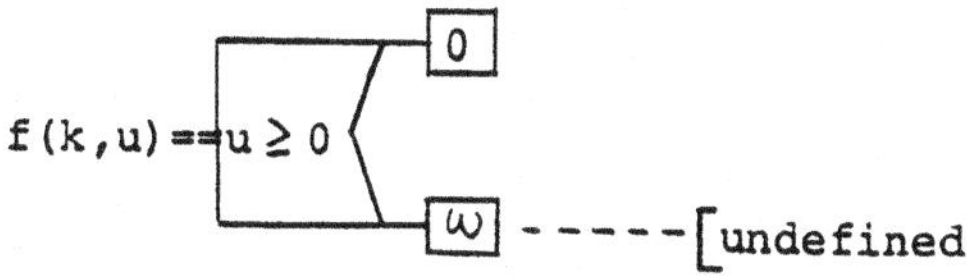

Following Examples 2 to 4 describe this same function in different forms.

Example 2:

Example 3:

Example 4:

$f^4(k,u) == \boxed{u \geq 0} \begin{cases} \boxed{0} \\ \boxed{f^4(u+1,u)} \end{cases}$

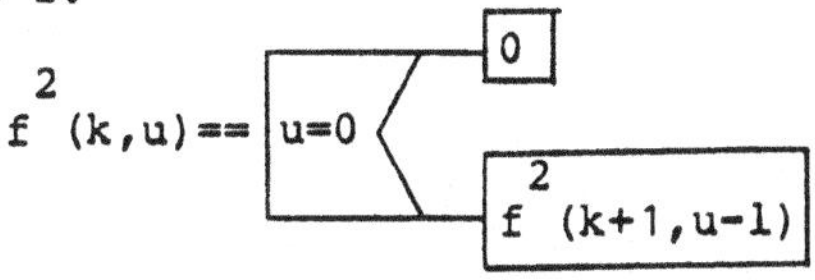

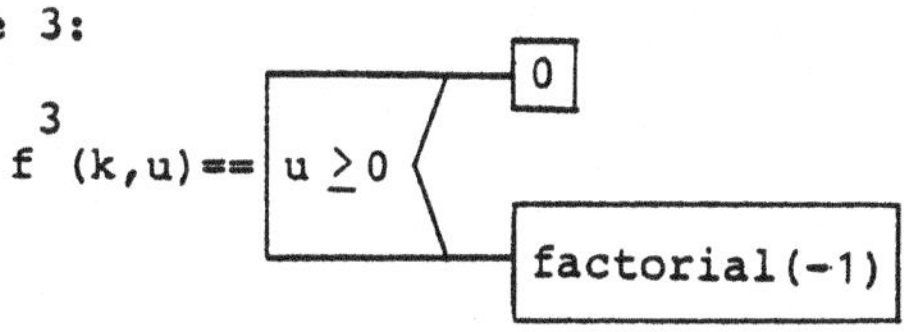

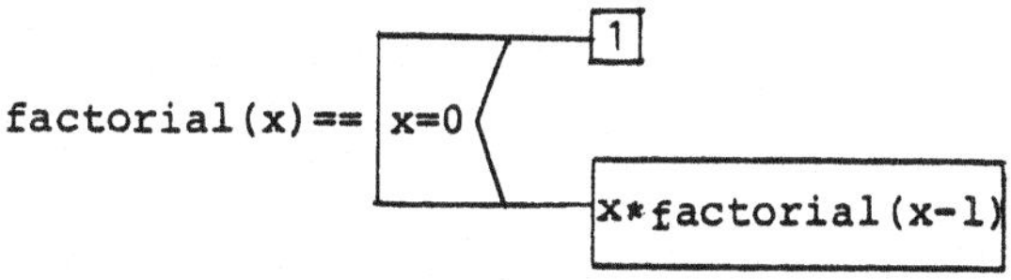

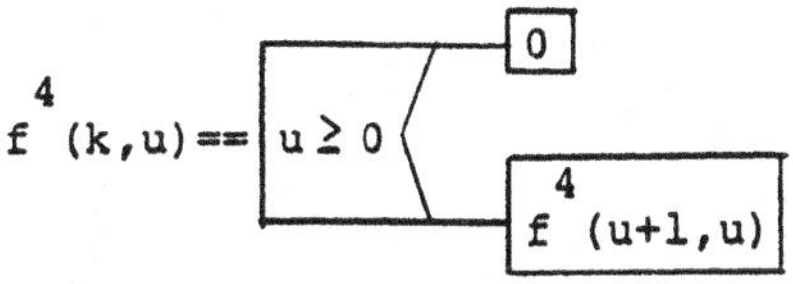

Example 5:

$$f^5(k,u) == \boxed{u=u} \quad \begin{cases} \boxed{0} \\ \boxed{f^5(k+1,u)} \end{cases}$$

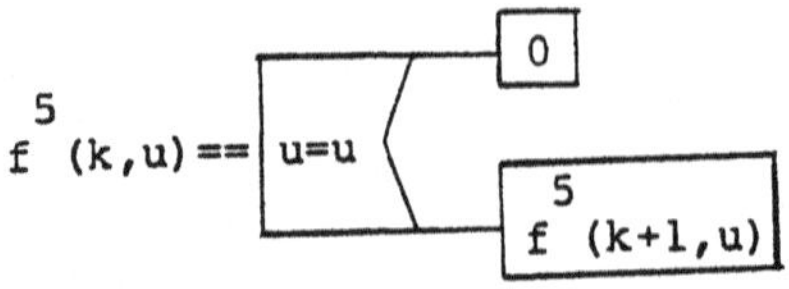

The value of f^5 is always 0.

Example 6:

$$f^6[u;k] == \boxed{null[u]} \quad \begin{cases} \boxed{k} \\ \boxed{cons[car[u];f^6[cdr[u];k]]} \end{cases}$$

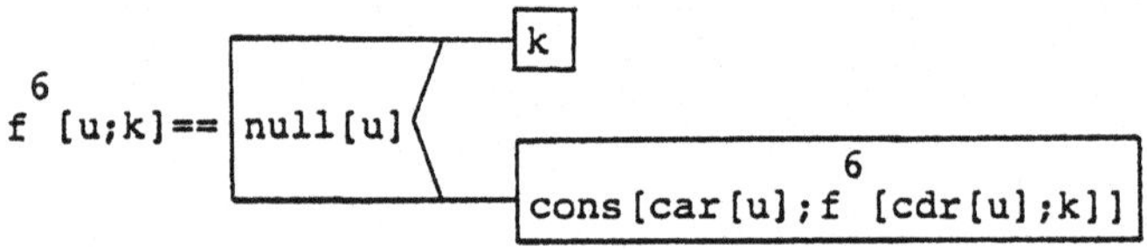

This is equal to append[u;k] in LISP.

As described in [42], a computation rule for a recursive program schema is repetition of (1) Rewriting and (2) Simplification. The computation terminates when there is no longer occasion to apply the rewriting rule.

Partial computation of f at k is performed by repetition of the following three rules:

(1) Rewriting: Replace f(k,u) by its definition body (This f(k,u) is called a semi-bound call by Ershov[13]). This rule is different from the rewriting rule for ordinary computation because it can only be applied when u is undefined.

(2) Simplification: Perform possible simplification to a program produced by the rewriting rule. This is not very different from the simplification rule for ordinary computation.

(3) Tabulation: Let E be the expression derived from the above two rules.

(3.1) If there are semi-bound calls, say f'(k',u'), in E, then each one is replaced by $f'_{k'}(u')$. Let the resulting expression be E'. Then define f_k as E', i.e.

$$f_k == E'.$$

(3.1.1) If there is an undefined $f'_{k'}$ in E', repeat (1), (2) and (3) for each f' and k' (i.e. partially compute f' at k').

(3.1.2) If there is no undefined $f'_{k'}$ in E', terminate the computation.

(3.2) If there is no semi-bound call in E', define f_k as E' and terminate the computation.

Thus, the discriminating characteristics of partial computation are the semi-bound call and tabulation. Six examples are shown to illustrate the above rules:

Example 1: Partially compute $f^1(4,u)$.

(1) Rewriting: Replace k by 4 in the $f^1(k,u)$ definition.

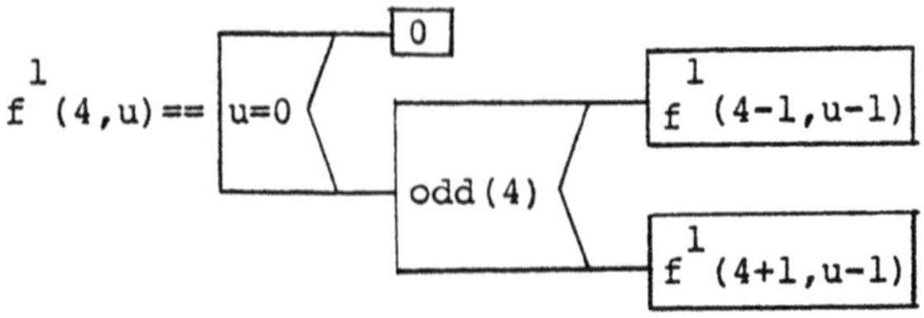

(2) Simplification: Perform computation and decide what can be done based upon the value of k, i.e. 4.

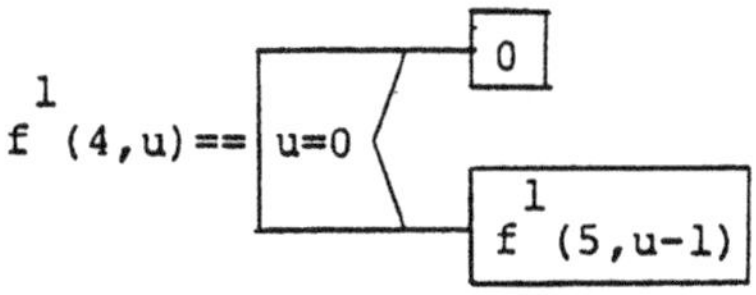

(3) Tabulation: Replace semi-bound call $f^1(5,u-1)$ by $f^1_5(u-1)$. The resulting expression is defined as $f^1_4(u)$.

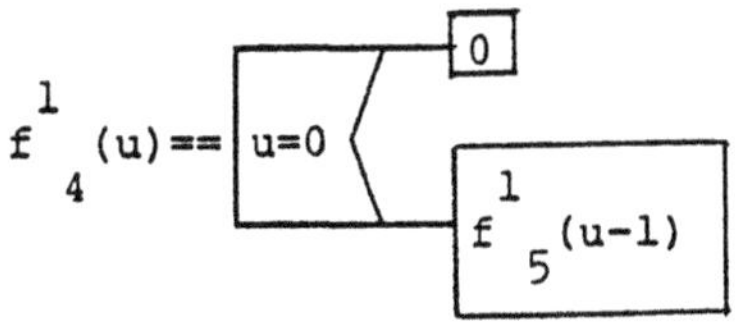

Since f^1_5 is undefined, define $f^1_5(u)$ through partial computation of $f^1(5,u)$, i.e.:

(4) Rewriting:

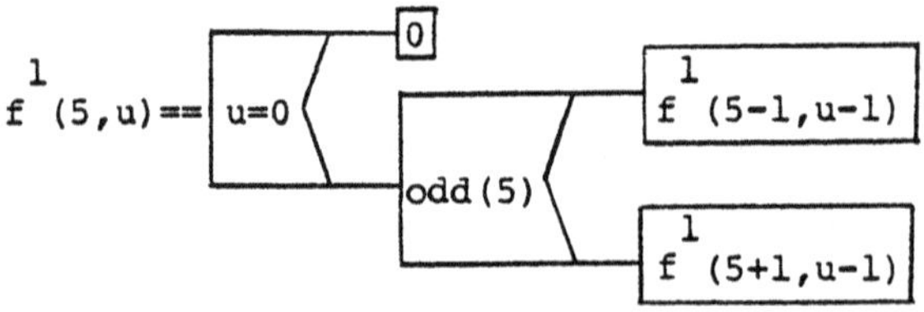

(5) Simplification:

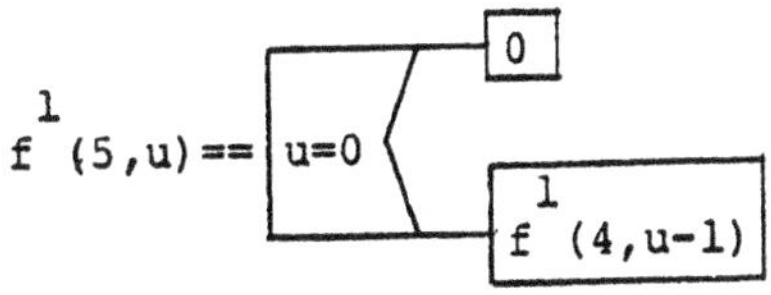

(6) Tabulation:

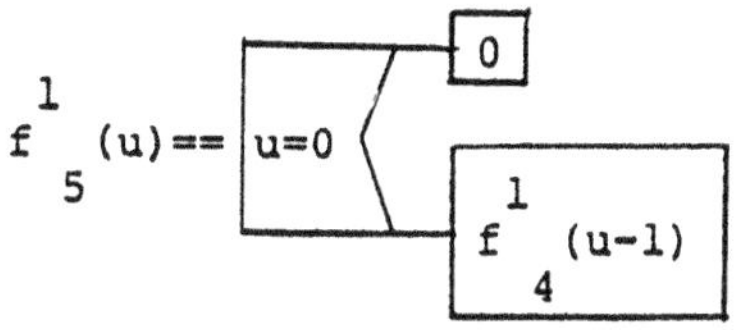

Since $f^1{}_4$ has been defined by previous tabulation, there is neither a semi-bound call nor an undefined function in the above expression. Thus, partial computation terminates. The results are the following two equations:

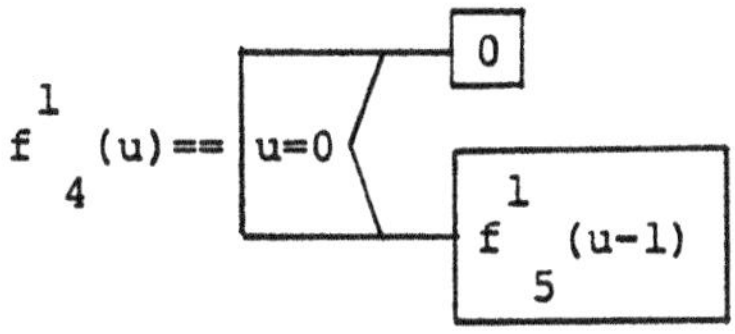

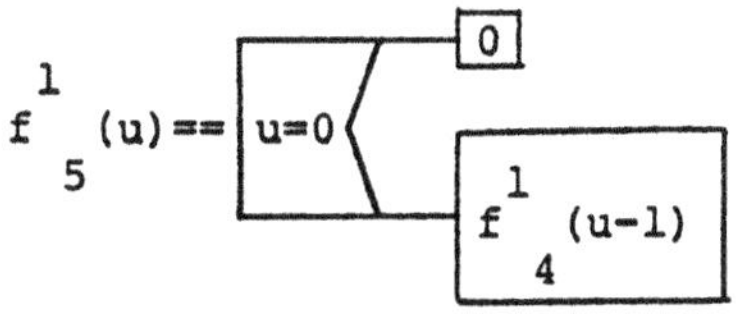

Example 2: Partially compute $f^2(4,u)$.

(1) Rewriting:

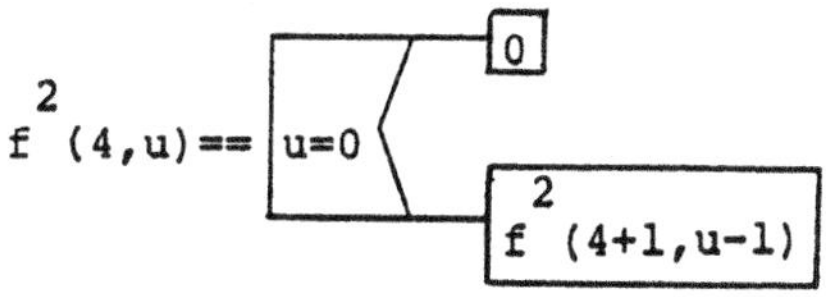

(2) Simplification:

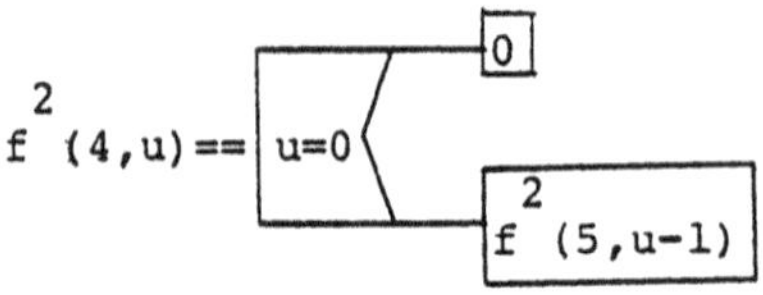

$$f^2(4,u) == \begin{cases} u=0 & \boxed{0} \\ & \boxed{f^2(5,u-1)} \end{cases}$$

(3) Tabulation:

$$f^2_4(u) == \begin{cases} u=0 & \boxed{0} \\ & \boxed{f^2_5(u-1)} \end{cases}$$

(4) Rewriting:

$$f^2(5,u) == \begin{cases} u=0 & \boxed{0} \\ & \boxed{f^2(5+1,u-1)} \end{cases}$$

(5) Simplification:

$$f^2(5,u) == \begin{cases} u=0 & \boxed{0} \\ & \boxed{f^2(6,u-1)} \end{cases}$$

(6) Tabulation:

$$f^2_5(u) == \begin{cases} u=0 & \boxed{0} \\ & \boxed{f^2_6(u-1)} \end{cases}$$

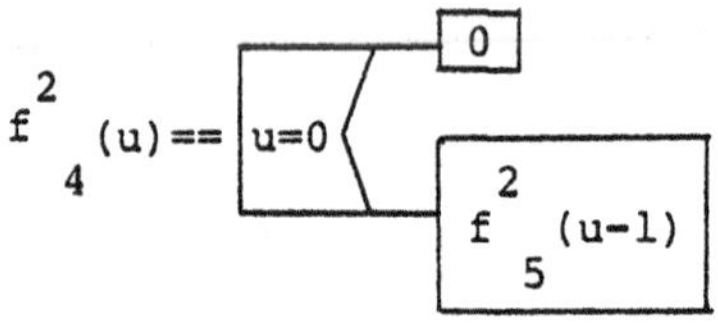

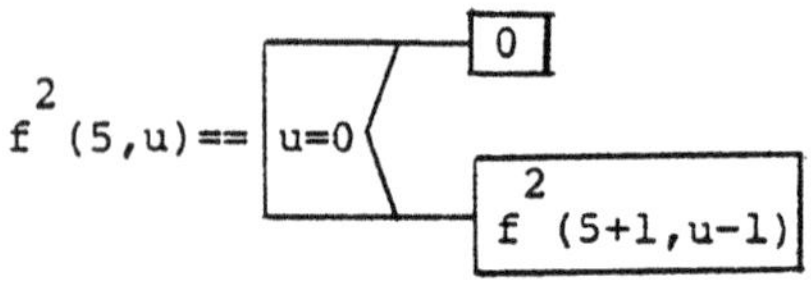

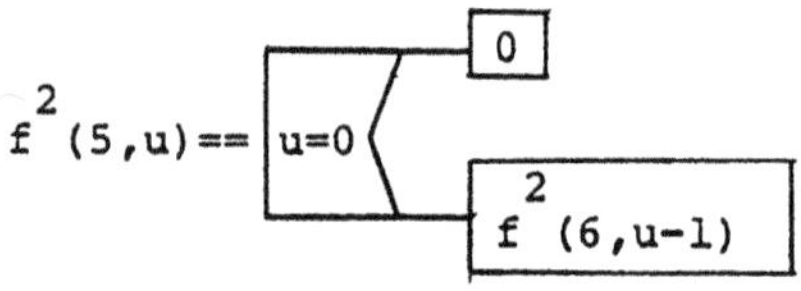

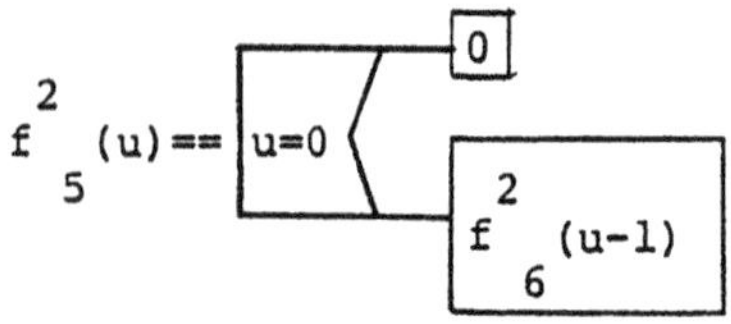

Undefined functions $f^2_6, f^2_7, \ldots$ appear indefinitely and partial computation does not terminate.   Thus, f^2_4 cannot be defined.

Example 3: Partially compute $f^3(4,u)$.

(1) Rewriting:

$$f^3 (4,u)== u \geq 0 \begin{cases} 0 \\ \text{factorial}(-1) \end{cases}$$

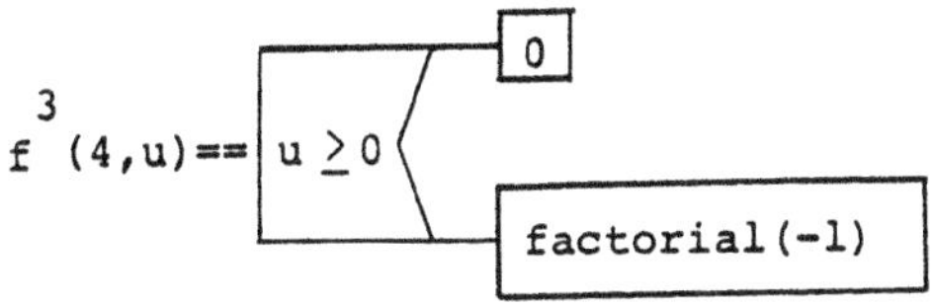

(2)   Simplification: According to the definition of the factorial function,   factorial(-1)   does   not   terminate.   Thus,   this   partial computation does not terminate.

Example 4: Partially compute $f^4(4,u)$.

(1) Rewriting:

$$f^4 (4,u)== u \geq 0 \begin{cases} 0 \\ f^4 (u+1,u) \end{cases}$$

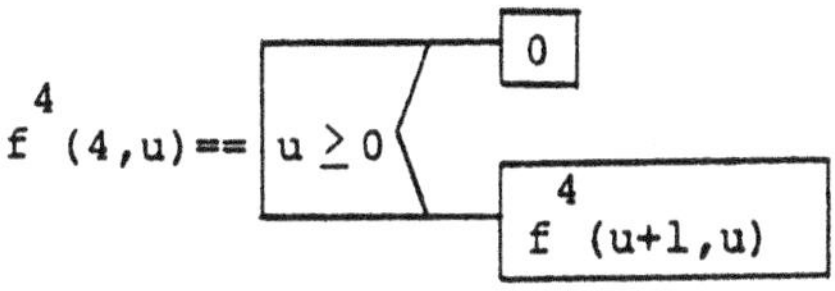

(2) Simplification:

The same expression as the above.

(3) Tabulation:

$$f^4_4 (u)== u \geq 0 \begin{cases} 0 \\ f^4 (u+1,u) \end{cases}$$

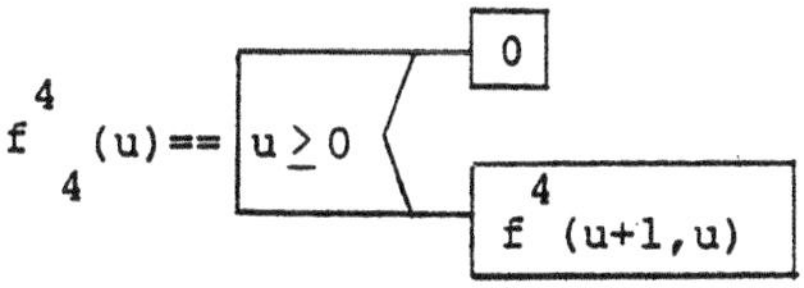

Since   neither   semi-bound   calls   nor   undefined   functions   are included in the expression, the partial computation terminates.

Example 5: Partially compute $f^5(4,u)$.

(1) Rewriting:

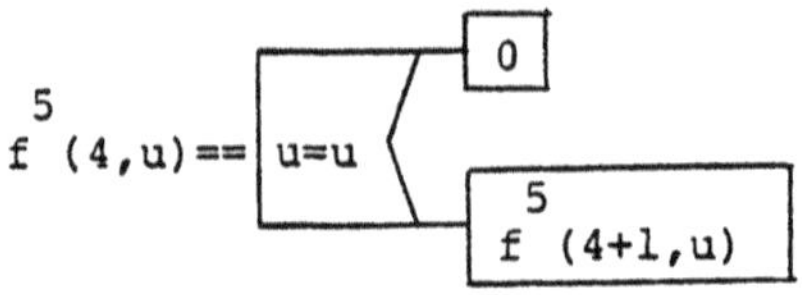

(2) Simplification:

If this rule is powerful enough to find that u=u always holds, then partial computation terminates and the result is 0 (i.e. $f^5(4,u)=0$). However, if boolean expression u=u is not evaluated because u is unknown, partial computation of f^5 at k=5 is performed. Thus, the same as in Example 2, partial computation will not terminate.

Example 6: Partially compute f^6[u;(A,B)].

(1) Rewriting:

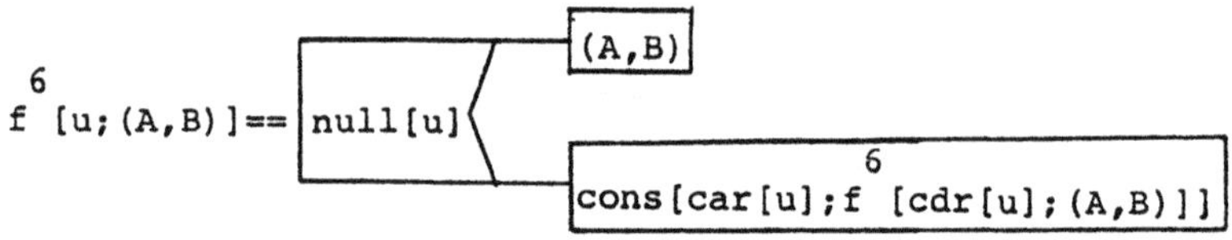

(2) Simplification:

The same as above.

(3) Tabulation:

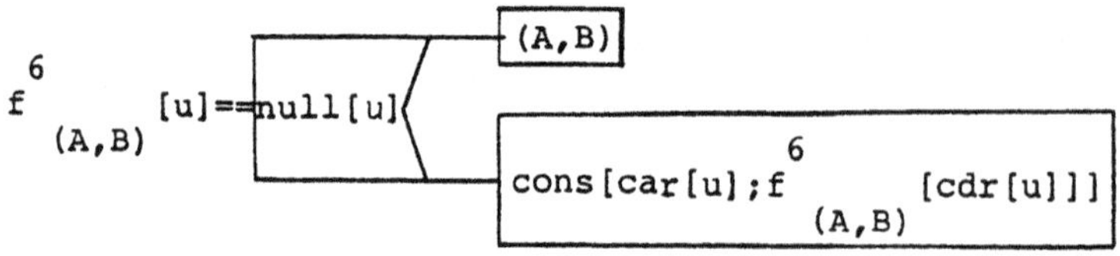

Partial computation is terminated and projection $f^6_{(A,B)}$ is produced.

In the above examples, partial computation of f^2 and f^3 do not terminate, but total computation terminates, at u>=0 (see Table 1). The reasons are as follows:

(1) The value of known variable k changes indefinitely (for f^2).

(2) Infinite computation that can be performed based upon known variable k and constants is included (for f^3).

Table 1: Summary of Examples.

	Total computation	Partial computation
Terminating	f^1, f^2, f^3, f^4, f^5, f^6 $(u \geq 0)$	f^1, f^4, f^5, f^6
Non-terminating	f^1, f^2, f^3, f^4 $(u<0)$	f^2, f^3, f^5

As described in Chapter 3, partial computation can be applied to such data of finite structure as a pattern, a set of axioms, a grammar or a program. Therefore, case (1) above will not happen in practical application. Furthermore, case (2) will not happen either, if the programs to be partially computed are written carefully. Thus, projection machine α described above seems powerful enough to be practical.

Tabulation is a unique feature of α (Tabulation technique for ALGOL-like programs was described in [18]). The first paper to mention the importance of tabulation in partial computation was probably [16]. Kahn[23] also mentioned a powerful tabulation method using Prolog.

Ershov[13] and REDFUN[6] devised mechanisms to terminate partial computation in such special cases as when the value of k does not change. For example, their α s can terminate for partial computation of f(k,u), where

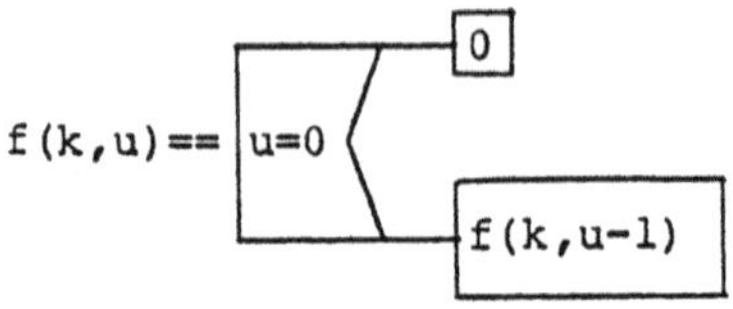

and the projection $f_k(u)$ is

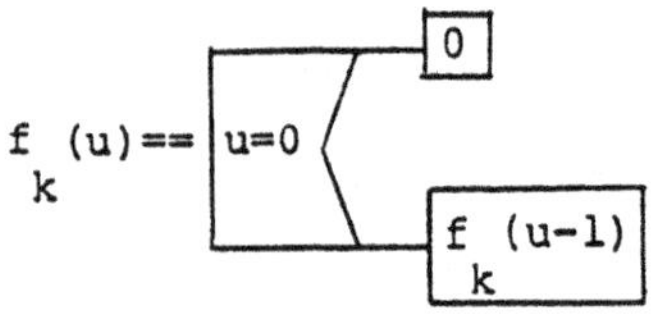

However, their α s do not terminate for partial computation of f^1 in Example 1 above.

5. Technical Problems in Making α Practical

One of the most energetic groups persuing how to make partial computation practical is Professor Sandewall's group in Sweden, to the best of the author's knowledge. In that group, Beckman[3], Haraldsson[20], Emanuelson[6], Komorowski[25] and Kahn[23,24] have been working on the subject for a long time. The recent achivements of Emanuelson[6] and Kahn[23] lead the author to believe that a practical projection machine can be implemented in the near future.

The group has been trying to partilly compute LISP programs. LISP is one of the best real-world languages for partial computation research because its interpreter is simple and clear, its programs can be easily manipulated as data, and its program structure is similar to a recursive program schema. Therefore, the Sandewall group's approach of establishing α for LISP as their first step seems very sound.

As described in the previous chapter, partial computation is a repetition of (1) rewriting, (2) simplification and (3) tabulation.

Rewriting is like so-called macro expansion and procedure integration[30] in the optimization techniques of a compiler. This operation is often performed in combination with simplification (see Fig.6).

ORIGINAL PL/I PROCEDURES

```
P:PROC(A);
  B=5;
  C=A*B;
  CALL Q(A,B);
RETURN(C);
END;

Q:PROC(X,Y);
  X=X*Y;
END;
```

AFTER INTEGRATION

```
P:PROC(A);
  B=5;
  C=A*B;
  A=A*B;
  RETURN(C);
END;
```

AFTER SIMPLIFICATION

```
P:PROC(A);
  A=A*5;
  RETURN(A);
END;
```

Fig. 6: Example of the integration and simplification [30].

Partial computation is useful because it automatically improves program efficiency. However, since α produces a projection by simple rules, there still remain redundant parts in the projection, such as car[cons[x;y]] for y with no side-effect. These redundancy should be avoided by program optimization, e.g. car[cons[x;y]] should be x. This operation is algebraic manipulation of a program that is similar to transformation of $\sin^2 x+\cos^2 x$ into 1 in symbol manipulation (A more complex example of program manipulation[23] is shown in Appendix 3).

Therefore, simplification is not a process unique to partial computation but a common one for program optimization[30]. This process is most important to the efficiency of a generated program (i.e. a projection). Program manipulation will become easier through advancement of Prolog-like languages.

Tabulation is a process used to associate program f and known data k to projection f_k. Essentially, it builds a table, like that shown below, and searches through it.

function	known data	projection
f	k1	f_{k1}
f	k2	f_{k2}
g	k3	g_{k3}
g	k4	g_{k4}
f	k5	f_{k5}
.	.	.
.	.	.
.	.	.

For example, when there is a semi-bound call f(k2,u+1), it is replaced by f_{k2}(u+1) and the table is searched for the entry of f, k2 and f_{k2}.

If the entry is found, partial computation of f(k2,u+1) is not performed again. If the entry is not found, f, k2 and the unfinished definition of f_{k2} are entered into the table and partial computation of f(k2,u+1) is performed. (A more detailed discussion of this is conducted in Chapter 4).

The tabulation technique is important not only in partial computation but also in ordinary computation. It has been studied for some time in the field of recursive programs[31,32]. Tabulation can be performed efficiently if a powerful hashing mechanism is implemented[33].

6. Conclusion

In the last decade, the author never believed that a practical projection machine could be implemented within 10 years. However, he is now more optimistic.

The achievements of Ershov, Emanuelson, Kahn and others establish the basis for the theory and practice of partial computation. Furthermore, the comming of a commercial LISP machine and inauguration of the 5th generation computer project in Japan will encourage research in this field.

7. References

Those papers not directly cited in this manuscript are also included.

[Theory and practice of partial computation]

[1] Babich,G.Kh., The DecAS algorithmic language for the solution of problems with incomplete information and its interpreting algorithms. Kibernetika, No., 1974, 61-71 (Russian).

[2] Babich,G.Kh. et al., The Incol algorithmic language for computations over incomplete information. Programmirovanie, No.4, 1976,24-32 (Russian).

[3] Beckman,L. et al., A partial evaluator and its use as a programming tool. Datalogilaboratoriet, Memo 74/34, Uppsala University, Uppsala, 1974 (also: Artificial Intelligence,vol.7, 1976, 319-357).

[4] Chang,C,-L. and Lee, R., Symbolic Logic and Mechanical Theorem -Proving, (Chapter 10.9) Academic Press, 1973.

[5] Dixon,J., The SPECIALIZER, a method of automatically writing computer programs. Div. of computer Res. and Technology, NIH: Bethesda, Md

[6] Emanuelson, P., Performance enhancement in a well-structured pattern matcher through partial evaluation. Linkoping Studies in Science and Technology Dissertations,No.55, Software Systems Research Center,Linkoping University, 1980.

[7] Ershov., A.P., Problems in many-language programming syatems. Language hierarchies and interfaces. LNCS, vol. 46, Springer-Verlag, Heidelberg, 1976,358-428.

[8]----------- , On the partial computation principle. Information Processing Letters, No. 2, 1977.

[9]----------- and Itkin, V.E., Correctness of mixed computation in Algol-like programs. LNCS,Springer-Verlag, Heidelberg, 1977.

[10]---------- and Grushetsky, V.V., An implementation-oriented method for describing algorithmic languages. Proc. of IFIP 77, 1977.

[11]-----------,A theoretical principle of system programming, Soviet Math. Dokl. Vol. 18 (1977), No. 2.

[12]-----------,On the essence of compilation. IFIP Working Conference on formal description of programming concepts, August 1977.

[13]-----------,Mixed computation in the class of recursive program schema. Acta Cyberneticca, Tom. 4, Fosc. 1, Szeged, 1978.

[14]------------,Mixed computation: Potential applications and problems for study, Theoretical Computer Science 18(1982) 41-67.

[15]------------,On Futamura projection. bit, Vol.12, No.14, 1980 (Japanese).

[16] Futamura, Y., Partial evaluation of computation process: an approach to a compiler-compiler. Systems Computers Controls, Vol. 2,

No. 5, 1971.
[17]------------,EL1 partial evaluator. Term paper manuscript, AM260, DEAP, Harvard University, 1972.
[18]------------,Compilation of Basic Transition Networks. Term paper manuscript, AM221, DEAP, Harvard University, 1972.
[19]------------,Partial computation of computer programs. AL78-80, Inst. Electronics Comm. Engrs. Japan, 1978 (Japanese).
[20] Haraldsson, A., A program manipulation system based on partial evaluation. Linkoping Studies in Science and Technology Dissertations, No.14, Department of Mathematics, Linkoping University, 1977.
[21]--------------, Experiences from a program manipulation systems. Informatic Laboratory, Linkoping University, 1980.
[22] Heuderson,P. and Morris,J.H., Jr., A lazy evaluator, Techn. Report No. 75, January 1976, Computing Laboratory, The University of Newcastle upon Tyne (also: 3rd ACM Symposium on principle of programming languages, January , 1976.
[23] Kahn, K., A partial evaluator of Lisp written in a Prolog written in Lisp intended to be applied to the Prolog and itself which in turn is intended to be given to itself together with the Prolog to produce a Prolog compiler. UPMAIL, Dept. of Computing Science, Uppsala University, March 1982.
[24]--------, The automatic translation of Prolog programs to Lisp via partial evaluation. UPMAIL, Dept. of computing Science , Uppsala University, P.O. Box 2059, Uppsala, Sweden.
[25] Komorowski, H.J., A specification of an abstract Prolog machine and its application to partial evaluation. Linkoping Studies in Science and Technology Dissertations, No. 69, Software Systems Research Center, Linkoping University, 1981.
[26] Lombardi,L.A. and Raphael, B., LISP as the language for an incremental computer. In E. Berkley and D. Bobrow (Eds), The programming language LISP: Its operation and application, MIT Press, Cambridge, 1964.
[27]-------------, Incremental computation. Advances in computers, Vol. 8, 1967.
[28] Ostrovsky, B.N., Obtaining language oriented parser systematically by means of mixed computation, in I.V. Pttosin, Ed., Translation and Program Models (Computing Center, Novosibirsk, 1980) 68-80 (in Russian).
[29] Turchin, V.F., Equivalent program transformation in REFAL. The automated system for construction control. Trans. of the CNIIPIASS institute, issue 6, M., 1974, 36-68 (Russian).

[Program optimization, recurtion removal and tabulation]

[30] Allen,F.E. et al., The experimental compiling system. IBM J ,RES DEVELOP. Vol. 24, No. 6, November 1980.
[31] Bird, R.S., Tabulation techniques for recursive programs. Computing Surveys, Vol. 12, No. 4, December 1980.
[32] Goto,E, Monocopy and associative algorithms in an extended Lisp. Report of the FLATS Project, Vol. 1, October 1978, Information Science Laboratory, The Institute of Physical and Chemical Research, Wako-Shi , Saitama 351, Japan.
[33] Keller, R.M. and Sleep, M.R., Applicative caching: Programmer control of object sharing and lifetime in distributed implementation of applicative languages. Proc. Functional Programming Language and Computer Architecture, 131-140, Dec. 1981.
[34] Rish, T., REMREC - A program for automatic recursion removal in LISP. Datalogilaboratoriet, Uppsala University, DLU 37/24, 1973.

[LISP, Prolog and PAD]

[35] Fuchi, K., Programming languages based on predicate logic. Inf. Processing journal of Japan, Vol. 22, No. 6, 588-591, June 1981 (Japanese).

[36] Futamura, Y. et al., Development of computer programs by PAD (Problem Analysis Diagram). Proc. of the Fifth International Conference on Software Engineering (New York: IEEE Computer Society, 1981), 325-332.

[37] Kahn, K., Unique Features of Lisp Machine Prolog. UPMAIL, Dept. of Computing Science, Uppsala University, March 1982.

[38] McCarthy, J. et al., LISP 1.5 Programmer's manual. M.I.T. Press, Cambridge, Massachusetts, 1962.

[39] Nakashima, H., Prolog/KR User's Manual for Version C-2. Wada Laboratory, Information Engineering Course, University of Tokyo, August 1981.

[40] Weinreb, D. and Moon, D., Lisp machine manual. MIT AI Laboratory, March 1981.

[Mathematical theory of computation]

[41] Kleene, S.C., Introduction to Meta-Mathematics. North-Holland Publishing Co., Amsterdam, 1952.

[42] Manna, Z., Mathematical Theory of Computation. McGraw-Hill, New York, 1974.

[43] Nakajima, R.,Introduction to Mathematical Information Science, Asakura-Shoten, Tokyo, 1982 (Japanese).

[44] Wegner, P., Programming, Languages, Information science and Computer Organization. McGraw-Hill, New York, 1968.

[45] Yasuhara, A., Recursive Function Theory and Logic. Academic Press, New York, 1971.

[5th generation computer project]

[46] Motooka, S., Overview of the 5th generation computer. Information Processing Journal of Japan, 426-432, Vol. 23, No. 5, May 1982 (Japanese).

Appendix 1: PAD (Problem Analysis Diagram) [36]

PAD is a new program schema for a substitute for both flow chart and recursive program schema. Correspondence between PAD, LISP and flow charts is described below.

	PAD	LISP	FLOW CHARTS
processing	E	E (LISP form)	E
selection	P E1 E2	[P->E1;T->E2]	P N Y E1 E2
definition	f(x)==E(f,x)	f[x]==E(f,x)	
sequencing	E1 E2	E1;E2 (in prog feature)	E1 E2
repetition	while P E		N P Y E
comment	-----[comment		-----[comment

Examples

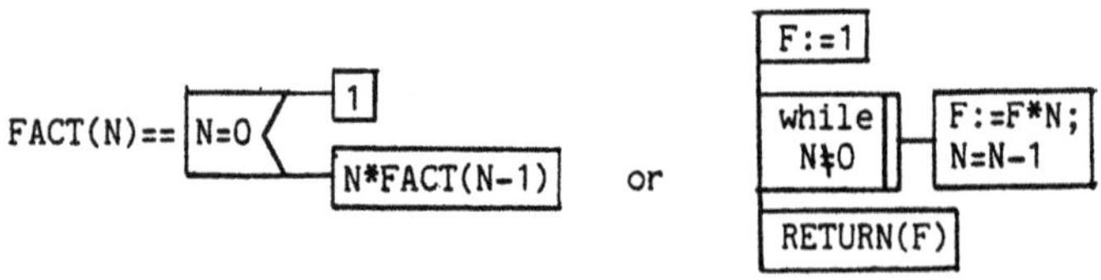
FACT(N)==
N=0
1
N*FACT(N-1)
or
F:=1
while N≠0
F:=F*N;
N=N-1
RETURN(F)

Appendix 2: Early Research in the Field (Cited from [3])

Partial evaluation has been used by several researchers and for a variety of purposes. We have identified the following applications (in an attemted chronological order):

Lombardi and Raphael used it as a modified LISP evaluation which attempted to do as much as it could with incomplete data[26].

Futamura studied such programs for the purpose of defining compilers from interpreters[16].

Dixon wrote such a program and used it to speed up resolution[5].

Sandewall reports the use of such a program for code generation in [S71].

Deutsch uses partial evaluation in his thesis[D73], but had written a partial evaluation program much earlier (personal communication, December 1969). His idea at that time was to use it as an alternative to the ordinary eval, in order to speed up some computations.

Boyer and Strother Moore use it in a specialized prover for theorems about LISP functions[B74]. Somewhat strangely, they call it an "eval" function.

Hardy uses it in a system for automating induction of LISP functions[H73].

Finally, the language ECL[W74] can evaluate, during compilation, a procedure call only containing "frozen" or constant arguments. Most of these authors seem to have arrived at the idea independently of each other, and at least number 2 through 5 at roughly the same time.

[B74] Boyer, R. and Moore, S., Proving theorems about LISP functions, Third International Joint Conference on Artificial Intelligence, Stanford Research Institute, 1974.

[D73] Deutch, P.,An interactive program verifier. Ph.D. thesis,

Xerox, Palo Alto Research Center, 1973.

[H73] Hardy,S., Automatic induction of LISP functions. Essex University, Dec. 1973.

[S71] Sandewall,E. A., Programming tool for management of predicate-calculus-oriented data bases. Proc. of Second International Joint Conference on Artificial Intelligence, British Computer Society, 1971.

[W74] Wegbreit, B., The treatment of data types in EL1. Communications for the Association of Computing Machinery, 5, 251-64, 1974.

Appendix 3: An example of program manipulation [23]

Example: Produce an efficient new program combining two programs.

Given programs:

l[x]: Compute the length of list x.

i[x;y]: Produce the intersection of lists x and y.

New program to be produced:

li[x;y]: An efficient program to compute the length of intersection of x and y.

Definitions:

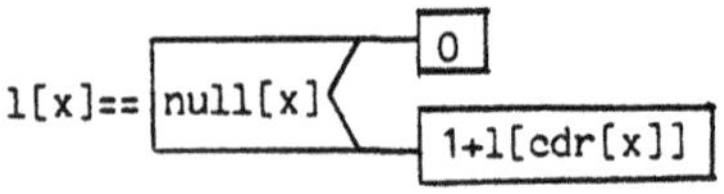

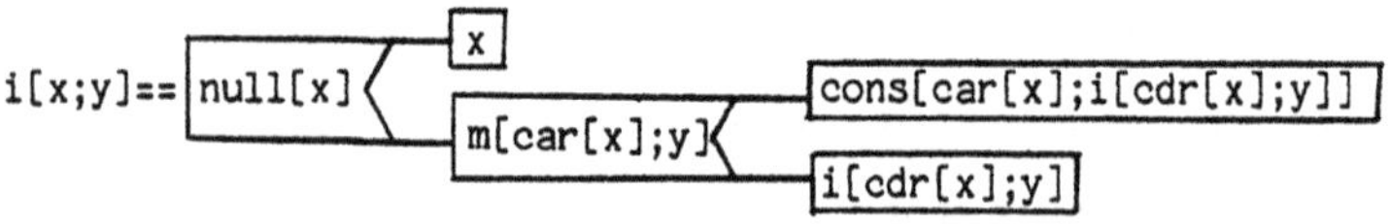

where m[x;y] is a predicate to give T if x is a member of y else it gives NIL, i.e. m[x;y]=member[x;y].

Algebraic manipulation:

li[x;y]==l[i[x;y]]

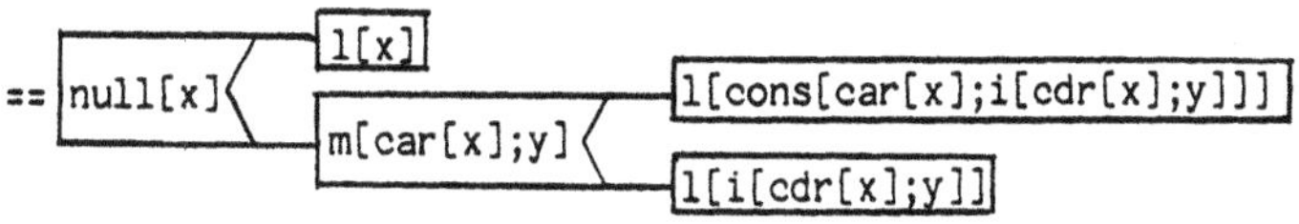

(From distribution over conditional expressions)

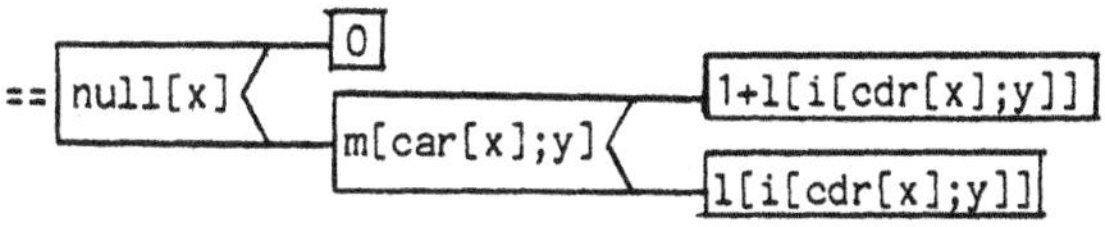

(From the definition of l)

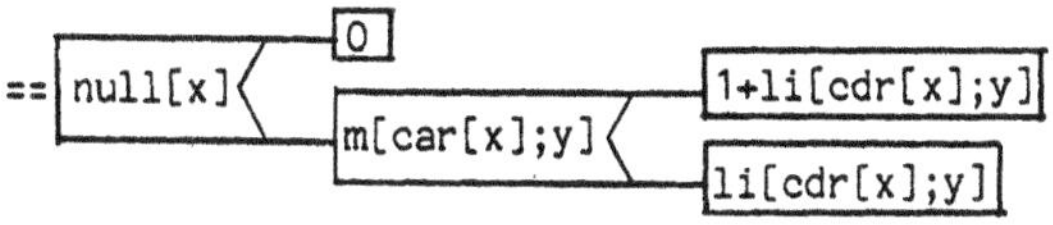

(From the definition of li)

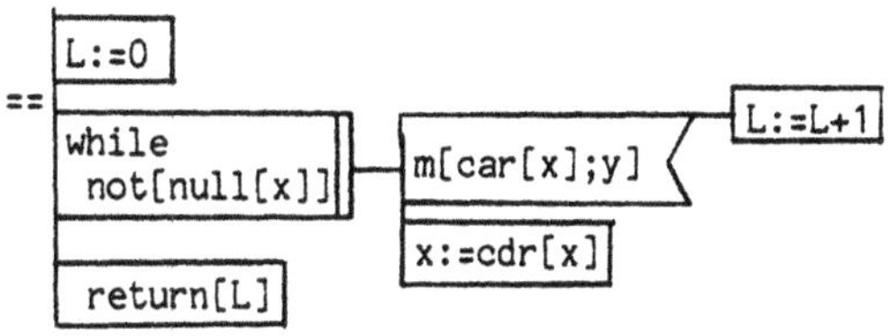

(From recursion removal)

Treatment of Big Values in an Applicative Language HFP

--- Translation from By-Value Access to By-Update Access ---

Takuya Katayama

Department of Computer Science
Tokyo Institute of Technology
2-12-1 Ookayama, Meguro-ku
Tokyo 152, Japan

1. Introduction

A method for treating big data in an applicative language HFP by converting the way of accessing data from by-value to by-update is stated. It is well-recognized that the applicative or functional style of programming is superior to the procedural one in its clarify, readability and verifiability. Its usage in the practical world of computing, however, is blocked by the lack of ways for efficiently executing programs in the style and it comes mainly from the fact that, in applicative or functional languages, every data is accessed through its value and there is no way of naming and updating it through the name. Although this problem may not be conspicuous when data is small in size, it is essential in dealing with big data such as files and databases.

Although there are cases where we can avoid computation involving big values by sharing structures or by delaying the operations on the values until they are really needed, i.e., by lazy evaluation technique [2], these techniques are not almighty and in many cases we have to do computations involving big values, such as updating files, large list structures or arrays.

This paper proposes a method of treating big data by converting by-value access to by-update access, which is used in the imlpementation of an applicative language HFP. HFP is an applicative language which admits hierarchical and applicative programming [3] and is based on the attribute grammar of Knuth [5]. It also has close relationship to Prolog. In the following, we first introduce HFP and discuss its implementation which solve the big data problem by using a simple file processing program.

2. HFP

HFP is an applicative language which supports hierarchical program design and verification. In HFP, a program is considered a module with inputs and outputs which are called attributes of the module. When a task to be performed by the module is complex, it is decomposed into submodules which perform corresponding subtasks and this fact is expressed by a set of equations which hold among attributes of the modules involved in the decomposition. Module decomposition proceeds until no more decompositions are possible.

In short, HFP is an extension of attribute grammar to write programs for general problems which are not necessarily language processing ones for which the attribute grammar has been originaly invented. In short, modules in HFP correspond to nonterminal symbols of attribute grammar and decompositions to production rules.

2.1 Formalism

HFP comprises (1) module, (2) module decomposition, and (3) equation.

(1) Module

Module is a black box with inputs and outputs termed attributes. The function of module is to transform its inputs to outputs. A module M with inputs $x_1, \ldots, x_n$ and outputs $y_1, \ldots, y_m$ is denoted diagramatically by

$$\boxed{M}\ \downarrow x_1, \ldots, \downarrow x_n, \uparrow y_1, \ldots, \uparrow y_m$$

and we write

$$IN[M] = \{x_1, \ldots, x_n\},\ OUT[M] = \{y_1, \ldots, y_m\}.$$

Module is completely specified by its input-output relationship and no side effect is permitted.

There are special modules: the initial module M_{init} and null module. The initial module corresponds to the main program and whose input-output relationship is what is to be realized by the HFP program under consideration. The null module, denoted by **null**, is used to terminate module decomposition.

(2) Module Decomposition

When a module M_0 has to perform a complex function and is decomposed into modules $M_1, M_2, \ldots, M_k$, we express this fact by writing module decomposition of the form

$$M_0 \Rightarrow M_1\ M_2 \ldots M_k$$

or

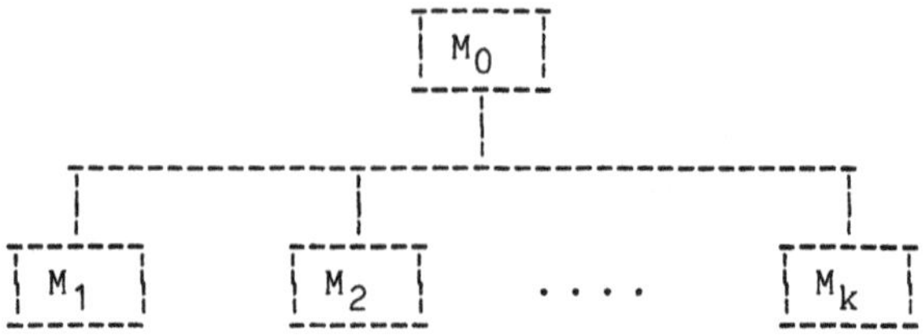

Decomposition process is terminated by applying a special decomposition with null module as its right side, which is denoted by

$M \Rightarrow$ **null** or simply | M |

It is usually the case that a module is decomposed in several ways. We attach the <u>decomposition condition</u> C to each decomposition d and write

$$d\colon M_0 \Rightarrow M_1\ M_2 \ldots M_k \textbf{ when } C.$$

The condition C is, in general, specified in terms of attributes, or more accurately, attribute occurrences of M_0, M_1, ..., M_k and it directs when this decomposition may apply. In this paper we only consider <u>determinitic</u> HFP in which (1) this condition C is specified in terms of input attributes of M_0 which are not dependent on other attributes and (2) there are no two decompositions with the same left side module whose decomposition conditions may become simultaneously true. When these conditions are not met, the HFP is called <u>nondeterministic</u>. In nonditerministic HFP, validity of applying a decomposition may be determined after it has been applied and several decompositions may be applicable simultaneously, so backtracking and nondeterministic branching can be very naturally expressed.

(3) Equations

We associate a set of <u>equations</u> to each decomposition d for specifying how the task performed by the main module M_0 is decomposed into subtasks performed by submodules M_1, ..., M_k. This is done by writing equations for (1) what data should be sent to submodules as their inputs and (2) how to combine their computation results to obtain outputs of the main module. Equations are of the form

$$v = F_v\ (v_1, \ldots, v_t)$$

where (1) $v, v_1, \ldots, v_t$ are attribute occurrences in d. We use the notation $M_i.a$ to denote an occurrence of attribute a of M_i in the decomposition d, (2) $v = M_i.a$ for a in $IN[M_i]$ $(i=1,\ldots,k)$ or $v = M_0.a$ for a in $OUT[M_0]$, (3) F_v is an attribute function for computing v from other attribute occurrences $v_1, \ldots, v_t$.

[Computation in HFP]

Given the values of input attributes of the initial module, computation in HFP is carried out in the following steps.

(1) Starting from the initial module, apply decompositions successively until terminated by null module. A tree which is virtually constructed in the above process and whose nodes are module occurrences of the decompositions applied is called a computation tree.

(2) Evaluate the values of attributes of the modules in the computation tree according to the equations associated with decompositions applied.

(3) The values of outputs of the initial module give the results of the computation activated by its input values.

2.2 Example : 91HFP

Let us consider a simple example. Although this example is too simple to verify the power of HFP it will be some aid in understanding it.

Let F be the McCarthy's 91 function. This function is defined by

$$F(n) = \textbf{if } n \leq 100 \textbf{ then } F(F(n+11)) \textbf{ else } n-10.$$

We introduce the module 'F' and associate input n and output v with it. The module F is decomposed into two ways as shown below with equations about its attributes n and v. (Note that '=' in the **with** clauses means 'equal to' and not a symbol for assignment.)

decomposition 1.

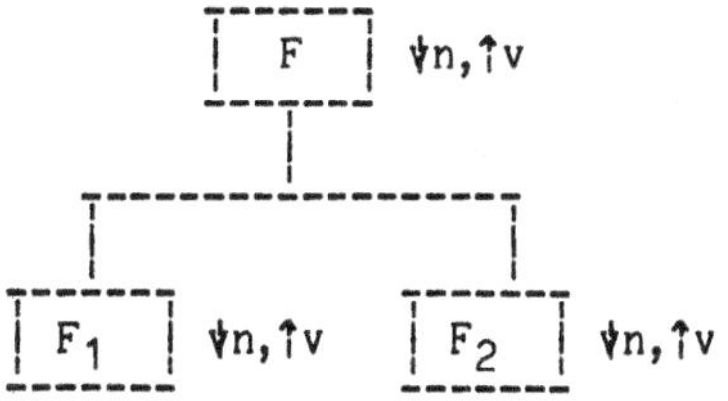

with $F_1.n = F.n+11$

$F_2.n = F_1.v$

$F.v = F_2.v$

when $F.n \leq 100$

decomposition 2.

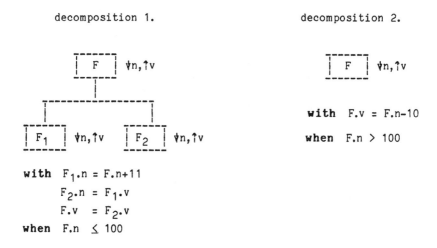

with $F.v = F.n-10$

when $F.n > 100$

In decomposition 1, module name F is indexed to distinguish different occurrences and no submodule exists in decomposition 2 to terminate decomposition. Attributes prefixed by '↓' and '↑' are inputs and outputs respectively. The **when** clause specifies the condition on which the decomposition may apply.

The tree below shows a computation tree for input n=99.

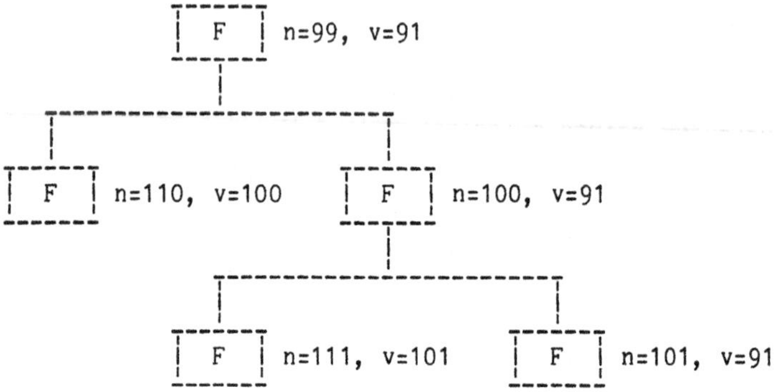

2.3 Comparison to Backus'FP and Prolog

[Backus'FP] Functionality (in the broad sense) of HFP consists in the facts: (1) relationship between attribute occurrences in every decomposition is specified by a set of equations which state that they are computed functionally from other attribute occurrences and (2) every module can be viewed as defining a set of functions which realize input-output relationship of the module. Difference between the Backus' functional programming (BFP) and ours is that BFP defines functions explicitly from primitive ones in terms of functional calculus, whereas HFP focuses on modules and their decompositions and the form of the functions realized by the module is scattered in the set of equations.

[Prolog] Nondeterministic HFP is very closely related to Prolog in which every variable is specialized for either input or output.

Let d be a decomposition in HFP.

$$d : M \Rightarrow M_1\ M_2 \dots M_k \textbf{ with } EQ \textbf{ when } C.$$

EQ is a set of equations associated with d and C a decomposition condition. This decompositions corresponds to the following (Horn) clause in Prolog.

$$\underline{M} \Leftarrow \underline{M}_1,\ \underline{M}_2,\ \dots,\ \underline{M}_k,\ [EQ],\ C$$

where $\underline{M}$ is the predicate for the module M and if M is represented, in the description of d, with names $x_1, \ldots, x_m$ for attribute occurrences, then $\underline{M} = M(x_1, \ldots, x_m)$. [EQ] is conjuncion of equations in EQ. For example, the clausal form which corresponds to the decomposition 1 of the 91HFP is given by

```
F(n,v) <= F(n1,v1), F(n2,v2),
          eq(n1,m), eq(n2,v1), eq(v,v2),
          plus(n,11,m), lesseq(n,1oo)
```

As we have just mentioned, every variable(attribute) in HFP is specialized for input or output, whereas it is not in Prolog. We, however, consider that this fact does not degrade HFP, rather distinction between input and output is essential in many cases and helps to write comprehensive programs. Furthermore, it enables to utilize data dependency analysis for implementing HFP far more efficiently than otherwise and to verify HFP program by the method similar to that for attribute grammar [3].

Another point HFP differs from Prolog is that backtracking in (nondeterministic) HFP is controlled by the explicitly declared decomposition condition and thus localized in a sense, on the other hand, every predicate has chance to cause costly backtracking in Prolog. In this sense, backtracking in HFP is considered more manageable than in Prolog. Also we want to emphasize that we can write any form of predicate as the decomposition condition and we do not have to introduce such sophisticated mechanism as cut symbol to represent 'not' as in Prolog.

3. Basic Implementation Technique for HFP

The basic idea for implementing HFP is to associate procedures $P_{M,y}$ to each module M and its output attribute y, and translate the given HFP into a procedural program which consists of these procedures. Here we only skech the technique and for the detail please refer to [3]. We only consider absolutely noncircular HFPs.

Let d_i's (i=1,2,...) be decompositions with left side module M and decomposition condition C_i. Then the procedure $P_{M,y}$ is of the form

procedure $P_{M,y}(x_1,\ldots,x_n;y)$
  if C_1 **then** H_1 **else**
  if C_2 **then** H_2 **else**
  .
  .
end

where $x_1,\ldots, x_n$ are input attributes of M on which y is dependent. $x_1,\ldots,x_n$ are decided by analyzing data dependency DG[M] in the attributes of M. H_i is sequences of assignment or procedure call statements to calculate the value of y from those of x_i's and their forms are determined from data dependency DG[d_i] among attribute occurrences of d_i and its associated equations.

The principle for constructing H_i is stated in the following way. First we prepare variables for attribute occurrences of d_i. H_i is a sequence of statements to assign values to these variables when the decomposition condition C_i holds. If a variable corresponds to the output attribute y of the main module M or to an input attribute of a submodule M' , the value assigned is computed from the defining equation for it. If it corresponds to an output attribute w of a submodule M', the value is obtained by calling a procedure $P_{M',w}$ which is associated with M' and w. These statements are listed in such order as determined by topologically sorting the dependency relation DG[d_i] among attribute occurrences of d_i so that the value of every attribute occurtence v is determined after valuation of attribute occurrences on which v is dependent has been completed.

The next program is obtained from the 91 HFP.

```
program   PROG91
     procedure F(n;v)
          if n > 100 then v := n-10
                      else n1 := n+11 ; call F(n1,v1) ; n2 := v1 ;
                             call F(n2,v2) ; v := v2
     end
     input(n) ; call F(n,v) ; output(v)
end
```

This program can be made simpler by folding assignment statements.

4. Implementing Access to Big Values in HFP

4.1 An Example : A Simple File Processing HFP Program

Let us consider the following simple file processing HFP program which, given an input file of records R_1, $R_2,\ldots$, R_k arranged in this order, produces an output file of records $F(R_1)$, $F(R_2),\ldots$, $F(R_k)$, where F is a function operated on input records.

attribute infile, outfile, outfile0 : **file of** record
inrec, outrec : record

module main **has** ↓infile ↑outfile
process-file **has** ↓infile ↓outfile0 ↑outfile
process-rec **has** ↓inrec ↑outrec

module decomposition

(1)

```
         ------
         |main|      ↓infile ↑outfile
         ------
           |
   --------------
   |process-file|  ↓infile ↓outfile0 ↑outfile
   --------------
```

with outfile0 = emptyfile

(2)

```
                 --------------
                 |process-file| ↓infile ↓outfile0 ↑outfile
                 --------------
                        |
      --------------------------------
      |                              |
   -------------               --------------
   |process-rec|               |process-file|
   -------------               --------------

   ↓inrec                        ↓infile$
   ↑outrec                       ↓outfile0$
                                 ↑outfile
```

with inrec = 1st(infile)
infile$ = rest(infile)
outfile0$ = appendr(outfile0, outrec)
when not eof(infile)

(3)

```
   --------------
   |process-file| ↓infile ↓outfile0 ↑outfile
   --------------
```

with outfile = outfile0
when eof(infile)

(4)

```
   -------------
   |process-rec| ↓inrec ↑outrec
   -------------
```

with outrec = F(inrec)
when always

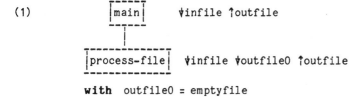

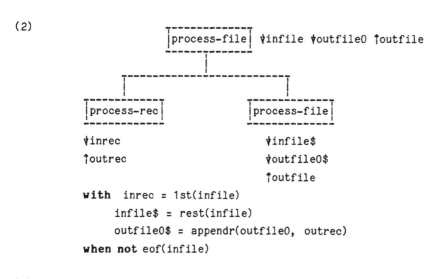

[1] The **attribute** clause defines data types of attributes. **file** is a data type with operations

```
emptyfile :                  => file
1st       : file             => record
rest      : file             => file
appendr   : file x record => file
eof       : file             => boolean
```

The meaning of these file operations are:

```
1st(f)         : first record of a file f,
rest(f)        : file obtained from f by deleting its first record,
appendr(f,r) : file resulted by writing the record r at the end of f,
eof(f)         : true iff f is emptyfile.
```

Details of the data type 'record' is irrelevant here.

[2] **module** clause declares module with their attributes. Input attributes are prefexed by ↓ and output attributes by ↑.

[3] **with** clause specifies equations associated with a decomposition and **when** clause defines its decomposition condition.

[4] Note that notation for attribute occurrence is different from what is given in section 2. In this HFP program, for the sake of clarity, attribute occurrence is not prefixed by module name and instead positional notation is used. Also note that when the same attribute name appears more that once in a decomposition this means that equations for copying the attribute values are omitted.

[Basic Step]

Suppose we apply the basic implementation technique to this HFP, we get the following procedural type program

```
program main(infile; outfile)
   procedure process-file(infile, outfile0; outfile)
       if not eof(infile) then
          inrec := 1st(infile)
          infile$ := rest(infile)
          call process-rec(inrec; outrec)
          outfile0$ := appendr(outfile0, outrec)
          call process-file(infile$, outfile0$; outfile)
       else
          outfile := outfile0
   end
```

```
    procedure process-rec(inrec; outrec)
        outrec := F(inrec)
    end
    outfile0 := emptyfile
    call process-file(infile, outfile0; outfile)
end
```

This program is, of course, unacceptable because big values of files are copied to variables and passed to and from procedures.

[Attribute Globalizaton]

Let's consider a computation tree for this HFP program, which illustrates how the input file 'infile' of the main module is transformed into the output file 'outfile'.

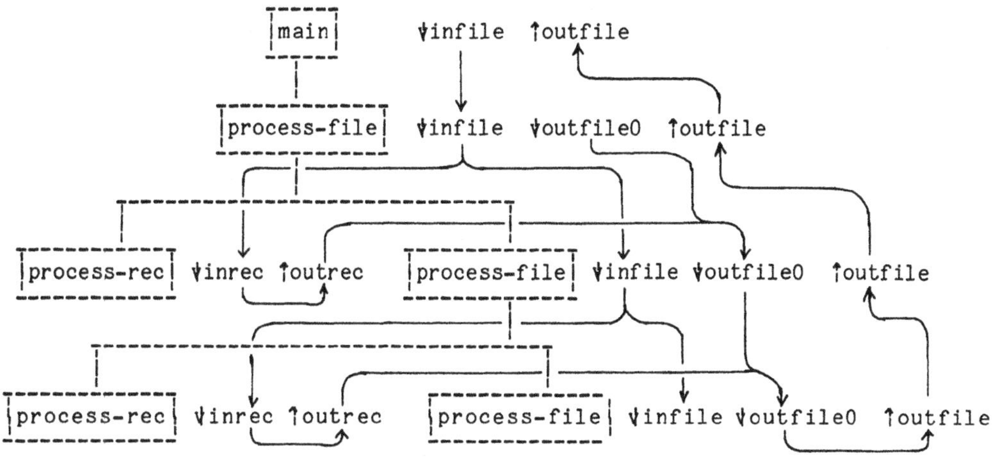

From this tree, we can see the following (1) and (2) hold.

(1) To all occurrences of the attribute 'infile' can be assigned a single global variable instead of a stack which is realized by activation records of the procedure 'process-file' in the above program. This is because after 'infile' is referred at a level to compute values of 'inrec' and 'infile' for the next level it is not referred any more.

(2) So are 'outfile' and 'outfile0'. Furthermore, we can assign the identical global variable to these attributes just by the similar reason.

Taking these facts into consideration, the generated program for our file processing HFP becomes as follows. The procedure 'process-file' has no parameters in this version.

```
program main(infile, outfile)
        procedure process-file
             if not eof(infile) then
                inrec := 1st(infile)
                infile := rest(infile)
                call process-rec(inrec; outrec)
                outfile := appendr(outfile, outrec)
                call process-file
        end
        procedure process-rec(inrec; outrec)
             outrec := F(inrec)
        end
        outfile := emptyfile
        call process-file
    end
```

Note that we omitted a useless assignment 'outfile := outfile' and changed the **if-then-else** statement to **if-then** statement.

[From By-Value Access To By-Update Access]

Now, what should be done next is to translate 'by-value' file access to 'by-update' access. This is accomplished by the help of translation axioms given below.

Axiom 1

```
inrec := 1st(infile)
infile := rest(infile)
            ===> read(infile, inrec)
```

Axiom 2

```
outfile := appendr(outfile, outrec)
            ===> write(outfile, outrec)
```

Axiom 3

```
outfile := emptyfile
            ===> rewrite(outfile)
```

The read, write, and rewrite statements are such that (1) read(f,r) transfers the record of f at the cursor position to r and move the cursor right, (2) write(f,r) writes the data in r at the cursor position and move it right and (3) rewrite(f) initializes f.

After applying these axioms to the generated program, i.e., replacing texts which matches the right side of any axiom by its left side, we have the following program in which files are accessed through the usual updating operations 'read' and 'write'. We also changed tail recursion to iteration.

```
program main(infile, outfile)
     rewrite(outfile)
     while not eof(infile) do
        read(infile, inrec)
        outrec := F(inrec)
        write(outfile, outrec)
end
```

4.2 General Scheme for Translation

Translating a given HFP program into a procedural type program in which big values are accessed through updating is performed in the following steps. Input to the translator is the given HFP program and a database of tranlation axioms and other optimization repertiores.

[1] Calculate data dependency relations DG[d] and DG[M] among (1) attribute occurrences of each decomposition d and (2) attributes of each module M respectively.

[2] Determine groups of attributes to which single global variables can be assigned. These groups can be determined from dependency relation DG[d] obtained in step 1. A simple and useful criterion that a group of attributes satisfies the above condition is that every dependency relation DG[d] does not contain 'branching' with respect to the attributes. More specifically, this is stated in the following way. Let DG[d]=(V,E), where V is a set of attribute occurrences of d and

$$E=\{(v1,v2) \mid \text{there is an attribute function } f_{v2} \text{ such that } v2=f_{v2}(\dots,v1,\dots)\}.$$

Let $A=\{a_1,\dots,a_k\}$ be a set of attributes which we want to globalize and V(A,d) be a set of occurrences of attributes in A with respect to the decomposition d. If there is no triple (v1,v2,v3) such that both of (v1,v2) and (v1,v3) are in E and v1, v2 and v3 are in V(A,d), then this criterion assures that these attributes $a_1,\dots,a_k$ can be globalized. Assign a global variable to each such group of attributes. Distinct local variables are assigned to other non-global attribute occurrences.

[3] For each module M and its output attribute y, construct a procedure $P_{M,y}$ by the method sketched in section 3.

[4] Inspect the body of $P_{M,y}$ and find a group of statements for big data access. If it matches to the left side of some conversion axiom, replace the group by its right side which is a sequence of statements for by-update data access.

[5] Apply, if possible, other optimization techniques such as folding of

assigment statements and recursion elimination.

5. Concluding Remarks.

We have proposed a method of treating big values which is used in an applicative language HFP. The underlying principle is to convert by-value data access to by-update data cccess with the aid of conversion axiom database. An examlpe is given to support this method. This principle may be applied to other applicative systems.

References

1. Backus, J. Can programming Be Liberated from the von Neumamm Style ? A Functional Style and Its Algebra. Comm. ACM, 21,8(Aug.1978), 613-641

2. Henderson,P. & Morris,J.H. A Lazy Evaluator. Proc. of 3rd ACM Smp. on Principle of Programming Languages (1976)

3. Katayama, T. HFP: A Hierarchical and Functional Programming Based on Attribue Grammar,Proc. 5th Int. Conf. On Software Engineering (1981)

4. Katayama,T., and Hoshino, Y. Verification of Attribute Grammars. Proc. 8th ACM Symposium on Principles of Programming Languages (1981)

5. Kunth, D.E. Semantics of context-free languages Math. Syst. Theoty J.2(1968),127-145

TOWARD THE DESIGN AND IMPLEMENTATION OF OBJECT ORIENTED ARCHITECTURE

Mario TOKORO
Department of E. E., Keio University
Yokohama 223 JAPAN

ABSTRACT

Object oriented programming languages and the computer architecture to support the reliable and efficient execution of programs written in these languages are important issues for providing better programming environment. The main purpose of this paper is to establish the foundation for the design and implementation of object oriented programming languages and object oriented architecture. First, various definitions for object in existing languages and systems are surveyed. Then a new model of object and computation on objects is introduced to establish the foundation. A new object oriented language is used both to rationalize the model and to exemplify how the model is applied to object oriented languages. Finally, issues in the design and implementation of an object oriented architecture which directly reflects this model is described.

1. INTRODUCTION

Recent requirements for high level programming languages can be summarized as follows:

(1) Easy to read/write programs,

(2) Flexible and yet efficient programming,

(3) Reliable execution, and

(4) Small software life-time cost.

Modular programming plays the most important role in satisfying these requirements, and object orientation, which is the ultimate form of data abstraction, is the paramount notion for achieving modular programming.

Briefly, object oriented languages/systems are languages/systems in which all the objectives of operations are called objects, and the operations permitted to operate on an object are described in or together with that object. Hydra [Wulf 75, Wulf 81], CLU [Liskov 79], and Smalltalk [Goldberg 76, Xerox 81] are systems/languages which adopted the notion of object orientation. Actor [Hewitt 73, Hewitt 77] is a computational model which is closely related to this notion. Although the notion of object orientation has become popular in recent years, detailed definitions for _object_ and related terms differ much from one system to another. Thus, it would be worthwhile to survey difinitions of _object_ and related terms in order to comprehend the reasons why such definitions are employed in their respective systems.

In order to establish the foundation for the design and implementation of object oriented programming languages and object oriented architecture, we propose a model of _object_ and computation on objects. The model is based on the preliminary discussion in [Tokoro 82]. The definitions of _object_ and related terms in this model are also described. A new object oriented programming language is used both to rationalize the model and to exemplify how the model is applied to object oriented programming languages. This object oriented language features the decomposition of types into _property_ and _attribute_, the notion of _multiple_ _representation_ in a class, and the notion of the _link_ of one name to another name to give a different _view_ of objects.

As the level of abstraction in programming languages becomes higher, architectural support for more efficient and reliable execution of programs becomes indispensable. A few object oriented machines have been proposed [Snyder 79] or implemented [Giloi 78] [Rattner 80]. An architecture which adopts our object oriented model and executes programs written in the new object oriented language efficiently and reliably is outlined. Issues in the design and implementation of the object oriented architecture, including program structure, cache for object memory, the context switching/parameter passing mechanism, variable length operations, and garbage collection, are also described. Rationale for the new object oriented language and the architecture are discussed.

2. A SURVEY OF DEFINITIONS OF OBJECT

In this section, we survey various definitions of _object_ and related terms in some existing systems.

2.1. Object in Hydra

Hydra [Wulf 81] is the kernel of an operating system. Object in Hydra is the analog of a variable in programming languages. An object is the abstraction of a typed cell. It has a value or state. More precisely, an object is defined as a 3-tuple:

(unique name, type, representation)

A unique name is a name that differs from all other objects. A type defines the nature of the resource represented by the object in terms of the operations provided for the object. The representation of an object contains its actual information content.

In Hydra, capability is the other important element for its conceptual framework. Capability is the analog of a pointer in programming languages; the main differences are 1) that a capability contains a list of permitted access rights in addition to pointing to an object and 2) that a capability can be manipulated only by the kernel.

Representation may contain capabilities for other objects. That is to say, representation consists of a data part which contains simple variables and a C-list part which is a list of capabilities. Objects are named by path routed from the current Local Name Space (LNS) of a program. Fig. 1 shows objects in Hydra.

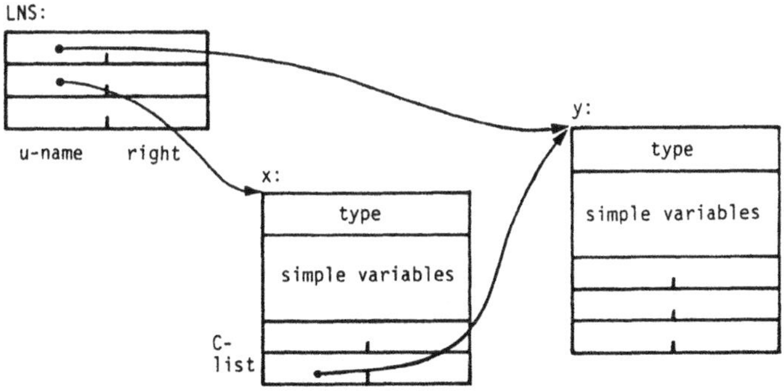

Fig. 1 Objects in Hydra

Objects in Hydra can be envisaged as the extension of the notion of storage resources in two directions: one is to incorporate type, i.e., the set of operations defined over the object, and the other is to include capabilities in an object. There is no notion of assignment at the object level. Assignement is performed by copying a value (but not an object) into a simple variable in the representation of an object. If we would like to incorporate *assignment* at the object level, we could ask the kernel to store in a slot of the current LNS the capability that points to an object. The slot is associated with the variable name to which the assignment has taken place. There is no type associated with a slot. Thus, we would do nothing with type checking for *assignment* at the object level. Objects in iAPX/iMAX 432 [Rattner 80, Kahn 81] follow the notion of objects in Hydra.

2.2. Objects in CLU

There are two basic elements in the CLU semantics [Liskov 79], *object* and *variable*. Objects are data entities that are created and manipulated by a program. Variables are just the names used in a program to denote objects.

Each object consists of a type and a value. A type is defined by a set of operations which create and manipulate objects of this type. An object may be created and manipulated only via the operations of its type. The operations of a type are defined by a module called *cluster* which describes the template (or skeleton) of the internal representation of objects created by that type and the procedures of the operations of that type. There is no *subtype* or *derived type*. *Invocation*, which is the procedure call in CLU, is specified as:

```
type $ operation ( parameters )
```

There are two categories of objects: *mutable object* and *immutable object*. A mutable object may change its *state* by certain operations without changing the identity of the object. Arrays and records are examples of mutable objects. If a mutable object m is *shared* by objects x and y, then a modification to m made via x will be visible from y. There are also copy operations for mutable objects. On the other hand, immutable objects do not exhibit time-varying behavior. Examples of immutable objects are integers, booleans, characters, and strings. There are immutable arrays, called *sequences*, and immutable records called *structure*. Since immutable objects do not change with time, there is no notion of *share* or *copy* for immutable objects.

A variable can have type *any* to denote an object of any type. In *assignment*, the object which results from the execution of a right-hand side expression must have the same type as the variable to be assigned. There are no implicit type conversions. There is an operation, force, which checks the type of the object denoted by the variable. CLU also has tagged discriminated union types oneof for an immutable labeled object and variant for a mutable labeled object. The tagcase statement is provided for decomposing oneof and variant objects.

Assignment y := z causes y to denote the object denoted by z (Fig. 2). The object is not copied; after the assignment is performed, the object will be shared by x and y. Assignment does not affect the state of any object.

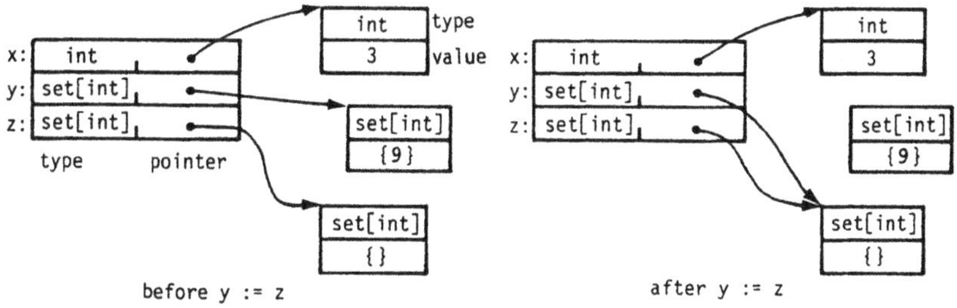

Fig. 2 Objects in CLU

2.3. Actor Model

In actor model [Hewitt 73, Hewitt 78, Yonezawa 81], actor is the unified entity of procedures, storage resources, and data. An actor is activated when it receives a message. A message conveys the requests of operations with or without data, or replies to requests or results of operations. Replies and results can be passed to an actor other than the requesting actor.

There are two kinds of actors: pure actors and impure actors. A pure actor is immutable while impure actor is mutable. The simplest impure actor is called cell and accepts two kinds of messages: one is "contents:" to reference its content and the other is "update: <value>" to update the content. Thus, variables in programming languages are implemented by using cell actors. There is no notion of class or type for an actor.

Fig. 3 shows an actor named 3 which accepts a message "+ 4" and returns "7" and a cell actor named x which first accepts "update: 4", then accepts "content:", and returns "4".

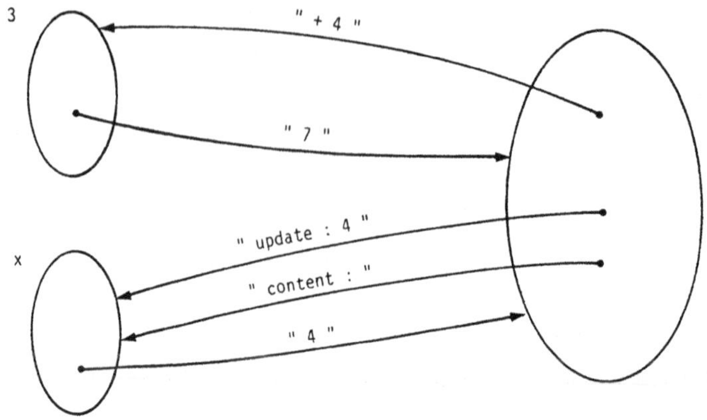

Fig. 3 Actors

2.4. Smalltalk

In Smalltalk [Goldberg 76, Xerox 81], there are two basic elements: object and variable. An object is a package of information and a description of its manipulation. A variable is a name in a program which refers to objects. An operation on an object is designated in terms of a message to the object. A message contains selectors (i.e. the operation names) and parameters. Fig. 4 shows variables and objects before and after an operation.

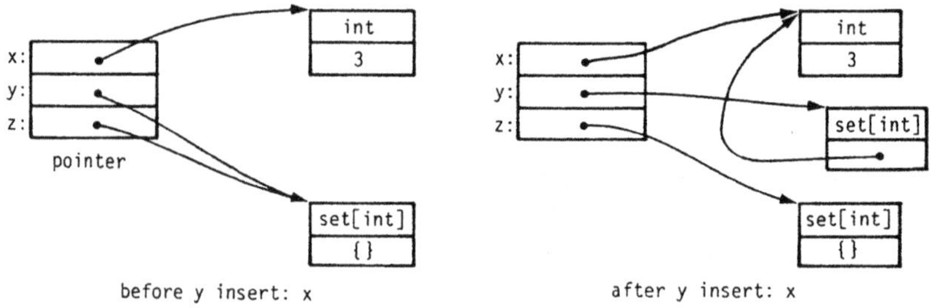

Fig. 4 Objects in Smalltalk

Class is a module which defines the operations to create an object and the operations to operate on objects created by the class. An object which is created by a class is called its instance. A class itself is an object. A class may contain its own variables (i.e., the state of the class). That is to say, a class is more than a template.

A class inherits methods (i.e. operations) directly from one (and only one) other class. In such a case, the class from which methods are inherited is called its _super class_. A super class itself may be subsumed under another super class again. There is a special class called OBJECT, which is the ultimate super class of any classes.

Smalltalk's objects are independent of the structure of storage resources. In paricular, there is no explicit declaration for the representation of the internal structure of an object.

In Smalltalk, there is no class (type) declaration for variables. Therefore, it is not easy for a compiler to determine classes and methods at compile-time. This impedes fast execution. It may possibly lead to less reliable execution, since no (static nor dynamic) type checking is performed for an assignment. There is research work being done on type inference at compile-time done without changing the language construct [Suzuki 80] and on the incorporation of explicit type declarations into variable declarations in order to increase reliability and efficiency of execution [Borning 81].

2.5. Discussion

Programming languages should be independent of the restrictions imposed by the existing machine architecture, especially from the structure and management schemes of storage resources. After we establish a firm foundation as to what high level programming languages should be, performance issues for the execution of programs should be discussed along with the design and evaluation of new computer architecture.

We think the notion of objects employed in Smalltalk seems to be very appropriate in the sense that it is independent of storage resources. The notion of class and the inheritance mechanism of Smalltalk also seems to be natural. We also think that the notion of variables in CLU is powerful in the sense that a variable specifies the type of objects which can be denoted by the variable.

In CLU, a _mutable object_ is effectively used when it is shared by objects which communicate through it. Communication, however, can also be achieved by sharing a _variable_, if such a mechanism is provided. Then, we need not distinguish mutable objects from immutable objects. A mutable object in CLU also contributes to reducing the number of memory claims and reclamation. This will, however, not always be true in our programming language that is proposed later, since objects of a class can vary in size. An efficient memory claim/reclamation scheme should be provided by object oriented memory architecture.

In contrast with CLU, the actor model employs a cell object which functions as a variable. A cell object can receive messages from two or more objects. This is equivalent to sharing in CLU.

3. THE MODEL OF OBJECT AND COMPUTATION

In this section we present the model of object and computation with definitions. Example programs are shown to explain and rationalize the model as well as to exemplify how this model is applied to object oriented programming languages.

3.1. Object and Name

Let us define name as an identifier which denotes objects. Assignment is defined as the association of an object with a name. A variable is a name in a program with which an object can be associated.

We have decided to construct our model with object and name. This decision has been made through the following reasoning:

(1) Let us define information as the accessible form of objects. The semantic structure of information is decomposed into two disjoint subfunctions: the access/visibility control and the entity which is the subject of operations. Name takes charge of the access/visibility control and object takes charge of the entity to be operated on.

(2) History sensitiveness can be attained only by a name. That is to say, a state change occurs only when an assignment is executed, and is explicit. Thus, there are no mutable objects. Every object is immutable.

(3) Sharing of an object is specified only through a name. There is no implicit sharing of an object. For this purpose, we introduce link which is a logical pointer from one name to another. A name which is linked by more than one name is shared by these names. Thus, sharing is controlled by name which performs the access/visibility control function. Therefore, an object can

remain an entity to be operated on.

3.2. *Reconsideration of Type*

In most of the strongly-typed languages such as CLU and ADA [ADA 80], *class* or *type* is specified for *variable*, although *type* is defined as a set of objects on which a set of operations is defined. To specify a *class* for a variable contributes giving information to compilers for determining the class of an operation in advance of its execution. However, when *subtype* (and/or *derived type*) was incorporated in strongly-typed languages, it became difficult to infer types of operations from the types of variables. Of course we could infer types if we defined all the operations for all the possible combinations of the domains and types of operands, and the ranges and types of the results. However, this would be tedious when the number of combinations increased. Thus, we have decided not to incorporate *subtype* in our language or architecture.

Remembering that the purpose of specifying a *class* to a variable is to check whether the assignment of an object to a variable is valid, we introduce the notions of *property* and *attribute* [Feldman 79, Williams 79]. *Property* is attached to a *name* to specify the set of objects which can be denoted by the *name*. Thus, *property* is the assertion to be validated for a name, which is described in the form of a proposition. *Property* is validated at compile-time if it is possible, or it is translated into object codes to verify when an association of an object with a name occurs at execution-time. *Attribute* is attached to an *object* to specify the *type* or *class* to which the object belongs. Thus, the class of an operation is determined by the operands' properties at compile-time if possible, and if not, by the operands' attributes at run-time.

Properties of names are declared as shown in the following examples.

```
i:  property {i: attributeof( i ) = integer};
j:  property {j: attributeof( j ) = integer & 0 < j < 100 };
k:  property {k: attributeof( k ) = integer & k mod 2 = 0};
m:  property {m: attributeof( m ) = integer & m > k};
x:  property {x: attributeof( x ) = real & x > 3.5 |
                 attributeof( x ) = integer  & x <= 0 };
```

In an assertion proposition, it is permissible to refer to other names and to invoke operations if necessary. *Attributeof*(x) is the operation of the most primitive class CLASS which will return the attribute of x.

In practice, it might be more amenable to programmers to declare *property* as:

<name>: <attribute> [*in* <representation>]
[*property* <assertion proposition>];

where <assertion proposition> describes a more detailed property which cannot be expressed in *attribute* or *representation*. A clause surrounded by "[" and "]" may be omitted. *Representation* is discussed below. If the same proposition is used for many names, it would be beneficial to use a compile-time facility to define property texts.

3.3. Reconsideration of Representation

In most abstract data type languages, an operation is strongly related to the representation of values. One representation is associated with and used in a class. An operation, however, should simply be the mapping of values to values and be independent of their representation. That is to say, we would like to unify semantically equivalent classes into one class, regardless of differences in representation.

In order to express a value, we need a bit-string which is just long enough to express that value. Integer numbers of any size should be operated on in a unified manner and the result should be generated in appropriate sizes. Binary integers and decimal integers should automatically be adjusted in the operations. Floating numbers of any size and format should also be operated on in a unified manner. The orthogonal and polar representations for a complex number should be used interchangeably. The list representation, dope vector representation, and linear representation for an array should also be used interchangeably. In most cases, representation should not be seen by programmers. For primitive representation such as length or radix, computer architecture should provide the functions for bit-strings interpretation/generation/conversion in accordance with the description of the representation of an object. For more complex representation, programming languages should provide a method of describing the interpretation/generation/conversion of representations.

Thus, we introduce _multiple representation_ in a class. That is to say, a class can have multiple representations with their operations' definitions, and the transformation rules between representation if possible. An example of a multiple-representation class is shown as follows:

```
class complex is
  create_xy,create_polar,add,sub,mul,div,
  x_cordinate,y_cordinate,abs,angle;
rep structure {r, i: real} as orthogonal;
rep structure {r, theta: real} as polar;

procedure create_xy (x, y: real) returns( orthogonal cvt)
  return rep${ r: x, i: y } as orthogonal
end create_xy
      .
      .
      .
procedure add (x: cvt, y: cvt) returns (cvt)
  repcase x of
    orthogonal: repcase y of
                    orthogonal: return complex$create_xy (x.r
                        + y.r, x.i + y.i);
                    polar: return complex$create_xy( x.r +
                         y.r * cos(theta),
                         x.i + y.r * sin(theta) );
                end repcase
    polar:      repcase y of
                    orthogonal: ..... ;
                    polar: ..... ;
                end repcase
  end repcase
end add
      .
      .
      .
end complex
```

Note that we do not need to have different classes for orthogonal and polar complex numbers, although the same effect could be expressed by _tagcase_ for the _oneof_ type in CLU. _Representation_ can neatly describe units, such as the class for length in meters, centimeters, inchs, miles, and so forth. Representation, however, is most powerful when it is used with architectural functions for the interpretation/generation/conversion of bit strings. Representation is attached to object in addition to attribute.

3.4. On the Visibility and View of Objects

Access to an object should be regulated for the purpose of protection. Capability, as used in Hydra, is one of the methods of regulating access. However, we do not feel secure in using capability, since a given capability is valid forever and is transferable. Thus, we introduce the notion of scope. Scope resembles the access list method of protection [Denning 76] in the sense that all the information to be protected is kept in the accessed entity, and permission to access it is determined by the accessed entity.

It should be noted, however, that scope is not specified for an object, but is specified for a name. Scope specifies the visibility of a name with respect to time and usage. Like a virtual circuit in a communication network, permission is given to an accessing entity, but can be revoked at any time. The permission is not transferable to others.

As for usage, there are three different modes: evaluate enable/disable, associate enable/disable, and link enable/disable. If evaluation is enable for a name which denotes an object, then the evaluation of the object returns the object as a whole. If evaluation for a name which denotes a procedure in the object is also enable, then the procedure can be evaluated. Associate enable is equivalent to write enable in many operating systems. In our object oriented programming language, variable must have at least associate enable for the surrounding context (i.e., local variable). Link enable is used to specify that a name can be shared (pointed) by other names.

Scope can specify permission for specific accessing entities. Thus, we can regulate access for each accessing entity in various modes. Scope is declared as:

```
    <name>: scope {<mode> [ / [except]
                   <specific accessing entity> ... / ] ... }
```

where <mode> can be any combination of association, evaluation, and link.

It is sometimes strongly demanded that different views be provided for the same object. A simple example of _view_ is the redefinition of a subarray as an array. More complex examples are found in database systems [Date 81, Baroody 81] and knowledge representation systems [Bobrow 76, Goldstein 80]. Basically, _view_ is achieved by creating an object and making links from the internal name space of the object to names of other objects. A detailed discussion of this will be given in a separate paper.

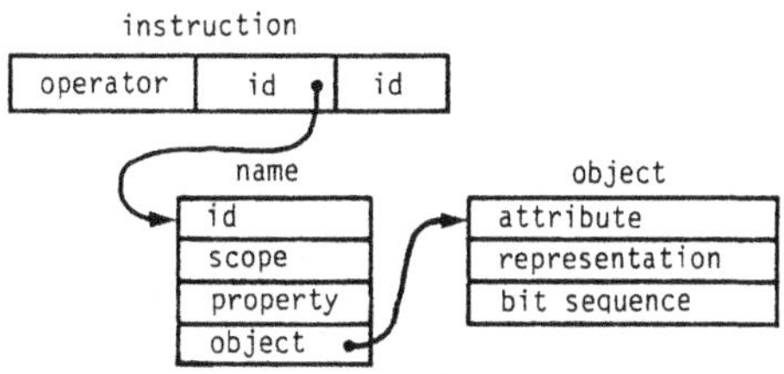

Fig. 5 Structure of Information

3.5. The Model

In this section, we will summarize the definitions of _object_ and related terms discussed above.

An access by an accessing entity to an accessed object is defined by the following four vertical relations (see Fig. 5):

Scope: _Scope_ specifies for a name the authorized accessing entities and access methods. The access methods include _evaluate_, _associate_, and _link_. Thus, _scope_ is used to regulate access to shared information.

Property: _Property_ defines the relation between a name and the set of objects which can be associated with the name. Thus, _property_ is the assertion of a name, which is validated at each association. Thus, _property_ describes an assertion proposition which is evaluated on association. An assertion proposition may specify _attribute_ and _representation_, may refer to other names, and may invoke operations.

Attribute: _Attribute_ defines the relation between an object and the set of operations to be performed on the object. That is to say, _attribute_ corresponds to specifying _class_, and is used to check the eligibility of an operation to be performed.

Representation: _Representation_ defines the relation between an object and its physical representation. For example, representation specifies radices and sizes for an integer and real number, the orthogonal or polar representation for a complex number, mapping methods for an array, and the length of a string. Representation can be transformed dynamically at the execution-time to meet the operation to be performed*.

Thus, *object*, *class*, and *name* are formally presented as follows:

Object: *Object* is a package of information and the description of its manipulation. An object is represented as a 3-tuple:

(<attribute>,<representation>,<bits>)

An object possesses its execution environment (or *context*), where <attribute> specifies a class and <representation> and <bits> are local data in the environment.

Class: *Class* is a module that defines the operations which create an object and the operations which operate on objects created by the class. An object which is created by a class is called its *instance*. A class itself is an object. A class may contain its own variables.

Name: *Name* is an identifier that denotes an object (i.e., the current object or *nil*). A name is represented by a quartuple:

(<identifier>,<scope>,<property>,<pointer to an object>)

Objects can be referenced only through names. Thus, a *variable* is a name in a program with *associate enable*.

3.6. Operations

We define three basic operations for the execution of programs: *association*, *evaluation*, and *link*. *Association*, which is used to assign a value to a variable, performs the association of a *name* with a new *object* which has just returned from an evaluation. This basic operation is expressed in the following form:

assoc (<name>,<object>)

Association frees the old object that has been associated with a name and then associates a new object with the name. In advance of this

* and the property of name to be associated, if necessary.

association, first the eligibility of the access right of *association* is checked according to the *scope* of the name, and then the eligibility of *association* of the object with the name is checked according to the *attribute* (and *representation*, if necessary,) of the object and *property* of the name.

The *evaluation* of a name or operation on objects returns an object. *Evaluation* is expressed in one of the following forms:

eval (<name>)
eval (<class name>$<procedure name><parameter list>)

where <parameter list> represents the list of names. The eligibility of *evaluation* of the name or operation is checked in advance according to the *scope* of the name. *Evaluation* of <name> simply returns the object denoted by that name. *Evaluation* of a procedure is performed as follows: after the necessary checking is performed, the context of the <procedure name> of the <class name> is created; after the parameters are passed, the procedure body is executed. The returned object has *attribute* and *representation*, but can have neither *property* nor *scope* until it is associated with a *name*.

Thus, for example, the statement z := x + y is compiled as either of the followings:

assoc(z, *eval*(number$add(x, y)))

which is the object program when the compiler knows that the attribute of x is type "number", or

assoc(z, *eval*(CLASS$attributeof(x)$add(x, y)))

when the compiler does not know the attribute of x. Computer architecture should provide functions for the fast execution of CLASS$attributeof(x) to support Smalltalk-like laugages.

Link makes a link from one name to another by a logical pointer. A link remains until it is unlinked. This basic operation is expressed in the following form:

link(<from name>, <to name>)

Assume this operation has been performed. Then, the *evaluation* of <from name> keeps on performing itself along these links until it reaches an object. Checking of the eligibility of access is done to every name it encounters. *Association* of an object to <from name>

associates the object to the ultimate name of <to name>.

<u>link</u>(<name>, <u>undefined</u>)

unlinks <name> and denotes <u>undefined</u>.

4. THE OBJECT ORIENTED ARCHITECTURE

4.1. Proposal

The internal structure of a computer can be envisaged as the pair of an operational unit and an informational unit. The operational unit controls and executes operations, while the informational unit preserves and supplies information. In conventional architecture, the operational unit and the informational unit correspond to the CPU and the memory, respectively. The interface between CPU and the memory is performed in terms of the physical or logical memory address.

In the object-oriented architecture proposed here, the operational unit performs the following four principal functions: 1) to check whether or not the attributes of objects are eligible for an operation to be performed, 2) to find an appropriate resolution for the representation of objects, 3) to perform the operation on the objects, 4) to request the informational unit to access/preserve the object with a name. The operational unit also controls the sequencing of instructions and process activities. The informational unit regulates all the access to objects and preserves objects. The informational unit supports the virtual memory system. The informational unit is also in charge of garbage collection.

The names which interface the operational unit and the informational unit in the object oriented architecture are similar to the names in capability based architecture [England 72] [Myers 80] [Houdek 81] [IBM 80] in each sense that they are unique in the system. In order to make a distinction between the local information environment of a program segment and the global information environment of the system, the object oriented architecture provides the local name space and the global name space. Fig. 6 shows the conceptual level description of a procedure being executed in the object oriented architecture.

4.2. Issues

Issues in the design and implementation of the object oriented

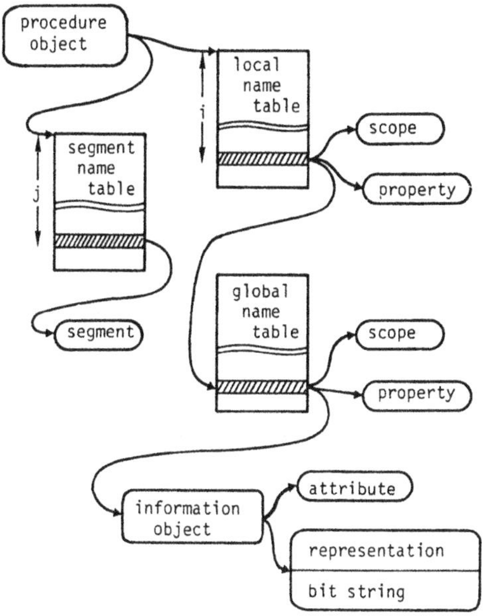

Fig. 6 Run-Time Environment

architecture are described here.

(1) Basic Machine Architecture

A stack architecture is being considered for the basic structure of our object oriented machine. In this machine, the stack(s) do not contain objects but names (i.e., pointers). The instruction set may adopt the canonical form [Hoevel 77] for efficient and compact representation of programs. It may also adopt the frequency based coding [Wilner 72] and the variable sized local names used in the Burroughs B1700 [Myers 81].

(2) Program Representation

A procedure object is represented by a tree of segments, each of which represents a block in a block level. There are no branch instructions that specify memory addresses. Without exception, control branchs to the top of a segment and returns from the bottom of the segment, and thus structured programming is attained at the architectural level.

(3) Name Table Structure

As seen in commercialized object oriented computers such as Intel

432 [Rattner 80] [Intel 81] and IBM System 38 [Houdek 81] [IBM 80], fast address transformation via a chain of tables is the most important issue for higher performance. Fast hashing hardware such as [Goto 77] and enough table lookaside buffers are indispensable.

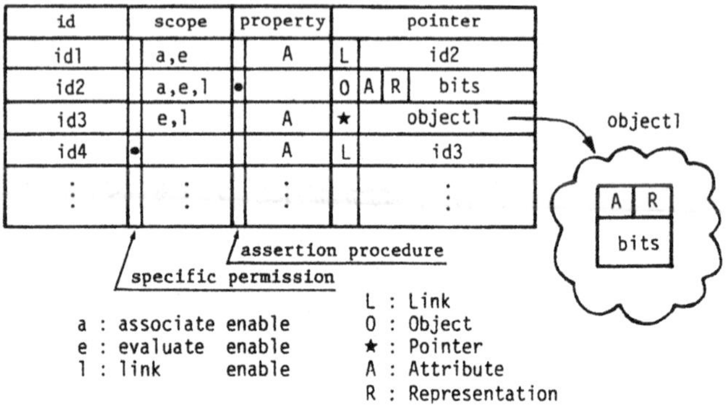

Fig. 7 Global Name Table

Fig. 7 shows the structure of the global name table. In our object oriented architecture, basic operations are performed in variable size fashion. Therefore, a simple object whose length is less than or equal to that of the pointer part is placed into the pointer part of the table. If the length of an object is larger, the object is created in the heap area of the system. Since the length of most of the primitive objects is less than the length of the object pointer, this strategy will result in a fast execution speed with variable length operations. Scopes without specifying permissions to specific accessing entities and properties without specifying assertion procedures are placed in the global name table. Other scopes and properties are placed in the separate tables.

(4) <u>Maximal Utilization of Cache</u>

Since it is expected that the instruction and data access patterns differ greatly from each other in the object oriented environment, to provide separate caches for instruction and data seems to be reasonable and effective. The instruction cache would be similar to those which are used in existing machines. However, the data cache would be peculiar to the object oriented architecture.

The data cache should be designed to provide for the very fast creation of objects. The object oriented architecture would try first to create an object in the data cache. If the cache has no free space, one or more objects which have copies in main memory and/or which are expected not to be referenced in the near future are rolled out. Thus, the cache acts as a fast local memory. The cache does not have to be a partial copy of the main memory addresss space, but can provide a separate address space. Garbage collection (possibly with a different scheme) separate from that of main memory might be effective for the data cache.

(5) Fast Context Switching and Parameter Passing

It is well known that a procedure call instruction appears on the average every 10 instructions in the object codes of a high level language program. In object oriented language programs, the frequency of the apparance is expected to be higher. Thus, fast context switching/parameter passing mechanisms are of particular significance. Since the main purpose of our object oriented architecture is to provide as much modularity as possible by the architecture, fast context switching/parameter passing mechanisms by optimized general register usage such as those employed in IBM 801 [Radin 82] or RISC [Patterson 81] cannot be employed. Object oriented context switching/parameter passing mechanisms are being investigated. The use of multiple stacks with stack caches has been of special interest.

(6) System Defined Attributes

An attribute is either system-defined or user-defined. A system-defined attribute can be either primitive or structured. Examples of primitive attributes are boolean, character, integer, and real. Class, procedure, and segment are other examples of primitive attributes. A structure object supplies a space for names and has an internal addressing mechanism so that an element of the structure object is accessed via an element name. An element of a structure object may be a structure or a primitive object. Examples of structure attributes are array, string, and record.

(7) Support for Dynamic Checking

We have decided on the policy that in order to generate efficient object codes the compiler checks the eligibility of access and operation as much as it can at compile-time, and leaves what it cannot check at compile-time for dynamic checking at execution time. We employ tags [Iliffe 68] [Feustel 72] for frequently used properties, attributes, and representations. Thus, we need high speed tag manipulation hardware.

(8) Variable Length Operations

One of the purposes of our object oriented architecture is to provide variable length operations by hardware. Variable size operations for binary and decimal integers, boolean, and character strings will be implemented by hardware. Variable size floating-point operations by means of recurring rationals [Yoshida 82] are being considered for hardware implementation.

(9) Garbage Collection

In object oriented architecture, efficient garbage collection schemes must be adopted. This is especially important in our architecture, since objects are not only dynamically created and freed but also the sizes of the bit strings vary in concert with their values. The sizes are indicated in their representation specification. Thus, we might have to incorporate compaction as well as garbage collection.

We employ the combination of the reference counter method and the marking method for garbage collection. In most cases, the reference counter method is effective and efficient. For the cases in which links make a cycle and in which a reference counter overflows, the marking method is effective. The employment of a garbage collection processor is being considered in the informational unit. It will run parallel with the other processor which manages information.

5. CONCLUSION

For the reliable and efficient execution as well as the reduction of the maintenance cost of programs, high-level languages with high descriptive capability and architecture to efficiently execute programs written in these languages are inevitable. In this paper, we focused on object oriented languages as such high descriptive languages due to their modular programming capability. We have attempted to establish the foundation for the design and implementation of object oriented languages and object oriented architecture. A new model of the structure of information was proposed, where names and objects are the essential elements. Scope and property are attached to a name while attribute and representation are attached to an object. A name functions to provide history sensitiveness. Thus, any <u>object</u> is immutable in this model. <u>Link</u> was introduced to share a name (and thus the object denoted by the name) by other names. A new object oriented language was used to rationalize the model and to exemplify how to apply the model to object oriented languages.

We outlined an object oriented architecture to execute programs in this language and other object oriented languages based on this model. Issues in the design and implementation of this architecture were described. This architecture brings us a novel feature: as in database systems, information can be independent of programs and programming languages. This independence enables sharing of information among programs written in different programming languages. We believe that the execution speed in this architecture is much higher than that of the conventional architecture when modular programming and reliable execution are required.

The final decision for the design of the object oriented language is being made by the object oriented programming/processing system (OOPS!) group at Keio University. The design of the architecture is also being done along with performance evaluation. In addition, applicability of our object oriented language and architecture to database systems such as [Codd 81, Baroody 81] and knowledge representation systems such as [Bobrow 76, Goldstein 81] is being studied.

ACKNOWLEDGEMENT

The author is grateful to Professor E. I. Organic of the University of Utah, Dr. J. R. Hamstra of Sperry Univac, and Dr. Dennis Frairy of Texas Instrument for their invaluable comments. The author is also thankful to Mr. Takashi Takizuka of KDD and Miss Kaoru Yoshida and Mr. Yutaka Ishikawa of Keio University for their earnest and keen discussions.

REFERENCES

[ADA 80] Reference Manual for the Ada Programming Language, United States Department of Defense, 1980.

[Baroody 81] Baroody Jr., A.J. and DeWitt, D.J., "An Object-Oriented Approach to Database System Implementation," ACM Trans. on Database Systems, Vol. 6, No. 4, Dec. 1981.

[Bobrow 76] Bobrow, D.G. and Winograd, T., "An Overview of KRL, a Knowledge Representation Language," CSL-76-4, Xerox PARC, July 1976.

[Borning 81] Borning, A.H. and Ingalls, D.H.H., "A Type Declaration and Inference System for Smalltalk," Proc. of Principle of Programming Languages, December 1981.

[Codd 81] Codd, E.F., "Relational Database: A Practical Foundation for Productivity," CACM, Vol. 25, No. 2, Feb. 1981.

[Date 81] Date, C.J., "An Introduction to Database Systems," Third Ed., Addison Wesley, 1981.

[Denning 76] Denning, P.J., "Fault Tolerant Operating Systems," Computing Surveys, Vol. 8, No. 4, pp.359-389, 1976.

[England 72] England, D.M., "Architectural Feature of System 250," Infotech State of the Art Report 14: Operating Systems, Berkshire, Infotech, England, pp.395-428, 1972.

[Feldman 79] Feldman, J.A., "High Level Programming for Distributed Computing," Communications of ACM, Vol.22, No.6, pp.353-368, 1979.

[Feustel 72] Feustel, E.A., "The Rice Research Computer - A tagged Architecture," AFIPS SJCC, pp.369-377, 1972.

[Giloi 78] Giloi, W.K. and Berg, H.K., "Data Structure Architecture - A Major Operational Principle", Proc. of 5th Annual Symposium on Computer Architecture, 1978.

[Goldberg 76] Goldberg, A. and Kay, A., ed., "Smalltalk-72 Instruction Manual," Xerox Palo Alto Research Center, March 1976.

[Goldstein 81] Goldstein, Ira and Bobrow, Daniel, "An Experimental Description-Based Programming Environment: Four Reports," Xerox PARC Research Report, CSL-81-3, March 1981.

[Goto 77] Goto, E, Ida, T, and Gunji, T, "Parallel Hashing Algorithms," Information Processing Letters, Vol. 6, No. 1, 1977.

[Hewitt 73] Hewitt, C., et al., "A Universal Modular Actor Formalism for Artificial Intelligence," Proc. of IJCAI, pp.235-245, 1973.

[Hewitt 77] Hewitt, C., "Viewing Control Structures as Patterns of Passing Messages," J. of Artificial Intelligence, Vol. 8, 1977.

[Hoevel 77] Hoevel, L.W. and Flynn, M.J., "The Structure of Directly Executed Languages: A New Theory of Interpretive System Design," TR-130, Digital Systems Laboratory, Stanford University, 1977.

[Houdek 81] Houdek, M.E, Soltis, F.G., and Hoffman, R.L, "IBM System/38 Support for Capability-Based Addressing," Proc. of the 8th Int'l Symp. on Computer Architecture, pp.341-348, May 1981.

[IBM 80] "IBM System/38 Technical Developments (G58060237)," IBM, 1980.

[Iliffe 68] Iliffe, J.K., "Basic Machine Principles," American Elsevier, New York, 1968.

[Intel 81] "Intel APX432 General Data Processor Architecture Reference Mannual," Intel, Aloha, Orengon, 1981.

[Kahn 81] Kahn, K.C., et al., "iMAX: A Multiprocessor Operating System for an Object-Based Computer," Proc. of the Eighth Symp. on Principles of Operating Systems, Dec. 1981.

[Liskov 79] Liskov, B., et al., "CLU Reference Manual," TR-225, Laboratory for Computer Science, MIT, Oct. 1979.

[Myers 80] Myers, G.J., and Buchinghan, B.R.S., "A Hardware Implementation of Capability-Based Addressing," Computer Arachitecture News, Vol. 8, No. 6, 1980.

[Myers 82] Myers, G.J., "Advances in Computer Architecture", Second Ed., John Wiley and Sons, 1982.

[Patterson 81] Patterson, D.A., and Sequin, C.H., "RISC-I: A Reduced Instruction Set VLSI Computer," Proc. of the 8th Int'l Symp. on Computer Architecture, pp.443-457, May 1981.

[Radin 82] Radin, G, "The 801 Minicomputer," Proc. of the Symp. on Architectural Support for Programming Languages and Operating Systems, pp.39-47, March 1982.

[Rattner 80] Rattner, J. and Cox, G., "Object-Based Computer Architecture," Computer Architecture News, Vol. 8, No. 6, 1980.

[Snyder 79] Snyder, A., "A Machine Architecture to Support an Object-Oriented Language," MIT TR-209, Laboratory for Computer Science, MIT, Marach 1979.

[Suzuki 80] Suzuki, N., "Inferring Types in Smalltalk," Proc. of Principle of Programming Languages, December 1980.

[Tokoro 82] Tokoro, M and Takizuka, T., "On the Semantic Structure of Information --- A Proposal of the Abstract Storage Architecture," Proc. of the 9th Int'l Symp. on Computer Architecture, pp.211-217, April 1982.

[Williams 79] Williams, G., "Program Checking", Sigplan Notice, Vol. 14, No. 9, pp.13-25, 1979.

[Wilner 72] Wilner, W.T., "Burroughs B1700 Memory Utilization," AFIPS FJCC, 1972.

[Wulf 75] Wulf, W.A., et al., "Overview of the Hydra Operating System," Proc. of the 5th Symposium on Operating System Principles, pp.122-131, Nov. 1975.

[Wulf 81] Wulf, W.A., Levin, R., and Harbison, S.P., "HYDRA/C.mmp: An Experimental Computer System," McGraw-Hill, New York, 1981.

[Xerox 81] Xerox Learing Research Group, "The Smalltalk-80 System," Special Issue on Smalltalk-80 System, Byte, Vol. 6, No. 8, August 1981.

[Yonezawa 81] Yonezawa, A., "Unified model of Algorithm Representation," in "Algorithm Representation," Iwanami Book Co., (in Japanese,) 1981.

[Yoshida 82] Yoshida, K., "A Research on Error Free Machines -- A proposal of New Floating-Point Arithmetic based on Recurring Rationals," Computer Science Report, Arithmetic-82-1, Department of E. E., Keio University, March 1982.

DURAL: an extended Prolog language

Shigeki Goto
Yokosuka Electrical Communication Laboratory
Nippon Telegraph and Telephone Public Corporation
1-2356 Take, Yokosuka-shi
Kanagawa 238 JAPAN

ABSTRACT

This paper proposes a new programming language DURAL which is an extension of the Prolog language. DURAL takes advantage of modal logic to classify clauses. The main features of DURAL are the following:

1) Modal symbols are introduced to discriminate between clauses.
2) The relative Horn clause represents the clause containing executable predicates.
3) The unit resolution as well as input resolution is adopted to facilitate debugging.

KEY WORDS & PHRASES

DURAL, extensional database, intensional database, modal logic, predicate calculus, program synthesis, programming language, Prolog, query language, relative Horn clause, resolution principle.

Introduction

Prolog is a new programming language proposed by Colmerauer [1]. It was implemented at both the University of Marseille [2] and the University of Edinburgh [3]. Prolog is essentially based on the first order predicate calculus. The logical foundation gives the Prolog program clear semantics. On the other hand, the rigorous syntax sometimes prevents ideas from being represented directly. For example, there is no priority between clauses. All clauses are equally applicable. This paper extends the underlying logic of Prolog in order to get more concise and flexible expressions in accordance with the logical formality.

Section 1 introduces the original Prolog language, which is based on the first order predicate calculus. Section 2 extends the underlying logic to modal logic by which a Prolog clause can be discriminated from others. Section 3 introduces another clause to treat the executable predicate. Section 4 discusses the resolution method which controls the Prolog interpreter. The final section is a conclusion of this paper.

1. Prolog Language

Each Prolog statement takes the form of a special logic formula, called a Horn clause, which can be interpreted operationally as a procedural declaration. Example 1 shows a simple program written in Prolog. This simple example is chosen for the convenience of explanation. Many actual programs have in fact been written in Prolog. They include natural language understanding systems, formula manipulation and symbolic integration systems, and a STRIP-like problem solver. [4]

Example 1. Addition in Prolog

```
1   +(ADD 0 *Y *Y)
2   +(ADD (S *X) *Y (S *Z)) -(ADD *X *Y *Z)
3   -(ADD 2 3 *Z)
```

Before examining example 1, it should be noted that Prolog's syntax is given in modified BNF notation by Definition 1. In this paper, the Prolog interpreter is written in Lisp and the Lisp "atom" is used as the primitive constituent.
In Definition 1, a pair of brackets [,] encloses the optional items. An ellipsis "..." indicates a list consisting of the preceding items, e.g., <A>... means <A> | <A><A> | <A><A><A> and so on. A vertical bar "|" means "or", and separates alternative items.

Def. 1 Syntax of Prolog

```
<program>   ::= <statement>...
<statement> ::= <Horn clause>
<Horn clause>::=[<positive literal>][<negative literal>...]
<positive literal> ::= +<atomic formula>
<negative literal> ::= -<atomic formula>
<atomic formula> ::= ( <predicate> <term>... )
<predicate> ::= <Lisp atom>
<term> ::= <Lisp atom> | <variable> | ( <function><term>...)
<variable> ::= *<Lisp atom>
<function> ::= <Lisp atom>
```

<Lisp atom> is the literal atom in the programming language Lisp, and is not defined here.

According to the above definition, a Horn clause has at most one positive literal and a finite number, possibly zero, of negative literals. Table 1 divides Horn clauses (Prolog statements) into four categories. In each category, the Horn clause can be interpreted operationally as a procedural declaration.

Table 1. Horn clause

neg / pos	0	not 0
0	[]	-B1-B2...-Bn
1	+A	+A-B1-B2...-Bm

Procedural interpretation:

[] STOP statement
i.e. procedure without a name or body
[] stands for an empty clause.

+A-B1-B2... PROCEDURE declaration
+A: procedure name and variable list
-B1-B2...: procedure body which calls B1, B2, ...

-B1-B2... EXECUTE B1, B2, ... (procedure without a name)
called a goal statement in Prolog

+A PROCEDURE without a body
This type of statement is not meaningless, since the variable list may contain various terms.

In Example 1, three types of statements appear. A normal program has only one goal statement. The goal statement changes its form during program execution, and takes the form of a STOP statement when the program terminates. Let's look at Example 1 again.

Example 1.

```
1   +(ADD 0 *Y *Y)
2   +(ADD (S *X) *Y (S *Z)) -(ADD *X *Y *Z)
3   -(ADD 2 3 *Z)
```

ADD is a predicate and (ADD a b c) means a+b=c; S is the successor function, namely (S x)=x+1. Program execution is performed by Algorithm 1 shown below.

Algorithm 1. (Prolog)

L means "line". The execution starts at line 1.

L	Action or Test	Next Line Succeed	Next Line Fail
1	Find a goal statement "G"	2	abort
2	G is [] (STOP)	stop	3
3	G:={Resolve G against another clause}	2	4
4	G:={The previous G}, backtracking	2	abort

The abortion in line 1 means there is no goal statement. The abortion in line 4 means G cannot be resolved against any clause in the program.

Backtracking is done by the stack mechanism.

The resolution in line 3 represents the input resolution. The adoption of the input resolution is supported by the following proposition. [5]

Prop. 1 If S is an unsatisfiable Horn set, then there is an input refutation of S.

The following example shows the execution of the above-mentioned program.

Example 2. Execution of Example 1

A number is automatically converted into the form (S (S ...0)), e.g. 2 -> (S (S 0)). L1, L2 and L3 represent the lines 1, 2 and 3 in Algorithm 1.

```
L1:  G:= -(ADD (S (S 0)) (S (S (S 0))) *Z)
L2:  G is not [].
L3:  G:= -(ADD (S 0) (S (S (S 0))) *Z)
L2:  G is not [].
L3:  G:= -(ADD 0 (S (S (S 0))) *Z)
L2:  G is not [].
L3:  G:= []
L2:  G is [].      normal termination
```

Although the execution is terminated normally, the answer 5 (=2+3) is lost, since there is no output statement. To implement the output statement, a built-in predicate "OUT" is introduced. It is assumed that the following statements (clauses) are built into the Prolog system.

```
+(OUT *X1)
+(OUT *X1 *X2)
    .
    .
```

```
     +(OUT *X1 *X2 ... *Xn)
          :
          :
```

If a goal statement contains a negative literal -(OUT T1 T2...Tm), where Ti is any term, the resolution is always successful due to the built-in clause +(OUT *X1 *X2...*Xm). The special OUT resolution has the side effect that T1, T2, ..., Tm are printed out on the terminal. Example 3 illustrates the OUT predicate.

Example 3. Subtraction using a built-in predicate OUT

```
  1    +(ADD 0 *Y *Y)
  2    +(ADD (S *X) *Y (S *Z)) -(ADD *X *Y *Z)
  3    -(ADD *X 3 5) -(OUT *X)                  <=== goal
```

A number is converted, e.g. 3 -> (S (S (S 0))).
A comma "," indicates a line continuation.

```
L1:    G:= -(ADD *X (S (S (S 0))) (S (S (S (S (S 0)))))) ,
           -(OUT *X)
L2:    G is not [].
L3:    G:= -(ADD *X (S (S (S 0))) (S (S (S (S 0))))) ,
           -(OUT (S *X))
L2:    G is not [].
L3:    G:= -(ADD *X (S (S (S 0))) (S (S (S 0)))) ,
           -(OUT (S (S *X)))
L2:    G is not [].
L3:    G:= -(OUT (S (S 0)))
L2:    G is not [].
L3:    G:= []         side effect: (S (S 0)) is printed out.
L2:    G is [].       normal termination
```

The output (S (S 0)) is also converted to 2.
The above example also demonstrates that subtraction can be performed with the same clauses 1 and 2 as for addition in Example 1. Many different goal statements can also be executed using clauses 1 and 2 (see below).

Example 4. A variety of goal statements

```
-(ADD 2 3 *Z) -(OUT *Z)                 addition:    *Z=5
-(ADD 2 *Y 5) -(OUT *Y)                 subtraction: *Y=3
-(ADD *X 3 5) -(OUT *X)                 subtraction: *X=2
-(ADD 2 3 5)                            normal termination
-(ADD 2 3 6)                            abnormal termination
-(ADD 2 *Y *Z) -(OUT *Y *Z)             *Y=*Y, *Z=(S (S *Y))
-(ADD *X 3 *Z) -(OUT *X *Z)             *X=0,  *Z=3
```

```
-(ADD *X *Y 5) -(OUT *X *Y)              *X=0,  *Z=5
-(ADD *X *Y *Z)-(OUT *X *Y *Z)           *X=0, *Y=*Y, *Z=*Y
```

The goal statement -(ADD 2 3 6) causes an abnormal termination, since 2+3 does not equal 6. -(ADD 2 *Y *Z) and -(ADD *X *Y *Z) have symbolic (not numerical) answers. It should be noted that even if the goal statement has more than one answer, only the one that is found first is printed. This is because of the character of the resolution principle, and will be discussed further in section 2.2.

2. Modal symbol

2.1 Modal logic and the modal symbol

The Prolog interpreter is considered to be a theorem prover in the first order logic. Each statement in a Prolog program constitutes an axiom, and the goal is a theorem (Figure 1A).

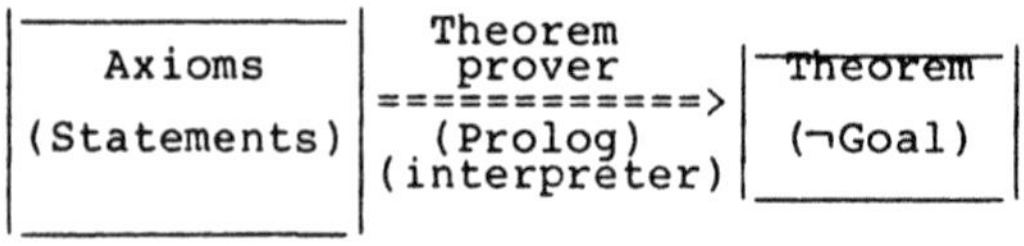

Figure 1A. Prolog interpreter as a prover

Prolog uses the resolution principle as a theorem prover, which proves a theorem by refutation (Figure 1B). The goal statement corresponds to the negation of a theorem.

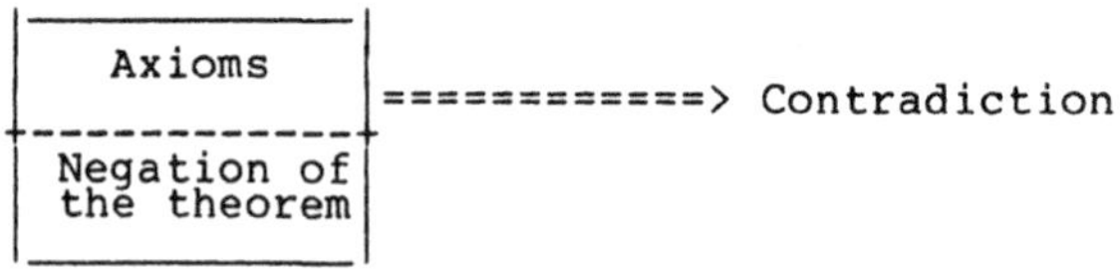

Figure 1B. Proof by refutation

Modal logic introduces modal symbols into the logic. The modal symbol is intuitively explained in figure 2, where L denotes a subset of the axioms. If a logic formula "q" is proved from L, we say that "Lq" is a theorem.

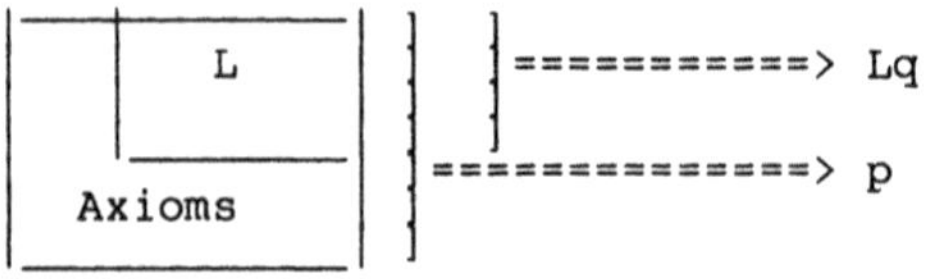

Figure 2. Subset of the axioms

The symbol L in the formula Lq is called a modal symbol and stands for strong provability. In modal logic, a logic formula may be preceded by modal symbols. Modal logic has a variety of applications. [6]

2.2 Clause discrimination

A faster version of the addition program (Example 1) can be written as follows, where "PLUS" is a Lisp function.

Example 5.

```
+(ADD *X *Y (PLUS *X *Y))
-(ADD 2 3 *Z) -(OUT *Z)        <=== goal
```

The goal statement is executed and terminated immediately. It prints out the answer *Z=(PLUS 2 3). If the answer is evaluated in Lisp, *Z is equal to 5. This paper shows that the fast version enable us to execute programs quickly in Prolog.
It is worth noting here, however, that the fast version clause is less flexible than the ordinary clause. Example 6 below shows that the fast clause +(ADD *X *Y (PLUS *X *Y)) cannot be used for subtraction, since (PLUS *X *Y) is a term and cannot be unified with the constant "5".

Example 6.

```
+(ADD *X *Y (PLUS *X *Y))
-(ADD 2 *Y 5) -(OUT *Y)      <=== goal, abnormal termination

+(ADD *X *Y (PLUS *X *Y))
-(ADD *X 3 5) -(OUT *X)      <=== goal, abnormal termination
```

There is no syntactical distinction between the fast version (Example 5) and the ordinary version (Example 1). Sometimes, however, it is necessary to distinguish them. For example, a goal statement -(ADD 2 *Y 5) -(OUT *Y) cannot be executed under the fast version clause +(ADD *X *Y (PLUS *X *Y)), but if the ordinary clauses +(ADD 0 *Y *Y) and +(ADD (S *X) *Y (S *Z)) -(ADD *X *Y *Z) coexist, the goal is achieved through the ordinary clauses. DURAL (Prolog) has an automatic backtrack facility, so it cannot be confirmed whether execution is performed only in the fast version environment. To distinguish the clauses, the modal symbol is introduced.

Def. 2. (modal symbol in DURAL)

```
<statement> ::= L(<positive literal>[<negative literal>...]) |
                <positive literal>[<negative literal>...]    |
```

```
                &lt;goal statement&gt;
<goal statement> ::= [] | <negative literal>[<negative literal>...]
<negative literal> ::= -<atomic formula> |
                        - L <atomic formula>
```

The symbol "L" is a modal symbol. A negative literal - L (ADD 2 *Y 5) must be resolved exclusively against a fast clause in the form L(<positive literal>[<negative literal>...], whereas -(ADD 2 *Y 5) may be resolved against any clause in the program. The use of this modal symbol "L" solves the distinction problem.

A further application of the modal symbol is explained in the database field. The resemblance between relational database models and Prolog-like languages is well known. [7] [8] As is often the case, the extensional database (elementary facts) is allowed to change over time, whereas the intensional database (general laws) tends to remain fixed. DURAL facilitates the distinction between the time-varying portion and the fixed portion. The modal symbol "L" can be regarded as indicating fixed clauses.
Example 7 defines path relation in Directed Graph G.

Example 7.

```
L1  L( +(PATH *X *X) )
L2  L( +(PATH *X *Z) -(ARC *X *Y) -(PATH *Y *Z) )
1   +(ARC 1 2)
2   +(ARC 1 4)
3   +(ARC 2 3)
4   +(ARC 4 2)
5   +(ARC 4 3)
6   -(PATH 4 *W) +(QUERY *W)          <=== goal
7   - L (PATH 4 *W) +(QUERY *W)       <=== goal
```

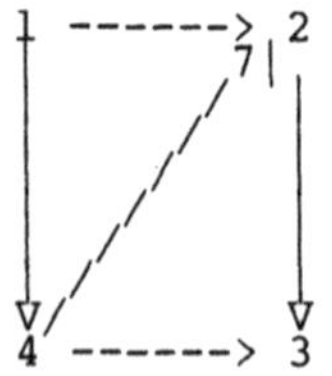

Fig.3 Directed Graph G

L1 and L2 are assumed to be fixed. These clauses have property "L", and are enclosed by "L( )". The time varying portion (1 to 5) represents arcs in G. A goal statement contains an executable predicate "QUERY" and has two side effects:

i) It prints out the answer;

ii) It assumes that the evaluation fails, and evokes a backtrack mechanism.

The goal statement 6 is answered by a set of nodes, {4, 2, 3}. The other goal, 7, must be carried out in the fixed portion, and the answer is only {4} (using L1).

The usage of the backtrack mechanism for plural answers has been discussed by Futo, Darvas and Szeredi. [8] DURAL realizes this backtrcking mechanism with the use of a relative Horn clause.

3. Relative Horn clause

3.1 Executable predicate

In section 2.2, a fast version program includes the Lisp function "PLUS". Lisp predicates, as well as Lisp functions, can be used in fast version programs. In Example 8, it is possible to replace clauses 3 and 4 with a predicate GE, if it is declared that GE is executable and the corresponding Lisp routines 3' and 3'' are defined.

Example 8.

```
 1    +(RESULT *X 1) -(GE *X 10)
 2    +(RESULT *X 0) -(GE 10 *X)
 3    +(GE (S *X) (S *Y)) -(GE *X *Y)     <--- to be replaced
 4    +(GE *X 0)                          <--- to be replaced
 3'   (DEFUN GE (X Y)                     <--- Lisp routine
          (OR (EQ X Y) (GREATERP X Y)) )
 3''  (DEFUN S (X) (ADD1 X))              <--- Lisp routine
```

If the declaration is issued, the negative literal -(GE *X *Y) stops searching for the corresponding positive literal, and evokes the evaluation in Lisp.

```
                          __ Error (*X or *Y      ===> Return to Prolog
                         |   is not a number)
Evaluate                 |
 (GE *X *Y)  ------------|    _ T    ===> Eliminate the literal
                         |   |            in the goal statement,
                         |___|            and proceed in Prolog
                             |
                             |_ NIL ===> Backtracking
```

The return to Prolog is carried out through the ERRSET facility in Lisp. When the value of (GE *X *Y) is T, the execution proceeds successfully. If the value is NIL, the value of -(GE *X *Y) becomes T. This means that the goal statement cannot be proved by resolution, and it brings the execution into the backtracking mode as in line 4 in

Algorithm 1.

The built-in predicate in section 1 has this function. It can be regarded as an executable predicate and its corresponding Lisp routine brings about the print out side effect.

3.2 Relative Horn clause

If the executable predicate is to be used freely, the Prolog system must check whether the predicate is built-in or not. The DURAL system saves checking by modifying the Prolog syntax. In the original Prolog syntax, a goal statement contains only negative literals. In DURAL, however, a positive literal may appear in a goal statement. When this occurs, it is treated as an executable literal. The syntax is extended to include positive literals in the goal statement, and is formally based on the relative Horn clause. [9]

Def. 3. (Relative Horn clause) A set S of clauses defines a set of Horn clauses relative to M, where M is a setting (defined below), if and only if each clause of S has at most one literal false in M.

A setting is in effect a consistent set of literals from the Herbrand universe of S.

Def. 4. (Setting) Given a set S of clauses, a <u>setting</u> M for S is a (possibly empty) set of literals satisfying the following conditions:

i) every literal of M is an S-instance of a literal of S or an S-instance of the complement of a literal of S;

ii) if $\tilde{l}$ is in M, then every S-instance of $\tilde{l}$ is in M;

iii) M does not contain a complementary pair of literals.

An S-instance of a literal $\tilde{l}$ is a literal $\tilde{l}$[tl/xl,...] where [tl/xl,...] denotes a substitution. The terms tl,... consist of the alphabet of S plus variable names. An S-generalization of a literal $\tilde{l}$ is an S-instance $\tilde{l}'$ of some literal of S such that $\tilde{l}$ is an S-instance of $\tilde{l}'$.

Def. 5. (Homogeneous setting; Partition)

i) A setting M is homogeneous if and only if whenever $\tilde{l}$ is in M then every S-generalization of $\tilde{l}$ is in M.

ii) A partition M is a setting M such that every S-instance of a literal of S is in M or has its complement in M.

A partition is by nature a homogeneous setting. The DURAL system utilizes only a subset of the relative Horn clause.

Def. 6. (Extended DURAL syntax)

<statement> ::= <positive literal>[<literal>...] |
<goal statement>

<goal statement> ::= [] | <negative literal>[<literal>...]

<literal> ::= <positive literal> | <negative literal>

Algorithm 2 (Dural)

L means "line". The execution starts at line 1.

L	Action or Test	Next Line	
		Succeed	Fail
1	Find a goal statement "G"	2	abort
2	G is [] (STOP)	stop	3
3	The left-most literal in G is positive	4	5
4	EXECUTE the literal in Lisp	2	5/6
5	G:={Resolve G against another clause}	2	6
6	G:={The previous G}, backtracking	2	abort

In line 4, three cases may occur:

```
                          __ Error in Lisp ===> (5) Return to DURAL
                         |                          resolution mode
Evaluate                 |
the positive  ---------- |        _ T ===> (6) The goal cannot be proved,
literal +P               |       |-              since the goal itself will
                         |_______|               be T.
                                 |
                                 |_ NIL ===> (2) Success
```

The resolution in line 5 represents the input resolution.

Prop. 2. If S is an unsatisfiable Horn set relative to a homogeneous partition M for S, then there exists an input refutation of S. [9]

EXAMPLE 9. Executable predicate ANS

```
1   -(ADD 2 3 *Z) -(OUT *Z)    built-in OUT
2   -(ADD 2 3 *Z) +(ANS *Z)    executable ANS
```

The predicate ANS performs the same job as OUT. However, clause 2 is not logically equivalent to clause 1. In fact, clause 1 is transformed into the formula -[(ADD 2 3 *Z) & (OUT *Z)], which is the negation of the formula [(ADD 2 3 *Z) & (OUT *Z)] to be proved. This type of goal statement is conventionally used, since the resolution principle proves the formula by refutation. On the other hand, clause 2 is equivalent to (ADD 2 3 *Z) -> (ANS *Z), which reads "whenever *Z satisfies (ADD 2 3 *Z), it is the answer". The answering predicate ANS is often applied to question answering systems. [5] DURAL expresses this answering predicate in the relative Horn clause.

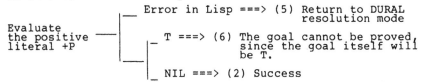

3.3 Program synthesis

Applying the program synthesis technique, a fast version program is constructed automatically from ordinary clauses. The theorem proving approach to program synthesis [10] is effective, since each DURAL statement takes the form of a logic formula. Moreover, as DURAL statements can be considered in intuitionistic logic, the synthesis method based can be used on Gődel's interpretation. [11] [12]

Example 10.

```
1   +(ADD 0 *Y *Y)
2   +(ADD (S *X) *Y (S *Z)) -(ADD *X *Y *Z)
```

From 1 and 2, the Lisp function ADD is synthesized.

```
3   (DEFUN ADD (*X *Y)
        (COND ((ZEROP *X) *Y)
              (T (S (ADD (SUB1 *X) *Y))) ))
```

Clauses 1 and 2 logically represent a mathematical induction which corresponds to recursive program 3. A more interesting example is shown below.

Example 11.

```
1   +(REVERSE NIL NIL)
2   +(REVERSE (CONS *X *Y) (NCONC1 *Z *X)) -(REVERSE *Y *Z)
3   (DEFUN REVERSE (*Y)
        (COND ((NULL *Y) NIL)
              (T (NCONC1 (REVERSE (CDR *Y)) (CAR *Y))) ))
```

In Example 11, a function REVERSE of LISP linear lists is synthesized. It is well-known that the linear list has a similar structure to the natural number. The table below shows the correspondence between them.

natural number	linear list
0	NIL
(ZEROP *X)	(NULL *X)
(S *Y)	(CONS *X *Y)
(SUB1 *X)	(CDR *X)

In addition, the synthesis algorithm itself can be represented in DURAL. In Example 12, GETC and PUTC are executable predicates. GETC searches for the clause which matches its argument, and PUTC makes a new clause containing the synthesized program. In production system terminology, the clause in Example 12 is a meta-rule, where the variable *P represents any predicate.

Example 12. Synthesis algorithm in DURAL

```
+(INDS *P) ,
    +(GETC (+(*P 0 *F *T0))) ,
    +(GETC (+(*P (S *E) *F *T2)-(*P *E *F *T1))) ,
    +(PUTC (+(*P *E *F (*P *E *F)))
           (*P (*E *F)
               (COND ((ZEROP *E) *T0)
                     (T subst[(*P (SUB1 *E) *F),*T1,*T2])))))
```

The goal statement -(INDS ADD) produces the function ADD in Example 10.

4. Unit resolution

An ordinary Prolog interpreter adopts the input resolution method. It has been proved that the unit resolution method is logically equivallent to the input resolution method. Therefore, it is possible to interpret a Prolog program using the unit resolution based system. [7] DURAL has two resolution methods (input and unit), and a user can "switch" the resolution mode. The two methods produce the same result when the program terminates normally. However, an abnormal program points out the difference between the two resolution methods. Example 13 is a rather tricky program which shows the usefulness of the unit resolution.

Example 13.

```
1   +(on a b)
2   +(on b c)
3   +(above *X *Y) -(on *X *Y)
4   +(above *X *Z) -(above *X *Y) -(on *Y *Z)
5   -(above a *V)                                <=== goal 1
6   -(above a d)                                 <=== goal 2
```

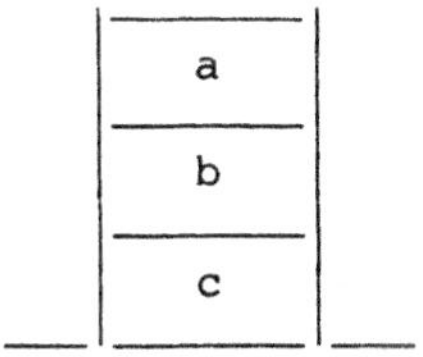

Fig.4 A is on b, and b is on c.

Goal 1 terminates normally and produces the answer *V=b in both methods. In the input mode, goal 2 never terminates and produces an infinite loop as shown below.

```
given goal   6   -(above a d)
from 4 & 6   7   -(above a *Y) -(on *Y d)
from 4 & 7   8   -(above a *Y') -(on *Y' *Y) -(on *Y d)
from 4 & 8   9   -(above a *Y'') -(on *Y'' *Y') -(on *Y' *Y),
                 -(on *Y d)
     .           .
     .           .
     .           .
```

Fig.5 Infinite Loop

On the other hand, goal 2 does not continue to expand in the unit mode. It is finitely determined to have no answer.

This difference between the two methods is useful in debugging a DURAL program. Some alert readers might have noticed that if clause 4 in Example 13 were written as:

4 +(above *X *Z) -(on *X *Y) -(above *Y *Z),

the situation would be changed and there would be no expansion. Indeed the example is chosen for the convenience of explanation. The usefulness of this difference is demonstrated in more complicated programs.

5. Conclusions

This paper proposes a new extension of the existing Prolog language. Since our extension is based on modal logic, DURAL still falls into the category of a logic programming language. The adoption of the relative Horn clause and the unit resolution method facilitates better program writing.

(1) DURAL is written in MACLISP, and runs on the DEC system 20. The size of the source program is about 660 lines when printed out neatly.
(2) Proposition 2 in section 3.2 can be strengthened by imposing a restriction on the resolution. The input resolution can be an ordered input resolution. The running DURAL described in (1) takes advantage of this ordering.

ACKNOWLEDGMENTS

The author wishes to express his sincere thanks to Dr. K.Fuchi at Institute for New Generation Computer Technology for his guidance in the Prolog language.

REFERENCES

[1] A.Colmerauer, H.Kanoui, R.Pasero & P.Roussel, Un systeme de

Communication Homme-machine en Francais, Rapport de recherche, Groupe d'Intelligence Artificielle, UER de Luminy, Universite d'Aix Marseille, 1973.

[2] G.Battani & H.Meloni, Interpreteur du langage de programmation PROLOG, Rapport de DEA, Groupe d'informatique appliquee, Groupe d'intelligence Artificielle, UER de Luminy, Universite d'Aix-Marseille, 1973.

[3] L.M.Pereira, F.C.N.Pereira & D.H.D.Warren, User's Guide to DEC System-10 PROLOG, Department of Artificial Intelligence, University of Edinburgh, 1978.

[4] M.H. van Emden, Programming with resolution logic, Machine Intelligence 8, pp.266-299, 1977.

[5] C-L.Chang and R.C-T.Lee, Symbolic Logic and Mechanical Theorem Proving, Academic Press, 1973.

[6] G.E.Hughes & M.J.Cresswell, An Introduction to Modal Logic, Methuen and Co., 1968.

[7] K.Fuchi, Predicate Logic Programming --- A Proposal of EPILOG, SM-1-2, Information Processing Society of Japan, 1977.

[8] I.Futo, F.Darvas and P.Szeredi, The Application of PROLOG to the Development of QA and DBM Systems, in LOGIC and DATABASES, pp.347-376, Plenum Press, 1978.

[9] D.W.Loveland, Automated Theorem Proving: A logical Basis, North-Holland, 1978.

[10] Z.Manna and R.J.Waldinger, Toward Automatic Program Synthesis, Comm. ACM, vol.14, no.3, pp.151-165, 1971.

[11] S.Goto, Program Synthesis from Natural Deduction Proofs, IJCAI-79, pp.339-341.

[12] M.Sato, Towards a Mathematical Theory of Program Synthesis, IJCAI-79, pp.757-762.

An Algorithm for Intelligent Backtracking

Taisuke Sato
Electrotechnical Laboratory

1. INTRODUCTION

The purpose of this paper is to propose a method to improve the backtracking behavior of Prolog. This method was derived from the one invented for OL(ordered linear) refutation. Therefore it takes consideration of failures by 'occur check' as well as failures by 'clash (different constant symbols)' unlike other intelligent backtracking methods [1][2][4][5]. Moreover we can be sure that it never destroys the completeness of proof search.

When we apply our backtracking method not to a prover like OL but to PROLOG, we have to treat the failures containning 'not' or 'cut'. Modifications needed for backtracking in case of those failures are given in the last section.

Currently, Prolog searches for an answer in a top-down and serial manner. If every choice has completely failed, Prolog starts backtracking to the most recent step where untried choices are left and selects an alternative to restart a proof search process.

The backtracking process described above simply goes back over the path in reverse order without analysing the cause of failure. Hence, it is called Naive Backtracking (NB for short).

However, NB is rather inefficient. Let us take a simple example(see fig. 1). For AND-goal A(x,y)B(x)C(y), A(x,y) is called first. A(z,a) ← unifies with A(x,y) and the resulting mgu(substitution, variable bindings) is {x\z, y\a}. Then the second goal B(x) is called. The actual goal is B(z) because the value of x is z. This goal immediately succeeds by unification with B(a) ←. mgu {z\a} is produced. The total substitution obtained up to this step is {x\z, y\a}*{z\a}(* denotes substitution composition). Thirdly C(y) is called. The actual goal is C(a). But since there is no input clause whose head is unifiable with C(a), backtrack occurs. According to NB, we return to the most recent step where B(a) ← was selected and retry the input clause B(b) ← as an alternative.

```
 ← A(x,y),B(x),C(y) ... AND-goal

     A(z,a) ←
     A(z,b) ←
     B(a)  ←
     B(b)  ←     x,y,z are variables.
     C(b)  ←     a,b are constants.
```

fig. 1 A simple example of a program

This selection is destined to fail. The reason is that since the cause of the backtrack -- the value <a> of the variable <y> -- is not eliminated, we will again reach the step of unifying C(a) with C(b) ← sooner or later. If we want to avoid the double occurrences of the same failure such as unifying C(a) which C(b) ←, we must retry the step where the vaiable <y> is bound to the constant <a>, i.e., where the goal A(x,y) is unfied with A(z,b) ←.

Based on these observatiuons, several Intelligent Backtracking methods (IB for short) have been proposed which analyze the cause of failures and decide what to do at the next step to avoid repeating the failure.

IB generally does not return step by step but skips over several steps at a time. In other words it cuts off OR branches of the proof search tree. Therefore the problem of IB is to insure that proof paths skipped over actually do not lead to a solution. If proof paths are cut off carelessly, we may lose the chance to find a solution.

Let us call such a backtracking method safe that cuts off only those paths that never lead to a solution. There are several literatures on IB[1][2][4][5], but they concentrate on the efficiency of IB, and have little consideration for safeness of their backtraking methods.

We pursure a backtracking method which is both intelligent and safe. We first define the proof search tree (search space) associated with the given Prolog program and goal clause.

2. PROOF SEARCH SPACE

In order to define IB without ambiguity, we need a description of the behavior of Prolog for the given program (Horn set) and goal clause.

The computation process of Prolog can be seen as a transition of an AND-goal. Every AND-goal in the process is represented by a pair, the template(skelton) of the goal and the substitution. For example, a goal A(a, f(z))B(b) is represented as <A(a, f(x))B(y), {x\z , y\b}>. A(a, f(x))B(y) is a template and {x\z, y\b} is a substitution.

In what follows, L,M,N... represent goals(literals). α, β, γ..., represent sequences of several goals, i.e., AND-goals. λ, μ, θ..., represent substitutions. $E\lambda$ represents the application of λ to an expression E.

If the initial AND-goal is α, then it is represented as $<\alpha, e>$(e denotes the null substitution). Suppose that the AND-goal becomes $<L\alpha, \theta_1*...*\theta_{i-1}>$ after i-1 resolution steps. The actual AND-goal is $L\alpha\theta_1*...*\theta_{i-1}$ and the next goal(conjunct) to solve is $L\theta_1*...*\theta_{i-1}$ because goals are solved in left-to-right order. Also suppose that an input clause $M \leftarrow \beta$ is selected and renamed whose head is unifiable with $L\theta_1*...*\theta_{i-1}$. Let the mgu(most general unifier) of M and $L\theta_1*...*\theta_{i-1}$ be θ_i. After the i-th resolution step, the AND-goal becomes $<\beta\alpha, \theta_1*...*\theta_i>$. $\theta_1,...,\theta_i$ are called substitution factors. If the template part of the AND-goal becomes empty, i.e., $<\square, \theta>$, the computation halts and the proof successes. Then the answer substitution is θ.

The proof process described so far corresponds to tracking some branch of a proof search tree(search space, see fig. 2). Each OR node si represents an AND-goal and is labeled by the template of the goal(conjunct) at si. The directed edge ei indicates the literal pair, the goal and the head of an input clause to resolve upon. The resulting mgu is represented by θ_i. The number of edges show the possibilities of resolution. A proof search tree like fig. 2 is easily obtained from the connection graph of the Prolog program and initial AND-goal.

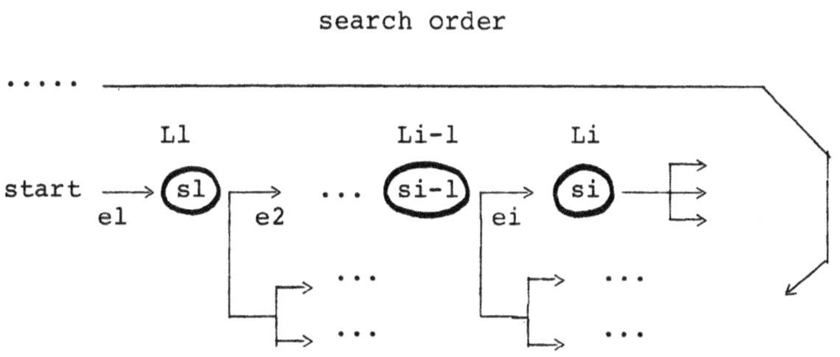

fig. 2 Proof search tree

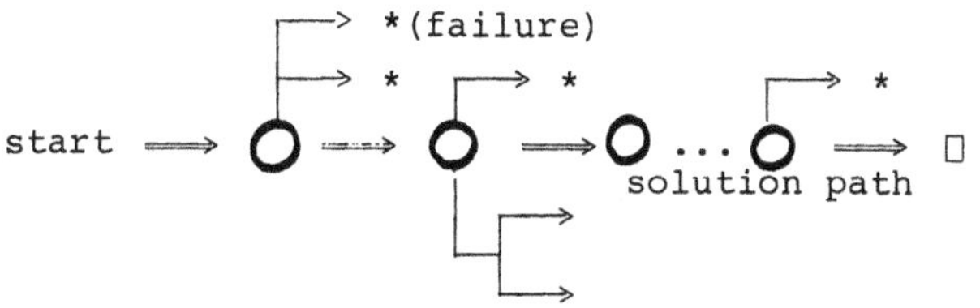

fig. 3 Solution path

If every path through a node s has failed, backtrack starts. We need some definitions before going into the details of the intelligent backtracking method.

We trace such a proof search tree in a left-to-right, up-down order. Resolutions(unifications) are performed when we go though edges. If a unification failed, we label the edge by an asterisk. A proof terminates when we reach the node where the AND-goal is empty. It is labeled by □ (null clause) and called an end node. The total path from a starting node to an end node is called a solution path or simply a solution.

3. DEFINITIONS

Def. 1 : introducing point of a goal template

For a goal template L to be resolved upon at some node s, there must be an ancestor node s' of s where L is introduced to the goal template for the first time. We call s' the introducing point of L and represent s' by intro(L).

We notice two facts.

(F1) For evry solution path through the introducing point of L, there must be a node labeled by L. This is obvious because any goal(template) of the AND-goal needs to be resolved upon until it succeeds.

(F2) Let t,t'... be descendent nodes of intro(L) labeled by L. The subtrees with top node t, t'... are the same since a search tree depends on only the template of an AND-goal. Suppose the AND-goal at intro(L) is $< \beta \angle \alpha , \theta_1 * \ldots * \theta_i >$. Then AND-goals have the form $< \angle \alpha, \theta_1 * \ldots * \theta_i * \mu >$ at t, t'....(μ is the composition of several substitution factors and depends on t, t'...)

Def. 2 : failure of edge

If the resolution(inification) indicated by an edge e does not success, we say that the edge e failed directly. If the resolution succeeded temporarily and a subsuquent search found no solution path through e, then we say that the edge e failed indirectly.

Def. 3 : failure of node

If all edges from a node s failed directly, we say that s failed directly. If there is at least one edge that indirectly failed and the other edges failed directly or indirectly, we say that s failed indirectly.

Note that the first backtracking is caused by the direct failure of some node. Fig.4 depicts the failures of a node.

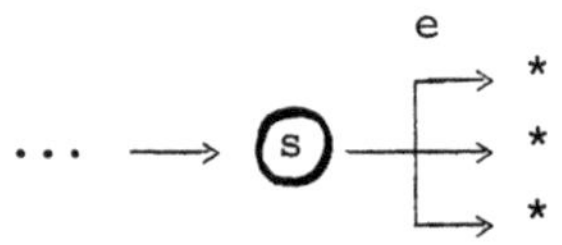

(direct failure of a node s and an edge e)

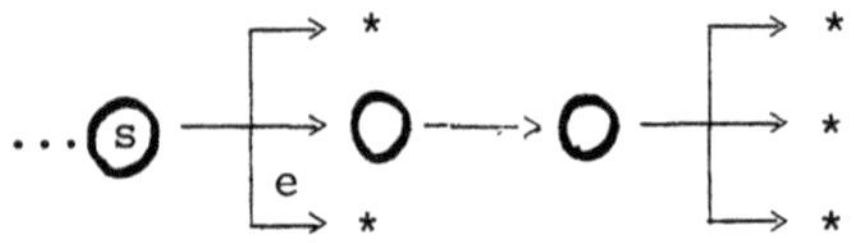

(indirect failure of a node s and an edge e)

fig. 4 Failed node

4. INTELLIGENT BACKTRACKING AT THE DIRECT FAILURE

Since the first backtracking is due to the direct failure of a node, we treat the case of direct failure first.

When an edge e fails directly, the failure often depends on the subset of the previous unifications. To put it differently, the edge e was destined to fail at some previous step.

Def. 4 : agent(e) of direct failure of an edge

Suppose that an edge e incident with a node s failed directly. Let $\{\lambda_1,\ldots,\lambda_k\ (k \geq 0)\}$ be all of the substitution factors referred to at s by UA(unification algorithm, we assume Robinson's unification algorithm[3]) to perform the unification indicated by the edge e. $\{\lambda_1,\ldots,\lambda_k\}$ contains the substitution factors referred to for 'occur check'. We call this set of substitution factors the agent of the direct failure of the edge e and represent it by agent(e).

Def. 5 : occurrence point of a prefix for a set of substitution factors

For a set of substitution factors $S = \{\lambda_1,\ldots,\lambda_k\ (k \geq 0)\}$ included in the substitution part of AND-goal at a node s, there is the farthest ancestor node s' where S is included in the substitution part $\langle \nu \rangle$ of the AND-goal at s'. We call $\langle \nu \rangle$ the prefix for S and s' the occurrence point of S. If k = 0 then s' is the top node of the proof search tree. Or else the AND-goal of s' has the form $\langle \alpha, \ldots \lambda_k \rangle$.

Def. 6 : det(e) of the directly failed edge e incident with a node s labeled by a goal template L

det(e) is defined as the most recent node of s in the occurrence point of agent(e) and intro(L). det(e) has the form $\langle \ldots L \ldots, \nu * \nu' \rangle$ where ν is the prefix for the agent(e).

At det(e) we can foresee the direct failure of an edge e. The reason is as follows:

Suppose that the edge e indicates the unification of L with the head M of some input clause. Let agent(e) be $\{\lambda_1,\ldots,\lambda_k\}$ and ν be the prefix for agent(e). The direct failure of e means that $L\lambda_1*\ldots*\lambda_k$ is not unifiable with M. Accordingly $L\nu$ is not unifiable with M either.

If we follow a path from det(e) or its descendants, we inevitably pass through e again by (F1) and (F2). At that time AND-goal must have the form $\langle L\alpha, \nu * \nu' \rangle$. Although the edge e indicates the unification of $\iota\nu^*\nu'$ with M, it necessarily fails because $\iota\nu$ is not unifiable with M.

From this observation, we can define det(s) for a directly failed node s where the failure of s is already inevitable.

Def. 7 : det(s) for a failed node s

Let {el,...,en} be all nodes incident with s and suppose that every ei failed. det(s) is defined as the most recent node of {det(el),..., det(en)}.

There is no solution path through det(s) or its descendants. For any solution path must go through one of failed edges {el,...,en} by (Fl) and (F2). Therefore our intelligent backtracking method should skip over det(s) or its descendants.

If we return back to the parent node of det(s), at least one of the bindings that caused the failure of s is eliminated. Therefore,

Def. 8 : btk(s) for a failed node s

When a node s failed, the backtracking destination btk(s) is the parent node of det(s).

Def. 9 : agent of a failed node s and det(s)

Suppose that edges el,...,en incident with a node s failed. We define agent(s) and agent(det(s)), by

agent(s) = agent(det(s))
= agent(el) $\cup$ ... $\cup$ agent(en)

Obviously det(s) is the most recent node of the occurrence step of agent(s) ane intro(L) of the goal template L at s. The next facts hold with respect to a failed node s and det(s) (see fig.5).

(F3) Let $\{\lambda_1,\ldots,\lambda_k\ (k \geq 0)\}$ be the agent of a failed node s $=<\alpha, \nu>$. If a node s' can be represented as $<\alpha, \nu'>$ and ν' includes $\{\lambda_1,\ldots,\lambda_k\}$, there is no solution path through s'. Therefore no solution path contains det(s) or its descendants.

(F4) Suppose that a node s failed and det(s), btk(s) are defined respectively. If the substitution factor λ obtained by the edge from btk(s) to det(s) is included in agent(det(s)), λ is the most recently obtained factor in agent(det(s)). The substitution part ν at det(s) is represented as $\nu = \nu'^* \lambda$ for some substitution ν'.

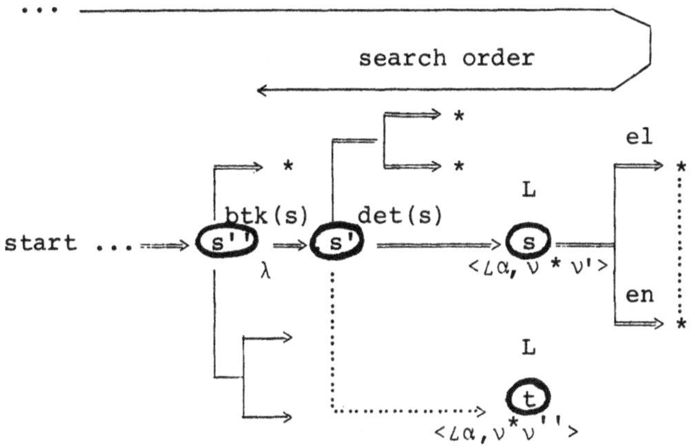

ν is a prefix for agent(s) = agent(s').
Dotted lines stand for OR branches skipped over by intelligent backtracking.

fig. 5 The relationship among directly failed node s, det(s), btk(s)

5. INTELLIGENT BACKTRACKING AT THE INDIRECT FAILURE

In this section we discuss indirect failures. We assume that F3 and F4 are true of every failed node previous to an indirectly failed node s. Note that F3 and F4 hold for the first failed node(direct failure node).

Def. 10 : agent(e) and det(e) for an indirectly failed edge e

Suppose that an edge e incident with a node s labeled by a goal template L failed indirectly. e connects s with its son s'. If we need to start backtracking at s, s' failed already and must be det(s'') for some previously failed node s'' and s is btk(s''). Since we are interested in the backtracking at s, we consider s and s' as such. Therefore agent(s') is already defined. F3, F4 are true of s'. Let θ be the substitution factor obtained by the unification indicated by the edge e.

(1) agent(s') = { } : In this case the casue of failure of s' is the existence of the AND-goal template of s'. We define,

agent(e) = { }
det(e) = intro(L)

(2) agent(s') $\neq$ {} and θ is not included in agent(s') : θ is not responsible for the failure of s'. Therefore,

agent(e) = agent(s')
det(e) = the most recent node in the occurrence node of det(s') and intro(L)

(3) agent(s') $\neq$ { } and θ is included in agent(s') : Let agent(s') be$\{\lambda_1,\ldots,\lambda_k\}$ and λ_k be the most recently produced substitution factor in agent(s'). θ is λ_k since F4 holds for s'. Let $\{\mu_1,\ldots,\mu_\ell\ (l \geq 0)\}$ be the substitution factors referred to by UA to produce λ_k. $\{\mu_1,\ldots,\mu_\ell\}$ excludes the substitution factors referred to only for 'occur check' by UA. This is because in the structure sharing implementation of Prolog UA produces λ_k irrespective of 'occur check' if $\{\mu_1,\ldots,\mu_\ell\}$ exists.

agent(e) = $\{\lambda_1,\ldots,\lambda_{k-1}, \mu_1,\ldots,\mu_\ell\}$
det(e) = the most recent node in the occurence node for agent(e) and intro(L)

We complete the definition of agent(e) and det(e) for an indirectly failed edge e. Every proof search process starting at det(e) or its decendant nodes must pass through the edge e if it succeeds. But the unification indicated by e always fails.(proof omitted. It is based on the assumption that F3, F4 is valid for every node that already failed directly or indirectly)

As for the directly failed edge e, agent(e) and det(e) are already given by def. 4 and def. 6. Thus agent(e) and det(e) are defined for every failed edge incident with an indirectly failed node s. Det(s) and btk(s) for an indirectly failed edge is given by def. 7 and def. 8 respectively. Agent(s) and agent(det(s)) for an indirectly failed node s is given by def. 9. Note that F3 holds for an indirectly failed node s and det(s) and F4 for det(s) and btk(s) again. Thus based on the induction on the number of failures, we can be sure of the safeness of our backtracking method given by from def. 1 to def. 9.

6. MISCELLANEOUS MATTERS

In order to apply our intelligent backtracking method to a real situation, some modifications are needed.

a) false failures :

Our backtracking method is not applicable to failures other than the ones caused entirely by variable bindings. Consider a backtrack containning 'cut'. When such backtrack occurs, our intelligent backtracking method is not applicable since the information about variable bindings expected to be obtained from the skipped paths is lost. Similarly when a user forces a failure to get another answer, this failure is not ascribed to variable bindings. We call such failure a false failure. When we start backtracking at a failed node s and at least one of the failures of s is a false failure, we are compelled to adopt NB(details omitted).

b) 'not' : Safeness of our backtracking method is not insured with respect to failures containning a goal 'not(P)' because even if a goal not(P) occurred and P succeeded, the further instantiated P may fail. But if P in 'not(P)' is ground, this is not the case. Therefore, when 'not(P)' failed and P is ground, our method is applicable to this case.

c) prevention of the same failure :

Our backtracking method redoes one variable binding at one time. Therefore if a failure of an edge e is due to multiple variable bindings, we are in danger of failing again at e. To avert such danger we have only to avoid passing through the edge e unless we return to the ancestor node of det(e).

d) implementation :

Records needed for intelligent backtracking are,

(1) intro(L) for a goal template L
(2) agent(e) for a failed edge
(3) the substitution factor(mgu) for a successful edge and the list of the substitution factors referred to by UA.

In order to implement our method, we add a step identification number to a variable cell when variable binding occurs. For example, if a variable <x> gets its value <a> at N-th step, the record is <x, N, a>. But since recording a step number in a variable cell consumes extra bits and the number of total steps of a proof is unpredictable, we will records the step number block by block. This means that we assign 1 to the first 5 steps and assign 2 to the next 5 steps and so on. Thus we can save bits for recording a step number to avoid "step number over flow".

Our backtrack method could be more intelligent if we built our backtracking theory based on 'substitution component' instead of 'substitution factor'[2]. MGU produced by Robinson's unification algorithm has the form {x1\t1}*...*{xn\tn}. Each {xi\ti} is called substitution component. Since failures depend not on a mgu as a whole but on some substitution components of the mgu, we can develop intelligent backtracking theory based on the dependencies of substitution components which is completely in parallel with our method presented here. The resulting backtracking method, however, will be impractical because of the expense of book keeping for dependencies of substitution compoennts. Therefore we did not discuss such backtracking methods.

7. CONCLUSION

We have proposed an intelligent backtracking method based on a search proof tree obtained from the connection graph of a Prolog program. Although our method is assured to be safe, i.e., never overlooks a solution path, it only points out the step from which the retrial of an altenative search may succeed. It can not indicate a promising step to return. Therefore further refinement of our intelligent backtracking method would have to consider possiblities of success as well as safe.

ACKOWLEDGEMENT : The auther is grateful to Dr. Tanaka, Cheif of Machine Inference Section of Electrotechnical Laboratory and other members of the section for helpful discussion.

REFERENCES :

[1] Bruynooghe, M. : "Analysis of Dependencies to Improve the Behavior Logic Prolog", 5th conf. on Automated Deduction, Lec. Note in Comp. Sci. Springer, 1980.

[2] Bruynooghe, M. and Pereira, L.M. : "Revision of Top-down Logical Reasoninig through Intelligent Backtracking", Departamento de Informatica, Universidade Nova de Lisboa, 1981.

[3] Chang, C.L. and Lee, R.C.T. : "Symbolic Logic and Mechanical Theorm Proving", Academic Press, 1973.

[4] Preira, L.M. and Porto, A. : "Selective Backtracking for Logic Programs", 5th conf. on Automated Deduction, Lec. Note in Comp. Sci. Springer, 1980.

[5] Lasserre, C. and Gallaire, H. : "Controlling Backtrack in Horn Clauses Programming", ACM Logic Programming Workshop, Budapest, 1980.

A pattern matching algorithm in binary trees

Keiji Kojima

Central Research Laboratory, Hitachi, Ltd.
Kokubunji, Tokyo 185, Japan

Abstract

An algorithm is presented which searches all occurrences of a given complete binary tree in another, in running time proportional to the sum of the numbers of their nodes. The algorithm is essentially an application of Knuth-Morris-Pratt's string matching algorithm. An extension to more general tree structures is also described.

1. Introduction

Pattern matching for typical data structures (string, tree, graph etc.) plays an important role in symbolic manipulations ([1]). For string matching problems, a number of algorithms have been devised ([2],[3],[4],[5]). Knuth-Morris-Pratt's string matching algorithm (KMP) ([3]) is noted among others as the first linear time algorithm. This paper introduces a linear time pattern matching algorithm for binary trees, whose basic idea is derived from KMP.

Tree matching algorithm have wide-ranged applications in information processing, e.g., in automatic algebraic simplification and logical deduction. Figure 1.1 illustrates an example of the use of the tree matching algorithm.

```
                *                                  *
T:            /   \                  P:          /   \
             +      1                           x      1
           /   \
          *      1
        /   \
       a      1
                                     P':        * x 1


     T':  * + * a 1 1 1
```

fig.1.1

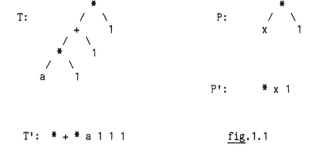

In fig.1.1, the binary tree T represents the algebraic expression (a*1+1)*1 which can be simplified to a+1 using the formula x*1=x. In order to perform this simplification, it is necessary to find all occurences of P in T regarding x as a special node that matches to any node of T. An ovious solution is to apply a string matching algorithm by transforming trees to strings. In fig.1.1, T' and P' are strings which are obtained from T and P by preorder traversing. In fact, *a1, which is an occurrence of P', will be found if we apply a string matching algorithm to T' and P'. Unfortunately, however, no other occurence of pattern P in T can be detected from P' or T'. Such a defect is not particular to preorder ordering. It can be easily proved that no matter what representation may be used, extra informations are required in order to disaround this defect([6]). Besides the obvious loss of time which is consumed by tree-string transformations, this fact shows that use of a string matching algorithm is not the best solution for the tree matching problem.

The tree matching algorithm which is presented in this paper works on tree structures directly. In KMP, string matching is considered as placing the pattern string (say x) over the text string (say y) and sliding x to the right. KMP slides x over y as far as possible making use of the information which is constructed by the analysis of x. The analysis of x requires $O(|x|)$ time (i.e., time proportional to the length of x) and the sliding x over y requires $O(|y|)$ time. Therefore KMP requires $O(|x|+|y|)$ time on a whole. In order to attempt to apply the idea of KMP to tree structures, following problems must be solved.

(i) What information is necessary in order to slide a pattern tree over a text tree efficiently ?

(ii) Is the information able to be constructed in linear time ?

The tree matching algorithm which is described in this paper solves these problems in the case that the pattern tree is complete binary. The algorithm is composed of two stages just as KMP : pattern analysis and matching between the text and the pattern.

In Section 2, we introduce basic data representations as a basis for the discussion in later sections. In Section 3, a linear time algorithm for the analysis of the pattern tree is described. In Section 4, we describe the matching algorithm whose running time is also linear. In Section 5, the matching algorithm is extended to cover text trees which are not completely binary.

2. Data Structures

In this section, we introduce the basic data structures and the operations. Given a complete binary tree with 2^h-1 (h>0) nodes, we represent it by an array T in the following manner :

(i) the root of the tree is stored in T[1],

(ii) the left son and right son of the node which is stored in T[i] are stored in T[2i] and T[2i+1] respectively.

That is, the nodes of the tree are stored in level-first order in array. So hereafter we denote the nodes of a complete binary tree by their level-first numbers.

We define "compact subtree" which is a suitable unit to our discussion.

Definition 2.1

Let T be a complete binary tree. A compact subtree T(i,j) is the tree whose nodes are descendants of i and numbered from up to j. The node i is the root of T(i,j). The node j is called the bottom of T(i,j). The |T(i,j)| represents the number of the nodes of T(i,j). (cf., fig.2.1)

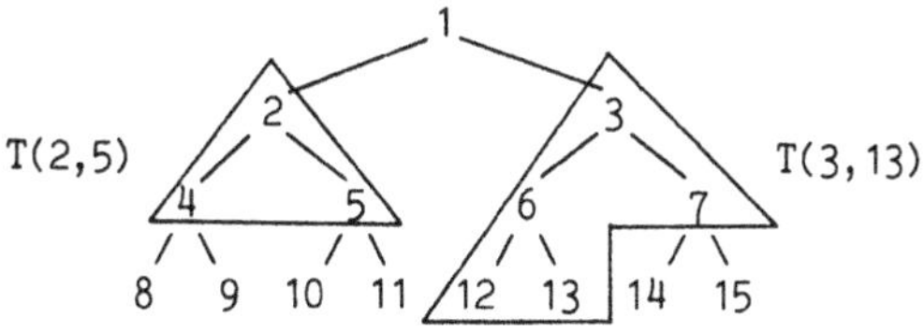

fig.2.1

Using def.2.1, our tree matching problem can be formally described as follows.

Problem

Let T and P be compelte binary trees. T is called a text tree and P is called a pattern tree. Construct an O(|T|+|P|) algorithm that searches all compact subtrees of T which is equal to P.

We define three operations left, right, fwd, which give characteristic positions on compact subtrees.

Definition 2.2

(i) $\mathrm{left}(i,j) = 2^{[\log j]-[\log i]}\, i \quad (i,j\neq 0)$

(ii) $\mathrm{right}(i,j) = 2^{[\log j]-[\log i]}(i+1) - 1 \quad (i,j\neq 0)$

(iii) fwd(i,j) = if left(i,j) $\leq$ j $<$ right(i,j)

then j + 1 else 2left(i,j) $(i,j\neq 0)$

These functions are also defined on i=0 as left(0,j)=j, right(0,j)=j and fwd(0,j)=j. (cf., fig.2.2)

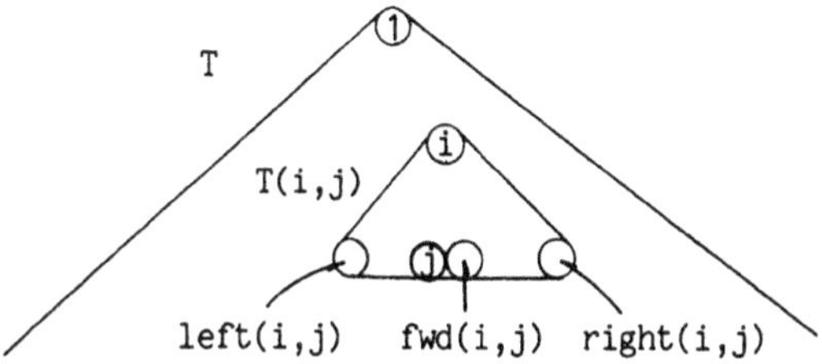

fig.2.2

Proposition 2.1

Let T be a complete binary tree and T(i,j) be a compact subtree of T.

(i) If T(i,k) is the minimal complete compact subtree of T which contains T(i,j), then left(i,j) is the leftmost leaf and right(i,j) is the rightmost leaf of T(i,k).

(ii) The (i,fwd(i,j)) is a compact subtree which contains T(i,j) and |T(i,fwd(i,j))|=|T(i,j)| + 1.

Corollary (i) left(i,j)=j iff T(i,k) is a complete compact subtree where j=fwd(i,k).

(ii) right(i,j)=j iff T(i,j) is a complete compact subtree.

Two auxiliary functions, trans and trans^{-1} are introduced by the following definition.

Definition 2.3

(i) $\text{trans}(i,j) = j - 2^{[\log j]-[\log i]}\ (i-1)\ (i,j\neq 0)$

(ii) $\text{trans}^{-1}(j,k) = 2^{-[\log i]}\ (k-j) + 1\ (j,k\neq 0)$

$\text{trans}(0,j)=\text{trans}^{-1}(0,j)=0$

It can be easily proved that trans(i,j)=|T(i,j)|. The trans^{-1} is an inverse of trans in the sense that $\text{trans}^{-1}(j,\text{trans}(i,j))=i$. The effects of trans and trans^{-1} are illustrated in fig.2.3.

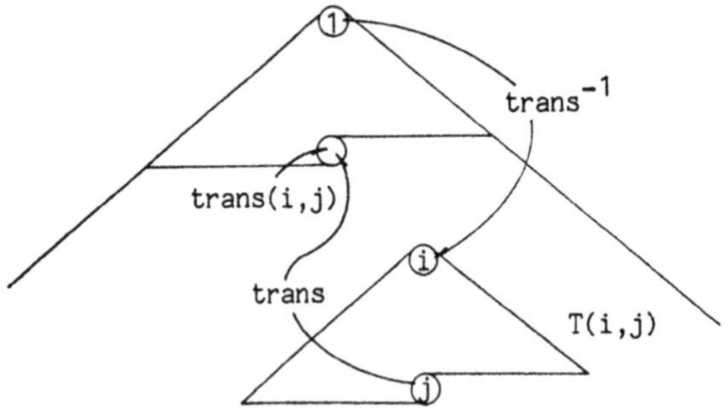

fig.2.3

3. Analysis of pattern tree

In this section, we describe an analyzing algorithm which produces information used to speed-up the matching procedure. This algorithm takes a linear time in the size (number of nodes) of the pattern tree. Throughout section 3,4 and 5, we assume that the pattern tree is stored in array P (size m) and that the text tree is stored in array T (size n). First of all, we define five characteristic compact subtrees. These trees supply all necessary informations for the linear matching.

Definition 3.1

A candidate tree (C-tree) of the node j is the maximal compact subtree P(i,k) that satisfies the following conditions :

(i) $i \neq 1$,

(ii) j = fwd(i,k),

(iii) P(i,k) = P(1,trans(i,k)).

If there is no such a compact subtree, C-tree for j th node of T is an empty tree. This also holds for defs 3.1 to 3.5. Notice that the C-tree for the node j is unique because there always exists the maximal tree among compact subtrees with the same bottom.

Definition 3.2

A success tree (S-tree) of the node j is the maximal compact subtree P(i,j) ($i \neq 1$) such that P(i,j) is equal to P(1,trans(i,j)).

Definition 3.3

A failure tree (F-tree) of the node j is the maximal compact subtree P(i,j) ($i \neq 1$) that satisfies the following conditions :

(i) P(i,k) = P(1,trans(i,k)) where j = fwd(i,k),

(ii) P[j] ≠ P[trans(i,j)].

Definition 3.4

A right tree (R-tree) of the node j is the maximal compact subtree P(i,k) ($i \neq 1$, j = fwd(i,k)) such that P(i,j) is complete and equal to P(1,trans(i,j)).

Definition 3.5

A left tree (L-tree) of the node j is the maximal comapct subtree P(i,j) ($i \neq 1$) such that P(i,j) is complete and equal to P(1,trans(i,j)).

We define the following four functions which return the root of C-,S-,F-,L- or R-tree of a given node.

Definition 3.6

Let X represent 'C','S','F','L' or 'R'. The function X-root is defined as follows :

X-root(j)=i iff i is the root of X-tree for j.

If the X-tree is empty, the value of the function is 0.

We define one more convenient function named 'next'.

definition 3.7

Let X be 'C','S','F','L' or 'R'. Then,
$next(X,j) = next^{(1)}(X,j) = trans(X\text{-}root(j),j)$,
$next^{(k)}(X,j) = next(X,next^{(k-1)}(X,j))$.

Our next task is to construct C-,S-,F-,L- and R-tree in a linear time. This can be achieved by using some inductive relations among these trees. Firstly, we show the inductive relations. The following five propositions show that we are able to find S-,F-,L- and R-tree of the node j if we know C-tree of the node j and these four trees of nodes whose numbers are less than j.

Proposition 3.1

Assume that the C-tree for j (j>1) is an empty tree (i.e., C-root(j)=0). Then,

(i) If P[j]=P[1], then S-root(j)=j, F-root(j)=0, L-root(j)=j and R-root(j)=0.

(ii) Otherwise, S-root(j)=0, F-root(j)=j, L-root(j)=0 and R-root(j)=0.

Proposition 3.2

Assume that C-root(j)=i_c ($i_c \neq 0$) and that P[j]=P[trans(i_c,j)]. Then,

(i) S-root(j)=i_c,

(ii) F-root(j)=$trans^{-1}$(j,next(F,trans(i_c,j))).

Proposition 3.3

Assume that C-root(j)=i_c ($i_c \neq 0$). If left(i_c,j)=j, then R-root(j)=i_c else R-root(j)=$trans^{-1}$(j,next(R,trans(i_c,j))).

Proposition 3.4

Let i_s=S-root(j). If right(i_s,j)=j then L-root(j)=i_s. Otherwise, L-root(j)=$trans^{-1}$(j,next(L,trans(i_s,j))).

Proposition 3.5

Assume that C-root(j)=i_c ($i_c \neq 0$) and that P[j]=P[trans(i_c,j)].

(i) Let m be the smallest integer which satisfies P[j]=P[$next^{(r)}$(F,j')] or $next^{(r)}$(F,j')=0, where j'=trans(i_c,j). Then S-root(j)=$trans^{-1}$(j,$next^{(r)}$(F,j')).

(ii) F-root(j)=i_c.

Proof of Proposition 3.5

(i) Let x=$next^{(r)}$(F,j'), y=$next^{(r-1)}$(F,j') and i_s=S-root(j).

Case 1 : Assume that r is the smallest integer such that P[j]=P[x] and that $next^{(k)}$(F,j')$\neq$0 for $0 \leq k \leq r$. In this case, P(1,x)=P($trans^{-1}$(j,x),j) from the definitions of C-tree and F-tree. Hence $0 \leq i_s \leq trans^{-1}(j,x)$ by def.3.2. On the other hand, by def.3.1 and def.3.2, P(1,j') includes P(1,trans(i_s,j)). If P(1,$next^{(k-1)}$(F,j')) includes P(1,trans(i_s,j)), then P(1,$next^{(k)}$(F,j')) must include P(1,trans(i_s,j)) since F-tree of $next^{(k-1)}$(F,j') would not be maximal otherwise. By simple induction, we can conclude that P(1,x) includes P(1,trans(i_s,j)), hence $trans^{-1}(j,x) \geq i_s$. Therfore $i_s = trans^{-1}(j,x)$.

Case 2 : Assume that r is the smallest integer such that x=0 and that $P[j] \neq P[next^{(k)}(F,j')]$ for $0 \leq k \leq r$. By the similar discussion in Case 1, it can be easily shown that P(1,y) is equal to $P(trans^{-1}(j,y),j)$ except at their bottoms and that P(1,y) includes P(1,trans(i_s,j)). Now suppose that $i_s \neq 0$. Then P(1,trans(i_s,j)) satisfies the condition of def.3.3 (i),(ii). By the maximality of F-tree, however, $y \geq trans(i_s,j) > 0$. It is a contradiction, therfore i_s must be zero. Hence S-root(j)=$trans^{-1}(j,x)$ since $trans^{-1}(j,0)=0$ by def.2.3.

(ii) Immedeately follows from def.3.1 and def.3.3.

If C-,S-,F-,L- and R-tree of nodes whose numbers are less than or equal to j are known, we are able to determine some C-trees of the nodes whose numbers are greater than j. The following two definitions and three propositions show this.

Definition 3.8

Let i_s=S-root(j) and i_c=C-root(j). The R^j is a set of nodes, which is constructed as follows.

(i) If right(i_s,j) < right(i_c,j), then right(i_s,j)+1 $\in R^j$.

(ii) If $x \in R^j$ and right(R-root(x),x) < right(i_c,j), then right(R-root(x),x)+1 $\in R^j$. (cf., fig.3.1)

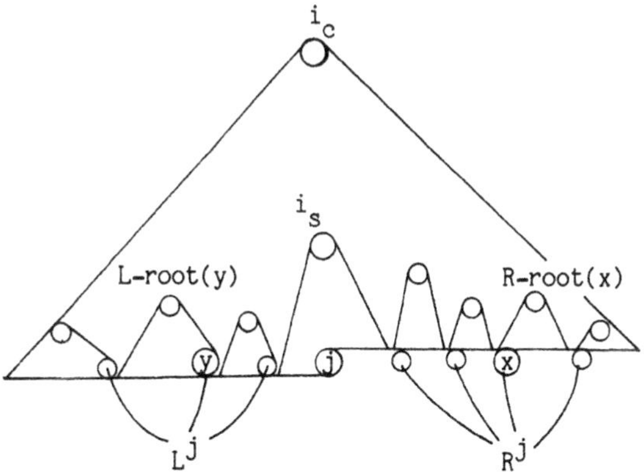

fig.3.1

Definition 3.9

Let i_s=S-root(j) and i_c=C-root(j). Then L^j is a set of nodes being constructed as follows :

(i) If left(i_s,j) > left(i_c,j), then left(i_s,j)-1 $\in L^j$.

(ii) If $x \in L^j$ and left(L-root(x),x) > left(i_c,j), then left(L-root(x),x)-1 $\in L^j$.

Proposition 3.6

Let i_s=S-root(j). If i_s=0, then C-root(2j)=0 and C-root(2j+1)=0. Otherwise, C-root(fwd(i_s,j))=i_s.

Proposition 3.7

Let $x \in R^j$ and i_c=C-root(j). Then, C-root(x)=$trans^{-1}$(x,next(R,trans(i_c,x))).

Proposition 3.8

Let $x \in L^j$. If L-root(x)=0, then C-root(2x)=0 and C-root(2x+1)=0. Otherwise, C-root(fwd(L-root(x),x))=L-root(x).

Now we are able to establish the pattern analysis algorithm which constructs C-,S-,F-,R- and L-tree for a complete binary tree T. The algorithm is based on the inductive relations of C,S,F,R,and L-tree which are shown by props.3.1 to 3.8. In fact, the algorithm is only a translation of props.3.1 to 3.8 into an algol-like program. The correspondence between the proposition and the algol-like statement is as follws :

prop.3.1 : lines 6,7,8,

prop.3.2 : line 11,

prop.3.3 : lines 16,17,

prop.3.4 : lines 18,19,20,

prop.3.5 : lines 12,13,14,15,

prop.3.6 : lines 22,23,

prop.3.7 : lines 24,25,26,27,

prop.3.8 : lines 28,29,30,31,32.

The algorithm uses five arrays c,s,f,l and r corresponding to the five trees. For example, f[j]=F-root(j) after the execution fo the algorithm.

Algorithm 3.1

```
1       (c[1],s[1],f[1],r[1],l[1]):=(0,0,0,0,0);
2       (c[2],c[3]):=(0,0);
3       for j from 2 to m do
4         ic:=c[j];
5         if ic=0
            then
6             if P[j]=P[1]
7               then (s[j],l[j]):=(j,j); (f[j],r[j]):=(0,0)
8               else f[j]:=j; (s[j],r[j],l[j]):=(0,0,0)
              end if
            else
9             j':=trans(ic,j);
10            if P[j]=P[j']
11              then s[j]:=ic; f[j]:=trans^-1(j,next(F,j'))
12              else x:=j'; f[j]:=ic;
```

```
13                until P[x]=P[j] or x=0 do
14                  x:=next(F,x)
                  end until;
15                s[j]:=trans^-1(j,x)
          end if;
16        if left(ic,j)=j then r[j]:=ic
17                          else r[j]:=trans^-1(j,next(R,j'))
         end if;
18        if right(ic,j)=j
19          then l[j]:=s[j]
20           else l[j]:=trans^-1(j,next(L,trans(s[j],j)))
          end if
      end if;
21    is:=s[j];
22    if is=0 then (c[2j],c[2j+1]):=(0,0)
23          else c[fwd(is,j)]:=is
      end if;
24    x:=right(is,j)+1;
25    while x<right(ic,j) do
26      c[x]:=trans^-1(x,next(R,trans(ic,x)));
27      x:=right(c[x],x])+1
      end while;
28    x:=left(is,j)-1;
29    while x>left(ic,j) do
30      if l[x]=0 then (c[2x],c[2x+1]):=(0,0)
31               else c[fwd(l[x],x)]:=l[x]
        end if;
32      x:=left(l[x],x)-1
      end while
    end for
```

The correctness and complexity of algorithm 3.1 are described in the following theorems.

<u>Theorem</u> 3.1

Algorithm 3.1 determines C-tree, S-tree, F-tree, R-tree and L-tree correctly.

<u>Proof of Theorem</u> 3.1

Assume that c[j]=C-root(j) at line 4 in algorithm 3.1. Then, s[j],f[j],l[j] and r[j] are correctly assigned using the inductive relations shown by props.3.1 to 3.5. Therefore, in order to prove this theorem, it is sufficient to show c[j]=C-tree(j) at line 4. To show this, we provide the loop invariant (*) for the major for-loop.

$$(*) \qquad \bigwedge_{k=1}^{p} c[j_k] = \text{C-root}(j_k)$$

where $j_1=j$, $j_k=\mathrm{right}(c[j_{k-1}],j_{k-1})+1$ $(1<k<p+1)$

and $j_{p+1}=2\mathrm{left}(c[j_1],j_1)$. (cf.,fig.3.2)

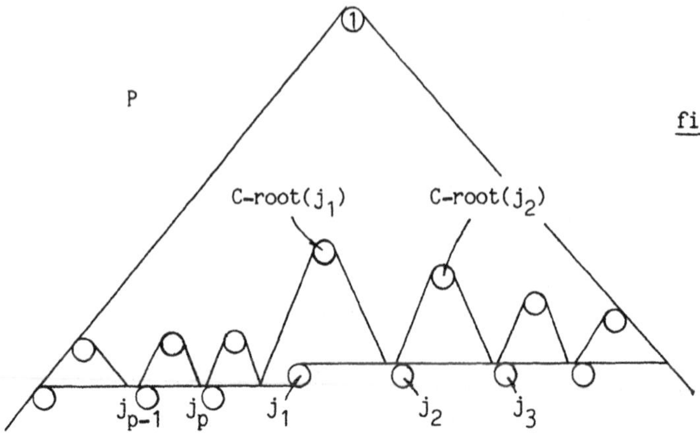

fig.3.2

Now we are going to verify that (*) is always true for each for-loop traversal by induction.

(Basis) When j=1, (*) is clearly true.

(Inductive step) Assume that (*) is true for $j \geq 1$. Then,

Case 1 : If C-root(j)=S-root(j)≠0 and j=right(C-root(j),j), then $j+1=j_2$ and c[fwd(C-root(j),j)]=C-root(j) by prop.3.6. Since $c[j_2]=$C-root(j_2) by hypothesis and c[fwd(C-root(j),j)] is correctly assigned at line 23, (*) is also true for j+1.

Case 2 : If C-root(j)=S-root(j)≠0 and j<right(C-root(j),j), then j+1=fwd(C-root(j),j) and C-root(j+1)=C-root(j) by prop.3.6. Since c[j+1] is correctly assigned at line 23, (*) is true for j+1.

Case 3 : If C-root(j)=S-root(j)=0, then $j+1=j_2$ and C-root(2j)=C-root(2j+1)=0 by prop.3.6. Since C-root(j+1)=c[j+1] by the inductive hypothesis and c[2j] and c[2j+1] are correctly assigned at line 22, (*) is true for j+1.

Case 4 : If C-root(j)≠S-root(j), then R^j and L^j can not be empty. From the constructions of R^j and L^j, which are shown in prop.3.7 and prop.3.8, (*) is clearly true for j+1.

Therfore we have established (*) as the for-loop invariant. Hence algorithm 3.1 correctly determines C,S,F,L and R-tree for the all nodes of the pattern tree.

Theorem 3.2

Algorithm 3.1 requires O(m) time.

Proof of Theorem 3.2

The algorithm has three loops at line 13, line 25 and line 29 on which the time complexity depends. Let us define two functions G and H in order to evaluate the cost of these three loops.

$$G(j) = \sum_{k=1}^{p} |\text{C-tree for } j_k|,$$

$$H(j) = p,$$

where $j_1=j$, $j_k=\text{right}(c[j_{k-1}],j_{k-1})+1$ $(1<k<p+1)$

$j_{p+1}=2\text{left}(c[j_1],j_1)$.

Notice that $c[j_k]=\text{C-root}(j_k)$ at line 4 by the loop invariant in the proof of theorem 3.1. Assume that x:=next(F,x) is executed u(j) times at line 14. Then G(j) decreases by u(j)-1 at least because $u(j)\leq\text{C-root}(j)-\text{S-root}(j)$ and j becomes an element of C-tree for fwd(S-root(j),j) if S-root(j)=0. Therfore $G(j+1)\leq G(j)-u(j)+1$. Hence $\sum_{j=1}^{m} u(j)\leq G(1)-G(m+1)+m\leq m$ since G(1)=0 and $G(m+1)\geq 0$. That is, line 14 is executed at most m times during the execution of algorithm 3.1. On the other hand, if the while loops at line 25 and 29 are repeated v(j) times, then $H(j+1)\geq H(j)+v(j)$ since at least v(j) new C-trees are added. Consequently, $\sum_{j=1}^{m} v(j)\leq H(m+1)-H(1)\leq m$ since $H(m+1)\leq m+1$ and H(1)=1. Therfore the total cost of the loop executions at lines 13, 25 and 29 is O(m). Hence algorithm 3.1 requires O(m) time.

4. Matching algorithm

We are going to construct the linear pattern matching algorithm for complete binary trees. The matching algorithm is quite similar to algorithm 3.1. In fact, the matching algorithm can be obtained from natural extensions of the ideas which we have developed in the previous section.

To begin with, we extend the definitions of C-tree and S-tree.

Definition 4.1

C-tree for j th node of T is the maximal compact subtree T(i,k) which satisfies the following conditions :

(i) j = fwd (i,k),

(ii) T(i,k) = P(1,trans(i,k)).

Definition 4.2

S-tree for j th node of T is the maximal comapct subtree T(i,j) which is equal to P(1,trans(i,j)) where trans(i,j)=m.

Since defs.3.1 to 3.2 are a special case that T=P and i=1 in def.4.1 and def.4.2, the revised versions of C-tree and S-tree are the extensions of old ones. From hereafter, we adopt defs.4.1 to 4.2 as the definitions of them. Because we do not utilize C-tree and S-tree of P which are constructed by algorithm 3.1 at all in this section, this causes no confusion.

By def.4.1, constructing C-tree for every node of T is equivalent to solving our pattern matching problem. That is, we are searching for C-tree T(i,k) such that |T(i,j)|=m and T[j]=P[m] with j=fwd(i,k). From this point of view, algorithm 3.1 is considered to be a matching algorithm between pattern tree and itself by regarding T as P in defs.3.1 to 3.2. Using def.4.1 and def.4.2, props 3.1,3.2 and 3.5 holds with minor changes as follows.

<u>Proposition</u> 4.1

Assume that C-root(j)=0. If T[j]=P[1], then S-root(j)=j, otherwise, S-root(j)=0.

<u>Proposition</u> 4.2

Assume that C-root(j)=i (i=0) and that T[j]=P[trans(i,j)]. If trans(i,j)$\neq$m, then S-root(j)=i, otherwise, S-root(j)=trans^{-1}(j,next(L,trans(i,j))).

<u>Proposition</u> 4.3

Assume that C-root(j)=i_c ($i_c \neq 0$) and that T[j]=P[trans(i_c,j)]. Then, S-root(j)=trans^{-1}(j,next$^{(r)}$(F,j')) where j'=trans(i_c,j) and r is the smallest integer which satisfies T[j]=P[next$^{(r)}$(F,j')] or next$^{(r)}$(F,j')=0.

Props. 3.6, 3.7 and 3.8 still true with no change.

Now we show the matching algorithm. This algorithm searches all T(i,j) which are equal to P and prints pair (i,j), constructing C-tree and S-tree for T and P. The construction of C- and S-tree are performed using prop.4.1 prop.4.3 and prop.3.6 prop.3.8 just as algorithm 3.1. As we have mentioned, the matching algorithm is essentially the same as algorithm 3.1. The correspondence between proposition and line is as follows :

prop.4.1 : line 5,
prop.4.2 : line 8,
prop.4.3 : lines 9, 10, 11,
prop.3.6 : lines 14, 16,
prop.3.7 : lines 17, 18, 19, 20,
prop.3.8 : lines 21, 22, 23, 24, 25, 26.

<u>Algorithm</u> 4.1

```
1     c[1]:=0;
2     for j from 1 to n do
3       ic:=c[j];
4       if ic=0
          then
5           if T[j]=P[1] then s[j]:=j else s[j]:=0 end if
          else
6           j'=trans(ic,j);
```

```
            if T[j]=P[j']
              then s[j]:=ic; if j'=m then print(ic,j)
              else x:=j';
                   until T[j]=P[x] or x=0 do
                     x:=next(F,x)
                   end until;
                   s[j]:=trans^-1(j,x)
            end if;
        end if;
        is:=s[j];
        if is=0 then (c[2j],c[2j+1]):=(0,0)
                else
                   if j'=m then is:=trans^-1(j,next(L,j')) end if;
                   c[fwd(is,j)]:=is
        end if;
        x:=right(is,j)+1;
        while x<right(ic,j) do
          c[x]:=trans^-1(x,next(R,trans(ic,x)));
          x:=right(c[x],x)+1
        end while;
        x:=left(is,j)-1;
        while x>left(ic,j) do
          y:=trans^-1(x,next(L,trans(ic,x)));
          if y=0 then (c[2x],c[2x+1]):=(0,0)
                 else c[fwd(y,x)]:=y
          end if;
          x:=left(c[x],x)-1
        end while
      end for
```

Theorem 4.1

Algorithm 4.1 searches all occurence of T(i,j) such that T(i,j)=P in O(n) time.

We are able to prove this theorem in quite a similar manner by supplying the same for-loop invariant and the cost evaluation functions as proofs of Theorem 3.1 and Theorem 3.2. We omit the detail.

5. Extension

Algorithm 4.1 can be extended to cover text trees which are not completely binary. Assume that a text tree T with n nodes is stored in array text in level-first order. The lson, rson and parent are arrays such taht lson[k], rson[k] and parent[k] are the left son, right son and parent of node k, respectively. (cf., fig.5.1)

```
            a
           / \
T         b   c
         / \
        d   a
           / \
          c   a
```

k	1	2	3	4	5	6	7
text	a	b	c	d	a	c	a
parent	0	1	1	2	2	5	5
lson	2	4	0	0	6	0	0
rson	3	5	0	0	7	0	0

fig.5.1

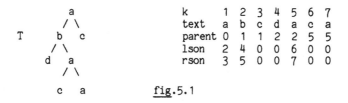

The T is embedded into a complete binary tree T' by two functions f:T->T' and f':T'->T. That is, for node k of T, f(k) gives the corresponding node of T' and f' is the inverse of f (cf., fig.5.2).

T

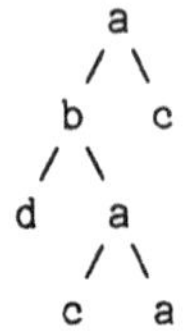

k	1	2	3	4	5	6	7
f	1	2	3	4	5	10	11

T'

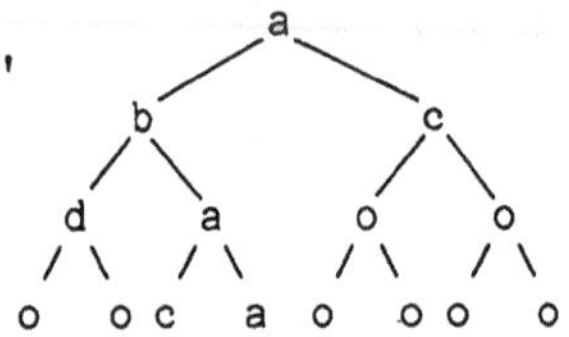

k	1	2	3	4	5	10	11
f'	1	2	3	4	5	6	7

fig.5.2

The table of f and f' can be constructed in O(n) time in an obvious manner. It is possible to calculate f' for nodes which do not appear in the table while their parents do.

f'(k) = if k is even then lson[f'(k/2)] else rson[f'(k-1/2)]

Note that the number of the nodes of T' for which f' is defined is O(n). Now assume that the construction of C-tree for T and P proceeds upto j th node of T and that C-root(j)=i_c. Then, using f, T(i_c,j) is translated to T'(f(i_c),f(j)) (cf., fig.5.3). That is, f(j) and f(i_c) are used instead of j and i_c, respectively, in algorithm 4.1.

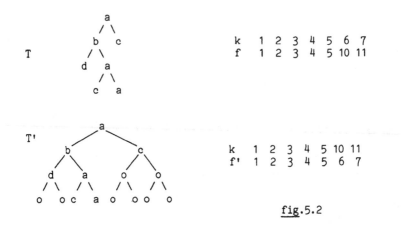

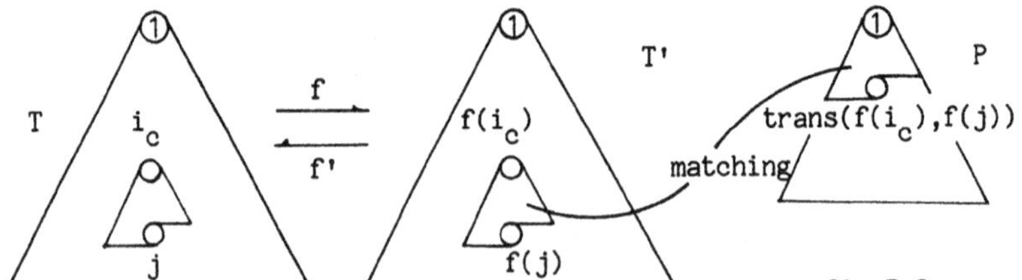

fig.5.3

Since T' is complete, algorithm 4.1 correctly works for T'(f(ic),f(j)) and decides C-trees for appropriate nodes of T'. Let us assume that C-root(x)=y for nodes x and y of T'. This time T'(y,x) is translated to T(f'(y), f'(x)). That is, let C-root(f'(x))=f'(y) in T.

There is a possibility, however, that the corresponding node to x may not exist in T. (That is, f'(x)=0. Notice that f' is defined for x since f'(parent[x]) is not zero in algorithm 4.1.) In this case the extended algorithm searches the node x' of T' such that $x<x'\leq right(y,x)$ and $f'(x')=0$. For such x', C-root(f'(x))=f'(y') in T where $y'=trans^{-1}(x',next(R,trans(y,x')))$. If there remains the node x'' such that $right(y',x')<x''\leq right(y,x)$ and $f'(x'')=0$, then C-root(f'(x''))=f'(y'') where $y''=trans^{-1}(x'',next(R,trans(y,x'')))$. This process continues until there remains no such x . A similar process is performed to the left of x using appropriate L-trees and R-trees.(cf., fig.5.4)

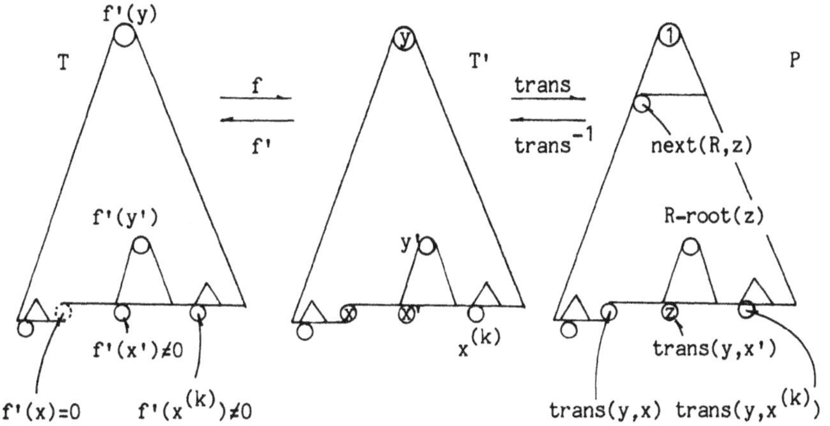

fig.5.4

After these processes, the construction of C-tree successfully proceeds to j+1 st node of T since all necessary C-trees have been decided and quite a similar loop invariant which is used in the proof of Theorem 3.1 also holds. The time complexity of the extended algorithm is O(n) since we have only to have access to the nodes of T' on which f' is defined. The space complexity is clearly O(n) from the above data structures.

Acknowledgements

The author wishes to express his deep appreciation to Professor Reiji Nakajima for his helpful advices. He also thanks to Etsuya Shibayama and Tatsuya Hagino for valuable discussions with them.

References

[1] Donald E.Knuth, Fundamental Algorithm, The art of Computer Programming, Vol.1, Addison-Wesley, Reading,Mass., 1968; 2nd edition 1973.

[2] Malcom C.Harrison, Implementation of the substring test by hashing, Comm.ACM,14 (1971), pp 777-779.

[3] D.E.Knuth, J.H.Morris,Jr., V.R.Pratt, Fast Pattern Matching in Strings, SIAM J. of Computer, Vol.6, No.2, June 1977, pp.323-350.

[4] Alfred .V.Aho and Margaret J.Corasick, Efficient string matching: An aid to

bibliographic search, Comm.ACM, 18 (1975), pp.333-340.
[5] R.S.Boyer and J.S.Moore, A fast string searching algorithm, Comm.ACM, Vol.20, No.2, Oct. 1977.
[6] C.M.Hoffmann and M.J.O'donnell, Pattern Matching in Trees, Journal of ACM, Vol.29, No.1, January 1982, pp.68-95.
[7] K.Kojima, A linear tree matching algorithm, Master Thesis, Kyoto University.

Polynomial Time Inference of Extended Regular Pattern Languages

Takeshi Shinohara

Computer Center, Kyushu University 91,
Fukuoka, 812 Japan

ABSTRACT

A pattern is a string of constant symbols and variable symbols. The language of a pattern p is the set of all strings obtained by substituting any non-empty constant string for each variable symbol in p. A regular pattern has at most one occurrence of each variable symbol. In this paper, we consider polynomial time inference from positive data for the class of extended regular pattern languages which are sets of all strings obtained by substituting any (possibly empty) constant string, instead of non-empty string. Our inference machine uses MINL calculation which finds a minimal language containing a given finite set of strings. The relation between MINL calculation for the class of extended regular pattern languages and the longest common subsequence problem is also discussed.

1. Introduction

Two kinds of inferences have been known, deductive inference and inductive inference. Many studies on deductive inference cover a wide range from theoretical problems to practical problems. Although some theories of inductive inference have been developed, few of them have reached practical applications to computer softwares. This study presents an approach to practical applications of inductive inference and gives its theoretical basis.

The direct motivation of this research is to develop a data entry system with learning function, proposed by Arikawa[4]. The following is a brief description of the learning system.

At the stage of data entry from a keyboard, a user types, for example, some bibliographic data:

$
Author: Angluin, D.
Title: Inductive Inference of Formal Languages from Positive Data
Journal: Inform. Contr. 45
Year: 1980
$
Author: Maier, D.
Title: The Complexity of Some Problems on Subsequences and Supersequences
Journal: JACM 25
Year: 1978
$
...,

where $ is a special symbol to mark off records. Halfway through this entry process, the system will learn or infer a structure of records in the form

Author: <w>Title: <x>Journal: <y>Year: <z>,

where each of <w>, <x>, <y>, and <z> stands for any string, and it will successively emit the constant parts "Author: ", "Title: ", "Journal: ", and "Year: " as prompts. Then the user may only type the data corresponding to <w>, <x>, <y>, and <z>.

The information, the system can use, is only the input data. Hence we should consider inference from positive data. Another important problem is computational complexity. The system should respond quickly. For these reasons we consider a polynomial time inference from positive data.

The inference from positive data has been considered of little interest to study, since Gold[4] proved that any class of languages over an alphabet is not inferrable from positive data if it contains all finite languages and at least one infinite languages, and hence that even the class of regular sets is not inferrable from positive data.

Recently Angluin[2,3] gave new life to the study of inference by characterizing the class of languages inferrable from positive data and presenting interesting classes. The class of pattern languages is one of her classes.

A pattern is a string of constant symbols and variable symbols, and the language of a pattern p is the set of all strings obtained by substituting any non-empty constant string for each variable by substituting any <u>non-empty</u> constant string for each variable symbol in p. A regular pattern is a pattern in which each variable symbol occurs at most once. Note that the structure obtained by the learning data entry system is a regular pattern. The class of regular pattern languages has been shown to be inferrable from positive data [Shinohara, 9].

In this paper we first point out some problems of our previous version of inference method and then we give a solution to them by considering polynomial time inferrability of extended regular pattern languages. Our extension is to allow the substitutions to erase some variable symbols. More precisely, an extended language of a pattern p is a set of all strings obtained by substituting any (<u>possibly</u> <u>empty</u>) constant string, instead of non-empty string, for each variable symbol in p. For example, the extended language of a pattern "0x1" can contain string "01" while the language by Angluin can not. The erasing variables requires a new discussion.

The inference, we deal with here, is carried out by using MINL calculation introduced by Angluin[2]. Hence our main attention is paid to the time complexity of MINL calculation for the class of extended regular pattern languages. MINL for extended regular pattern languages finds a regular pattern which represents a minimal extended regular pattern language containing a given non-empty finite set of strings. We also refer to the relation between MINL calculation for extended regular pattern languages and the longest common subsequence problem. We propose an algorithm which calculates MINL for extended regular pattern languages in polynomial time. By using this fact, the class of extended regular pattern languages is shown to be polynomial time inferrable from positive data.

2. Preliminaries

We begin with a brief review of previous results.

2.1. Patterns and Their Languages

Let Σ be a finite set of symbols containing at least two symbols and let $X = \{x_1, x_2, \dots\}$ be a countable set of symbols disjoint from Σ. Elements in Σ are called <u>constants</u> and elements in X are called <u>variables</u>. A <u>pattern</u> is any string over $\Sigma \cup X$. The set $(\Sigma \cup X)^*$ of all patterns is denoted by P.

We say that a pattern p is <u>regular</u> if each variable in p occurs exactly once in p.

Let f be a non-erasing homomorphism from P to P. If f(a) = a for any constant a, then f is called a substitution. If f is a substitution, f(x) is in X, and f(x) = f(y) implies x = y for any variables x and y, then f is called a renaming of variables. We use a notation $[a_1/v_1, \ldots, a_k/v_k]$ for the substitution which maps each variable symbol v_i to a_i and every other symbol to itself. We define two binary relations on P as follows:

1) $p \equiv' q$ iff $p = f(q)$ for some renaming of variables f,
2) $p \leq' q$ iff $p = f(q)$ for some substitution f.

The language of a pattern p, denoted by L(p), is the set $\{ w \in \Sigma^* \mid w \leq' p\}$. These syntactic relations $\equiv'$ and $\leq'$ are characterized by the following lemma.

Lemma 1. [2]

1) For all patterns p and q, $p \equiv' q$ iff $L(p) = L(q)$.

2) For all patterns p and q, if $p \leq' q$ then $L(p) \subseteq L(q)$, but the converse is not true in general.

3) If p and q are patterns such that $|p| = |q|$, then $p \leq' q$ iff $L(p) \subseteq L(q)$.

2.2 Polynomial Time Inference from Positive Data

Inference machine is an effective procedure which requires inputs from time to time and produces outputs from time to time. Let $s = s_1, s_2, \ldots$ be an arbitrary infinite sequence, and let $g = g_1, g_2, \ldots$ be a sequence of outputs produced by an inference machine M when inputs in s are successively given to M on request. Then we say that M on input s converges to g_0 iff g is a finite sequence ending with g_0 or all but finitely many elements of g are equal to g_0.

Let $\mathcal{L} = L_1, L_2, \ldots$ be a class of recursive languages, and let $s = s_1, s_2, \ldots$ be an arbitrary enumeration of some language L_i. Then we say that a machine M infers $\mathcal{L}$ from positive data if M on input s converges to an index j with $L_j = L_i$. We say that a class $\mathcal{L}$ is inferrable from positive data if there exists a machine which infers $\mathcal{L}$ from positive data.

Theorem 1. [2] If a class $\mathcal{L} = L_1, L_2, \ldots$ satisfies the following condition, then $\mathcal{L}$ is inferrable from positive data.

Condition: For any non-empty finite set S of strings, the set $\{ L \mid S \subseteq L, L = L_i \text{ for some index } i\}$ has finite cardinality.

Lemma 2. [2] The class of pattern languages satisfies Condition of Theorem 1.

Hereafter we omit the phrase "from positive data", hence for example "inference" means "inference from positive data."

An inference by a machine M is consistent iff a language L_{g_i} contains all inputs given so far whenever M produces output g_i. An inference is conservative iff an output g_i from M is never changed unless L_{g_i} fails to contain some of the inputs. These two properties are natural and valuable in inference problem. It is, however, known that inferrability does not always mean consistency and conservativeness [2].

A class $\mathcal{L}$ is polynomial time inferrable iff there exists an inference machine M which infers $\mathcal{L}$ consistently and conservatively, and requests a new input in polynomial time (with respect to the length of the inputs read so far) after the last input has been received.

MINL calculation for a class $\mathcal{L} = L_1, L_2, \dots$ is defined by Angluin[2] as follows:

MINL(S) = "Given non-empty finite set S of strings, find an index i such that $S \subseteq L_i$ and for no index j, $S \subseteq L_j \subsetneq L_i$."

The following theorem shows the importance of MINL calculation.

Theorem 2. [2] If a class $\mathcal{L} = L_1, L_2, \dots$ satisfies Condition of Theorem 1 and MINL for $\mathcal{L}$ is computable, then the procedure Q below infers $\mathcal{L}$ consistently and conservatively.

```
procedure Q;
  begin
    g1 := "none" ;  S := Ø ;
    for each input si do
       begin
          S := S ∪ {si} ;
          if si ∈ Lgi then
                gi+1 := gi
          else
             begin
                gi+1 := MINL(S) ;
                output gi+1
             end
       end
  end
```

Corollary 1. If a class $\mathcal{L} = L_1, L_2, \ldots$ satisfies Condition of Theorem 1, and the membership decision and MINL calculation for $\mathcal{L}$ are computable in polynomial time, then the class $\mathcal{L}$ is polynomial time inferrable from positive data.

Angluin [2] showed that the membership decision of pattern languages is NP-complete and ℓ-MINL calculation, a special case of MINL for pattern languages, is NP-hard.

ℓ-MINL(S) = "Given non-empty finite set S of strings, find a pattern of maximum possible length which represents a minimal pattern language containing S."

The following summarize the results of our previous study [9].

Lemma 3. For any regular pattern p and any string w, whether $w \in L(p)$ or not is decidable in $O(|p|+|w|)$ time.

Theorem 3. The following procedure computes ℓ-MINL(S) for regular pattern languages in $O(m^2 n)$ time, where $m = \max\{|w|;\ w \in S\}$, $n = card(S)$, and $w = a_1 \ldots a_k$ ($a_i \in \Sigma$) is one of the shortest strings in S.

```
begin
   p1 := x1...xk ;
   for i := 1 to k do
      begin
         q := pi[ai/xi] ;
         if S ⊆ L(q) then pi+1 := q
                     else pi+1 := pi
      end ;
   return pk+1
end
```

Theorem 4. The class of regular pattern languages is polynomial time inferrable from positive data.

3. Some Problems on ℓ-MINL Calculation for Regular Pattern Languages

There are some difficulties in the ℓ-MINL calculation when the polynomial time inference of regular pattern languages is applied to practical use. The main reasons are

1) restriction on the length of pattern, and
2) prohibition against substituting empty string for any variable.

We present some examples to explain these problems.

Example 1. Let S be the set {ABCdeFGh, ABCiFGjk}. Then every answer of ℓ-MINL is eight symbols long because the length of the shortest words in S is eight. Let p be any pattern of the form p_1FGp_2, where p_1 and p_2 are any regular patterns. Assume $S \subseteq L(p)$. Then, clearly, $ABCi \in L(p_1)$ and $h \in L(p_2)$, therefore

$$|p| = |p_1| + |FG| + |p_2| \leq 4 + 2 + 1 = 7 < 8.$$

Hence the string "FG" does not appear in any answer of ℓ-MINL(S). However the pattern $q = ABCx_1FGx_2$ is a possible answer of MINL(S).

Thus MINL(S) may have an answer which contains more constant symbols than any answer of ℓ-MINL(S).

Example 2. Let S = {aBcdf, GHcdBiii}. Then both patterns

$$p_1 = x_1Bx_2x_3x_4 \text{ and}$$

$$p_2 = x_1x_2cdx_3$$

are correct answers of ℓ-MINL(S). Our ℓ-MINL algorithm of Theorem 2 returns p_1 for S. If we change the order of substitutions in the algorithm, we can get p_2 as the answer of ℓ-MINL(S).

Example 3. Let S = {ABC, AC}. Then MINL(S) does not have any answer containing both symbols A and C because we can not substitute empty string for any variable.

To solve these problems, we extend the definition of pattern languages to allow erasing substitutions.

4. Extension of Pattern Languages

We give new definitions on pattern languages to allow substitutions to erase variables and we show some of their properties. The definitions of patterns and regular patterns are the same ones as in Section 2.

A substitution is any (possibly erasing) homomorphism from P to P which maps each constant symbol to itself. A special substitution which maps each variable to empty string is denoted by c. For example, if $\Sigma = \{0, 1, 2\}$ and $X = \{x, y, \dots\}$, then $c(0x1y2) = 012$. We define two binary relations $\leq'$ and $\equiv'$ as follows:

1) $p \leq' q$ iff $p = f(q)$ for some substitution f,
2) $p \equiv' q$ iff $p \leq' q$ and $q \leq' p$.

The <u>language of a pattern</u> p, denoted by L(p), is the set $\{ w \in \Sigma^* \mid w \le' p \}$. Hereafter we use the term "pattern languages" in the sense just defined above.

<u>Proposition 1</u>.

1) $p \le' q \implies L(p) \subseteq L(q)$.
2) $p \equiv' q \implies L(p) = L(q)$.

We say that a pattern $\hat{p}$ is in <u>canonical form</u> iff

$\hat{p} \equiv' q \implies |\hat{p}| \le |q|$ for any pattern q, and

$\hat{p}$ contains exactly k variables $x_1, x_2, \ldots, x_k$ for some integer k and the leftmost occurrence of x_i is to the left of the leftmost occurrence of x_{i+1} for i = 1, ..., k-1.

Note that the condition $|\hat{p}| \le |q|$ in the definition above is neccessary. For example, a pattern $0x_1x_21$ is not in canonical form because $0x_1x_21 \le' 0x_11 \le' 0x_1x_21$.

<u>Theorem 5</u>. There exists a unique canonical pattern $\hat{p}$ equivalent ($\equiv'$) to p for any regular pattern p.

<u>Proof</u>. Let $p = w_0v_1w_1v_2\ldots w_{n-1}v_nw_n$ ($w_0, w_n \in \Sigma^*$, $w_i \in \Sigma^+$ $(i=1,\ldots,n-1)$, $v_j \in X^+$ $(j=1,\ldots,n)$) be any regular pattern. Then $\hat{p} = w_0x_1w_1x_2\ldots w_{n-1}x_nw_n$ is in canonical form and $\hat{p} \equiv' p$. Any regular pattern equivalent to p is in the form $w_0u_1w_1u_2\ldots w_{n-1}u_nw_n$ $(u_i \in X^+)$. Therefore the uniqueness of such canonical pattern is obvious.

<u>Lemma 4</u>. If p is a canonical regular pattern, then
$|p| \le 2|c(p)| + 1$.

<u>Theorem 6</u>. The class of extended regular pattern languages satisfies Condition of Theorem 1 and it is inferrable from positive data.

<u>Proof</u>. Let $S \subseteq \Sigma^*$ be any non-empty finite set of strings and let w be one of the shortest strings in S. Assume $S \subseteq L(p)$, where p is any canonical regular pattern. Then $|w| \ge |c(p)|$ because $w \in L(p)$. By Lemma 4, $|p| \le 2|c(p)| + 1 \le 2|w| + 1$. Therefore the number of such patterns p is finite.

<u>Theorem 7</u>. For any regular pattern p and any string w, whether $w \in L(p)$ or not is decided in $O(|p|+|w|)$ time.

<u>Proof</u>. We can construct a deterministic finite automaton

recognizing L(p) in O(|p|) time by using the method of pattern matching machines [Aho, et al., 1].

5. MINL Calculation for Regular Pattern Languages

To show polynomial time inferrability of regular pattern languages, we first discuss MINL calculation. We also refer to the relation between MINL calculation and the longest common subsequence (LCS for short) problem.

We start with the definitions on subsequences:

1) For any strings $w = a_1 \ldots a_k$ ($a_i \in \Sigma$) and $s \in \Sigma^*$, $s \le w$ (or $w \ge s$) iff $s = a_{i_1} \ldots a_{i_m}$ $(1 \le i_1 < \ldots < i_m \le k)$.
We say that s is a <u>subsequence</u> of w (or w is a <u>supersequence</u> of s) if $s \le w$ (or $w \ge s$).

2) The set of <u>common subsequences</u> of a set S of strings is $CS(S) = \{ s \in \Sigma^* \mid s \le w \text{ for any string } w \in S \}$.

3) The set of <u>maximal common subsequences</u> of S is $MCS(S) = \{ s \in CS(S) \mid s = s' \text{ or } s \not\le s' \text{ for any } s' \in CS(S) \}$.

4) The set of the <u>longest common subsequences</u> of S is $LCS(S) = \{ s \in CS(S) \mid |s| \ge |s'| \text{ for any } s' \in CS(S) \}$.

<u>Proposition 2</u>.

1) $w \in L(p) \implies w \ge c(p)$
2) $L(p) \subseteq L(q) \implies c(p) \ge c(q)$
3) $L(p) = L(q) \implies c(p) = c(q)$
4) $S \subseteq L(p) \implies c(p) \in CS(S)$

We need three notations in the discussions below:

1) For any string $w = a_1 \ldots a_n$ and any integers i and j,

$$w\langle i:j \rangle = \begin{cases} a_i \ldots a_j & (\text{if } 1 \le i \le j \le |w|) \\ \varepsilon & (\text{otherwise}), \text{ and} \end{cases}$$

$w\langle i \rangle = a_i$.

2) For any symbol a and any integer i,

$$a^i = \begin{cases} \varepsilon & (\text{if } i=0) \\ aa^{i-1} & (\text{otherwise}). \end{cases}$$

3) For any variables $v_1, \ldots, v_k \in X$ and any constant strings $w_1, \ldots, w_k \in \Sigma^*$, $[w_1/v_1, \ldots, w_k/v_k]$ denotes the substitution which maps each variable v_i to w_i and every other variable to itself.

Theorem 8. Let p and q be any regular patterns and $\mathrm{card}(\Sigma) \geq 3$. Then $L(p) \subseteq L(q)$ implies $p \leq' q$.

Proof. We may assume, without loss of generality, that p and q are canonical regular patterns. We also assume $\mathrm{card}(\Sigma) \geq 3$, $L(p) \subseteq L(q)$, but $p \not\leq' q$. Let $q = w_0x_1w_1 \ldots w_{n-1}x_nw_n$, where $w_0, w_n \in \Sigma^*$, and $w_i \in \Sigma^+$ $(i=1,\ldots,n-1)$. Since c(p) is a supersequence of c(q) and $p \not\leq' q$, there exist integers i, j, and k such that

$0 \leq i \leq n$, $1 \leq j < k \leq |p|$, and
$p = p\langle 1:j\rangle\, p\langle j+1:k-1\rangle\, p\langle k:|p|\rangle$, where
$c(p\langle 1:j\rangle) \in L(w_0x_1 \ldots x_{i-1}w_{i-1})$,
$c(p\langle 1:j'\rangle) \notin L(w_0x_1 \ldots x_{i-1}w_{i-1})$ for any integer $j' < j$,
$p\langle j+1:k-1\rangle \neq rw_ir'$ for any patterns r and r',
$c(p\langle k:|p|\rangle) \in L(w_{i+1}x_{i+2} \ldots x_nw_n)$, and
$c(p\langle k':|p|\rangle) \notin L(w_{i+1}x_{i+2} \ldots x_nw_n)$ for any integer $k' > k$.

Let $p_1 = p\langle 1:j\rangle$, $p_2 = p\langle j+1:k-1\rangle$, and $p_3 = p\langle k:|p|\rangle$.
Then $L(p_2) \subseteq L(x_iw_ix_{i+1})$ because $L(p) \subseteq L(q)$ and $c(p_1)p_2c(p_3) \leq' p$. Let a be any constant symbol except $w_i\langle 1\rangle$ and $w_i\langle |w_i|\rangle$ and let $v_1, \ldots, v_m$ be all variables in p_2. Then $p_2[a^{|w_i|}/v_1,\ldots,a^{|w_i|}/v_m] \in L(p_2) - L(x_iw_ix_{i+1})$. This contradicts $L(p_2) \subseteq L(x_iw_ix_{i+1})$.

The following lemma says that the condition $\mathrm{card}(\Sigma) \geq 3$ is necessary in Theorem 8.

Lemma 5. When $\mathrm{card}(\Sigma) = 2$, there exist regular patterns p and q such that $L(p) \subseteq L(q)$, $p \not\leq' q$, and $q \not\leq' p$.

Proof. Let $\Sigma = \{0,1\}$, $p = x_101x_20x_3$, and $q = x_10x_210x_3$. Then, clearly, $p \not\leq' q$, $q \not\leq' p$, but $L(p) = L(q)$.

Hereafter we assume that the constants alphabet Σ contains at least three symbols.

Theorem 9. For any maximal common subsequence $s \in MCS(S)$, there exists an answer p of MINL(S) for regular pattern languages such that $c(p) = s$.

Proof. Let $s = a_1 \ldots a_k \in MCS(S)$. Then the pattern q_{k+1} defined as follows is an answer of MINL(S):

$$q_i := \begin{cases} x_1a_1 \ldots a_kx_{k+1} & (i=0) \\ \text{if } S \subseteq L(q_{i-1}[\varepsilon/x_i]) \text{ then } q_{i-1}[\varepsilon/x_i] \text{ else } q_{i-1} & (i=1,\ldots,k+1). \end{cases}$$

We must show that $L(q_{k+1})$ is a minimal regular pattern language containing S. Assume that there exists a regular pattern q' such that $S \subseteq L(q') \subsetneq L(q_{k+1})$. Then $c(q') \geq c(q_{k+1}) = s$. Since s is a maximal common subsequence of S, $c(q') = c(q_{k+1}) = s$. By Theorem 8, $q' \leq' q_{k+1}$ and $q' \not\equiv' q_{k+1}$. There exists a substitution f which maps q_{k+1} to q'. The substitution f maps at least one variables to empty string because $q' \not\equiv' q_{k+1}$. Let j be an integer such that x_j appears in q_{k+1}, $f(x_j) = \varepsilon$, and $f(x_{j'}) = x_{j'}$ for any integer $j' < j$. Then $q' \leq q_{j-1}[\varepsilon/x_j]$. Therefore $S \subseteq L(q_{j-1}[\varepsilon/x_j])$ and $q_j = q_{j-1}[\varepsilon/x_j]$. Hence the variable x_j can not appear in q_{k+1}. This contradicts the selection of j.

Here we should note that we can get an answer of MINL(S) in $O(m^2n)$ time from any maximal common subsequence of a set S of strings, where $m = \max\{|w|;\ w \in S\}$ and $n = card(S)$.

We may prefer the longest common subsequences to the maximal common subsequence. However the problem to find one of the longest common subsequences of a set of strings is known to be NP-complete [Maier, 8]. Therefore finding an answer of MINL(S), containing maximum number of constants, does not seem to be done in polynomial time. To find an answer of MINL(S) for regular pattern languages, is it necessary to select one of the maximal common subsequences of S? The following theorem asserts that it is not the case.

Theorem 10. There exists an answer p of MINL(S) for regular pattern languages such that $c(p) \notin MCS(S)$ for some set S of strings.

Proof. Let S = {01020, 0212}. Then $02 \notin MCS(S)$ because $012 \in CS(S)$. However the pattern $p = x_1 02 x_2$ represents a minimal regular pattern language containing S.

In the proof of Theorem 10, the pattern $q = 0x_1 1x_2 2x_3$ is a possible answer of MINL(S), and $c(q) = 012 \in LCS(S)$. In some cases q is not always better answer of MINL(S) than p because the pattern p contains a longer constant string "02" than q. Finally, from this obsevation, we get a MINL algorithm by using a method to find common strings in length decreasing order. The correctness is easily shown by Theorem 8 and the computing time is $O(m^4n)$, where $m = \max\{|w|;\ w \in S\}$ and $n = card(S)$.

In our MINL algorithm we use some notations for simplicity:

Let $\sigma = w_1, \ldots, w_n$ be a sequence of strings. The notation $L(\sigma)$ denotes the regular pattern language $L(x_1 w_1 x_2 \ldots w_n x_{n+1})$, $|\sigma|$ denotes the number of strings in σ, and $\|\sigma\|$ denotes the sum of lengths of strings in σ.

```
Procedure MINL(S);

  (* Input S: non-empty finite set of strings *)
  (* Output p: a pattern representing a minimal
              extended regular pattern language containing S *)
  begin
    s := one of the shortest strings in S ;
    σ := ε;          (* sequence of common strings *)
    n := |s| ;       (* length of candidate string *)
    while n > 0 do begin
      for i := 1 to |s| - n + 1 do
more:   for j := 0 to |σ| do
          if S ⊆ L(σ<1:j>,s<i:i+n-1>,σ<j+1:|σ|>)
            then begin
                   σ := σ<1:j>,s<i:i+n-1>,σ<j+1:|σ|>) ;
                   go to more
            end ;
      n := min(|s|-||σ||, n-1)
    end ;
    p := x_1 σ<1> x_2 ... σ<|σ|> x_{|σ|+1} ;
    if S ⊆ L(p[ε/x_1]) then p := p[ε/x_1] ;
    if S ⊆ L(p[ε/x_{|σ|+1}]) then p := p[ε/x_{|σ|+1}] ;
    return p ;
  end
```

Theorem 11. The class of extended regular pattern languages is polynomial time inferrable from positive data.

6. Concluding Remarks

We have discussed polynomial time inference for the class of the extended regular pattern languages and we have seen that MINL calculation for the class plays an important role in inference from positive data. We have also seen that finding a longest common subsequence is a special case of the MINL calculation for extended regular pattern languages. We believe that our discussion gives a new viewpoint of the LCS problem.

It should be noticed that our MINL algorithm for the extended regular pattern languages is consistent to the NP-completeness of the LCS problem. The MINL algorithm finds common strings to a set in length decreasing order. It should also be noticed that our method in the algorithm is natural.

Since our evaluation of the time complexity is not so acute, the exponent of the maximum length of strings might be reduced. The MINL algorithm is originally designed for the learning data entry system, and it should have other practical applications. A little modification may be needed for some problems.

ACKNOWLEDGMENTS

The author wishes to acknowledge Professor S. Arikawa for his helpful suggestions and encouragement. He would also like to thank Mr. S. Miyano for his useful comments in the course of starting this study.

REFERENCES

[1] Aho, A.V., Hopcroft, J.E. and Ullman, J.D. (1974), "The Design and Analysis of Computer Algorithms," Addison-Wesley, Reading, Mass.

[2] Angluin, D. (1979), Finding Patterns Common to a Set of Strings, in "Proceedings, 11th Annual ACM Symposium on Theory of Computing," pp. 130-141.

[3] Angluin, D. (1980), Inductive Inference of Formal Languages from Positive Data, Inform. Contr. 45, 117-135.

[4] Arikawa, S. (1981), A personal communication.

[5] Gold, E.M. (1967), Language Identification in the Limit, Inform. Contr. 10, 447-474.

[6] Hirschberg, D.S. (1977), Algorithms for the Longest Common Subsequence Problem, JACM 24, 664-675

[7] Hopcroft, J.E. and Ullman, J.D. (1969), "Formal Languages and their Relation to Automata," Addison-Wesley, Reading, Mass.

[8] Maier, D. (1978), The Complexity of Some Problems on Subsequences and Supersequences, JACM 25, 322-336.

[9] Shinohara, T. (1982), Polynomial Time Inference of Pattern Languages and its Application, in "Proceedings, 7th IBM Symposium on Mathematical Foundation of Computer Science."

[10] Wagner, R.A., and Fischer, M.J. (1974), The string-to-string Correction Problem, JACM 21, 168-73.

EFFECTS OF PRACTICAL ASSUMPTION IN AREA COMPLEXITY OF VLSI COMPUTATION

Ken'ichi HAGIHARA Kouichi WADA and Nobuki TOKURA

Department of Information and Computer Sciences
Faculty of Engineering Science
Osaka University
Toyonaka, Osaka 560 JAPAN

1. Introduction

Brent, Kung and Thompson have presented suitable VLSI models [1,12], and discussed area-time complexity of various computations such as discrete Fourier transform[12], and multiplication[1]. Following their pioneering works, several researchers have presented additional results[5,11,13,14,15].

Although the VLSI models by Brent-Kung and Thompson are suitable for analyzing VLSI circuits theoretically, their models are not yet sufficiently practical from the viewpoint of the current VLSI technology. Thus, it is important to add new assumptions to their original models so that the modified model may become more suitable for the current technology, and it is also important to obtain better lower bounds on the new model. In this paper, effects of the following assumptions on bounds of the area complexity are discussed.

BOUNDARY LAYOUT ASSUMPTION: All input/output (I/O) ports of a circuit must be located on the boundary of a region on which the circuit is embedded.

RESTRICTED LOCATION ASSUMPTION: In addition, the relative positions of I/O ports on the boundary must satisfy a certain restriction.

The boundary layout assumption is one of the practical assumptions and technologically important. A VLSI circuit is hierarchically composed of several subcircuits called "blocks " These blocks communicate with each other by the wires which connect the blocks through their boundaries. In this case, the inputs and the outputs of each block are performed on the boundary. The boundary layout assumption reflects such situation.

It has been shown that the boundary layout assumption affects lower and/or upper bounds of complexity[2,16,18]. For example, the area A necessary to embed the complete binary tree with n leaves under the present VLSI model satisfies

$A = \Theta(n)$ without the boundary layout assumption, and

$A = \Theta(n \cdot \log n)$ with the boundary layout assumption[2].
Another example is the area-time complexity AT^α for nontrivial n-input m-output functions such as decoder and encoder, where A and T denote the area and the computation time of a circuit to compute these functions respectively. It has been shown that the lower bound on AT^α ($\alpha \geq 2$) for these functions satisfies

$$AT^\alpha = \Omega(\max(n,m) \cdot [\max(\log N, \log M]^{\alpha-1})$$

without the boundary layout assumption, and

$$AT^\alpha = \Omega(\max(n,m) \cdot \max(\log^\alpha N/\log\log N, \log^\alpha M/\log\log M))$$

with the boundary layout assumption[16],
where N is the maximum of $N_1,\ldots,N_m$ (N_i $(1 \leq i \leq m)$ is the number of input variables on which the i-th output variable essentially depends), and where M is the maximum of $M_1,\ldots,M_n$ (M_j $(1 \leq j \leq n)$ is the number of output variables which essentially depend on the j-th input variable). In this case, the boundary layout assumption can reinforce the lower bound on AT^α measure by $\max(\log N/\log\log N, \log M/\log\log M)$.

The restricted location assumption is often encountered in practical case. That is, we can not arrange the I/O ports arbitrarily on the boundary. For example, it is well known that a combinational adder circuit of two n-bit integers can be constructed with $O(n)$ area. The construction of the adder requires that input ports of the addend and the augend are located alternatively on the boundary. The two operands usually come from different blocks and in this case it requires $\Omega(n^2)$ area to shuffle the two sets of wires outside the adder (Fig. 1). In order to reduce the wire area to connect the blocks, the two groups of input ports must be separated as shown in Fig. 2. However, it will be shown that it requires $\Omega(n^2)$ area to perform the function of addition itself.

In this paper, lower bounds on area of combinational circuits to perform addition, multiplication, division and sorting are derived on a VLSI model with the boundary layout assumption. In Section 3, a relationship between relative positions of I/O ports of a circuit and the circuit area is shown. By using the result, it is shown that a combinational circuit to compute the addition or the multiplication requires $\Omega(n^2)$ area, if some I/O port locations are specified, where n is the input bit-size. A similar result is shown by Savage[11]. But the result in this paper properly contains his result and is considered to be a generalized one.

In Section 4, lower bounds on area of combinational circuits to perform the multiplication, the division and the sorting are derived. It is shown that the combinational circuits to perform these functions require $\Omega(n^2)$ area under the boundary layout assumption. These results

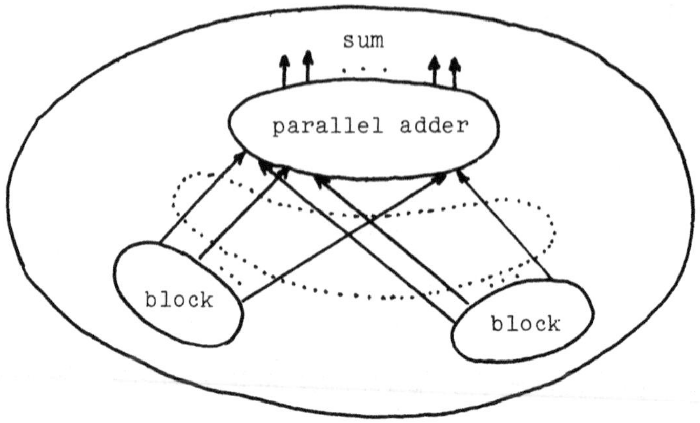

Fig. 1 $\Omega(n^2)$ area to shuffle two groups of wires.
The area enclosed with the dotted line is $\Omega(n^2)$.

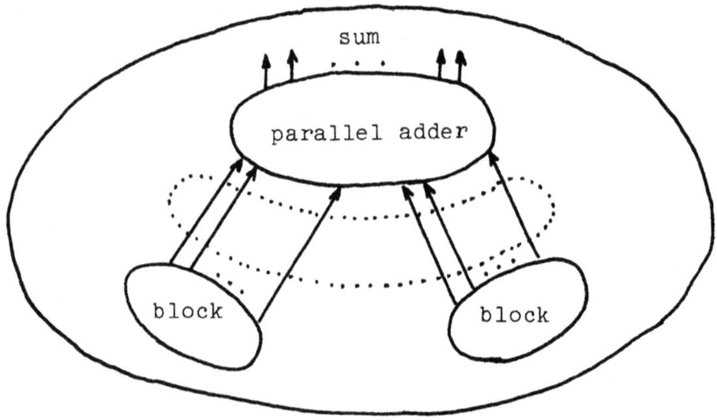

Fig. 2 Efficient connection of blocks.
The area enclosed with the dotted line is $O(n)$.

are obtained by using the relationship between the I/O port locations and the circuit area shown in Section 3. It should be noted that the lower bounds are independent of the I/O port locations and hold for any combinational circuit with the boundary layout assumption. These lower bounds are best possible for the multiplication and the division, and are optimal within a logarithmic factor for the sorting.

2. VLSI Model

In this section, a model of VLSI circuits is described and is used as a basis for deriving area bounds.

VLSI model

(A-1) A VLSI circuit is constructed from processing elements (PEs for short) and wire segments. A PE corresponds to a gate, a (1-bit) storage element and an input/output (I/O port for short).
A VLSI circuit is embedded on a closed planar region R.
(A-2) Wires have width of λ (> 0). Separation and length of wires are at least λ. Each PE occupies area of at least λ^2[12].
(A-3) Wire and PE, or PE and PE cannot overlap each other. At most ν ($\geq$ 2) wires can overlap at any point in the circuit.
(A-4) The number of the fanin of each PE and the number of the fanout of each PE are unbounded. It is assumed that the fanin of each input port and the fanout of each output port are zero.
(A-5) I/O operations are performed at I/O ports of a circuit. All I/O ports are located on the boundary of R. This assumption is called boundary layout assumption.

This model essentially the same as the model by Brent and Kung[1] except the nonconvexity of a circuit region (A-1) and the boundary layout assumption (A-5). Although Brent, Kung and others assume the convexity of a circuit region[1,10], the result in this paper does not require the convexity. The boundary layout is assumed by Chazelle-Monier[3] and Yasuura-Yajima[18].

In this paper, the area complexity of combinational circuits is discussed. A combinational circuit is an acyclic circuit without storage elements. With each PE of the circuit we associate a boolean function which the PE computes, defined as follows. With each of the input ports v_i we associate a variable x_i and an identity function $f_{v_i}(x_i) = x_i$. With each PE w of fanin d having $u_1,\ldots,u_d$ we associate the function $f_w=b_w(f_{u_1},\ldots,f_{u_d})$. The circuit computes the set of functions associated with its output ports. In what follows, it is assumed that a combinational circuit is embedded on a closed region and satisfies the boundary layout assumption, unless otherwise stated. And through this paper, for a combinational circuit C, let A(C) denote the area of the circuit.

For a VLSI circuit C, let V be the set of PEs in C. Let W be the set of wires connecting PEs in C, and an element of W is represented by <a,b>, where a and b are PEs and data flow from a to b.

The circuit graph corresponding to C (denoted by G(C)) is a directed graph $(G_p(V), G_w(W))$, satisfying the following conditions:

(1) The node in G(C) corresponds to each PE in C. The set of nodes in G(C) is denoted by $G_p(V)$, where G_p is a bijective mapping from the set of PEs to the set of nodes.

(2) The directed edge in G(C) corresponds to each wire connecting PEs in C. The set of directed edges in G(C) is denoted by $G_w(W)$, where G_w is a bijective mapping from the set of wires to the set of directed edges. When a wire <a,b> is in W, the directed edge $<G_p(a), G_p(b)>$ is included in $G_w(W)$, that is, the direction of the edge corresponds to the flow of data in C.

The circuit graph G(C) is used to analyze topological or graph theoretical properties for C.

3. Relationship between Circuit Area and I/O Port Location Restriction

In this section, a lower bound of the area complexity of a combinational circuit is discussed, which is embedded on a closed region and has some I/O port location restrictions.

The situations with I/O port location restriction are often encountered. For example, n input ports (or output ports) corresponding to an n-bit integer are usually located with preserving the bit order (Fig. 3). Another example is that the location of an operand X, the location of another operand W and the location of a result Y are separated one another (Fig. 4).

The results in this section insist that such constraints about the order of I/O port location possiblly require larger area than the complexity of the function itself. For example, a combinational adder circuit of two n-bit integers can be constructed with O(n) area by locating input ports of the addend and the augend alternatively on the boundary. However, separated I/O ports as shown in Fig. 4 must require $\Omega(n^2)$ area to perform the addition itself. It is shown that usual restriction on the location of I/O ports (e.g., Fig. 3 or Fig. 4) require large area, say $\Omega(n^2)$, when the multiplication or the division is computed by combinational circuits.

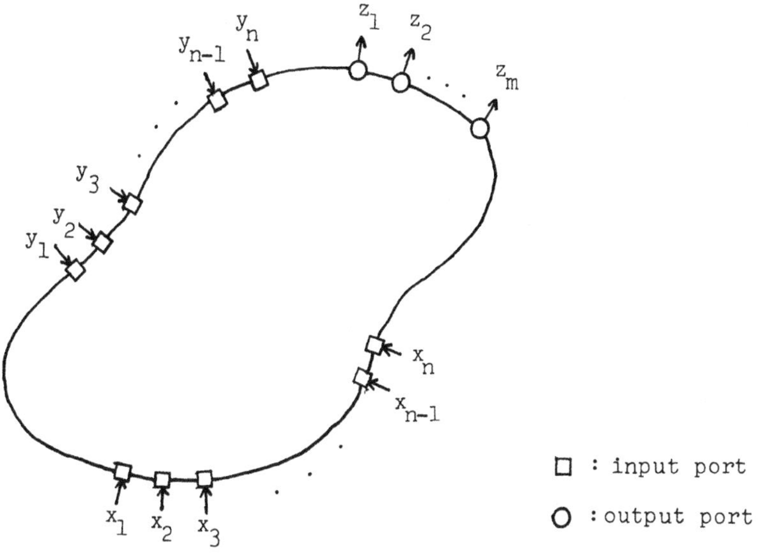

Fig. 3 I/O port locations with preserving the bit order.

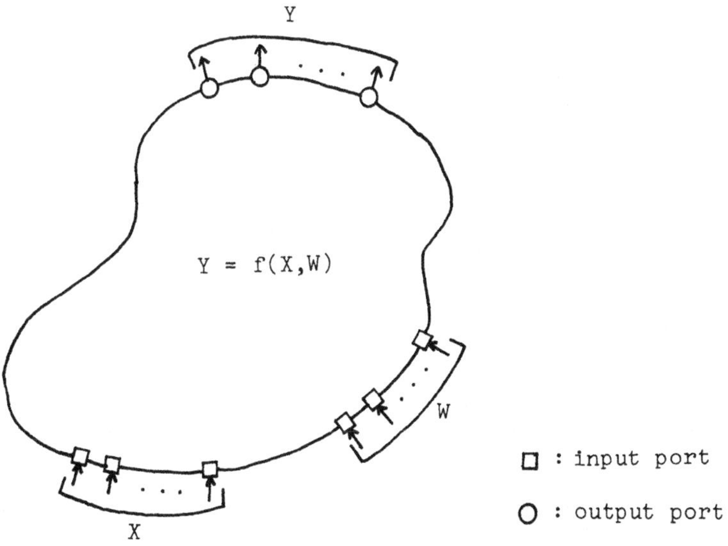

Fig. 4 Separated I/O ports.

It should be noted that the result in this paper is based on the following assumption: the amount of information which each logic gate can output is only one bit. That is, a logic gate may have some fanouts, but the values on them are identical.

Definition 1 Let $G = (V,E)$ be a directed graph. A path in the graph is represented by the sequence $(v_1,\ldots,v_n)$ of the nodes, where $(v_i,v_{i+1}) \in E$ for each i $(1 \leq i \leq n-1)$. A pair of paths $p = (v_1,\ldots, v_n)$ and $q = (u_1,\ldots,u_m)$ is called node-disjoint if p and q have no common nodes. A set P of paths is called node-disjoint if each pair of paths in P is node-disjoint.

Let V_1 and V_2 be subsets of the node set V such that $V_1 \cap V_2 = \phi$. A directed path $(v_1,\ldots,v_n)$ is called (V_1, V_2)-connecting, if it has the following properties:

1) $(v_1 \in V_1$ and $v_n \in V_2)$ or $(v_1 \in V_2$ and $v_n \in V_1)$,
2) for each i $(2 \leq i \leq n-1)$, $v_i \in (V-V_1-V_2)$. □

The following two lemmas demonstrate the relationship between a restriction of I/O port locations and the circuit area. Let R be a closed region on which a circuit is embedded. The boundary B of R forms a closed curve. A segment of the closed curve is called a contiguous subboundary of B.

Lemma 1 For a combinational circuit C, let $G(C) = (V,E)$ be the circuit graph of C, and let IO denote the set of I/O nodes of G(C). If there exist subset V_1, V_2 and V_3 of IO which satisfy the following conditions.

1) $V_i \cap V_j = \phi$ $(1 \leq i < j \leq 3)$.
2) G(C) has a node-disjoint set P_1 of (V_1, V_3)-connecting paths.
3) G(C) has a node-disjoint set P_2 of (V_2, V_3)-connecting paths.
4) $|P_1| = |P_2| = |V_3|$.
5) There exist two disjoint contiguous subboundaries B_1 and B_3 such that
   i) $V_1 \subseteq IO_1$ and $(V_2 \cup V_3) \cap IO_1 = \phi$,
   ii) $V_3 \subseteq IO_3$ and $(V_1 \cup V_2) \cap IO_3 = \phi$,
   where IO_i $(i = 1,3)$ denotes the set of I/O nodes located on B_i.

Then, it follows that

$$A(C) = \Omega(|V_3|^2).$$

(proof) For the nodes of V_3 on B_3, number the nodes from v_1 to v_k in the order of the location on B_3, where $k = |V_3|$. For each $v_i \in V_3$ $(1 \leq i \leq k)$, let p_i and q_i be the paths in P_1 and P_2 which have v_i as an endpoint respectively.

Since the endpoint of q_i, which is different from v_i, is located

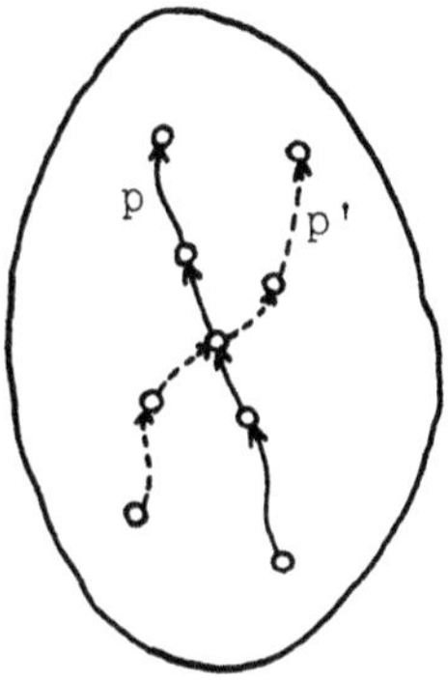

crossing with a common vertex

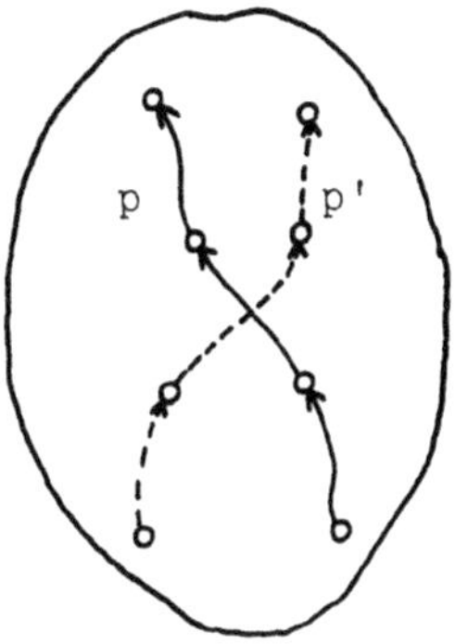

multi-level crossing

Fig. 5 Two kinds of crossings.

on neither B_1 nor B_3 by the condition 5), each q_i crosses each $p_j(j<i)$ on R at least once, or crosses each $p_h(h>i)$ on R at least once. Therefore, each q_i must cross at least min(i-1,k-i) paths in P_1 on R. Note that the expression "a path p crosses a path p' " has two meanings. One is that p and p' join at a common node and branch from the node. Another is that an edge in p and an edge in p' cross each other(Fig. 5).

Since a unit of area has at most ν crossing wires, a unit of area has at most $\binom{\nu}{2}$ crossing points. Thus A(C) has at least $(1/\binom{\nu}{2})\cdot\sum_{i=1}^{k}\min(i-1,k-i)$ units. So we have

$$A(C) \geq (1/\binom{\nu}{2})\cdot\sum_{i=1}^{k}\min(i-1,\ k-i)$$
$$\begin{cases} = (1/4\cdot\binom{\nu}{2})\cdot k(k-2) & \text{(if k is even)} \\ = (1/4\cdot\binom{\nu}{2})\cdot(k-1)^2 & \text{(if k is odd)} \end{cases}$$
$$= \Omega(|V_3|^2). \square$$

The next lemma is proved similarly to Lemma 1.

<u>Lemma 2</u> For a combinational circuit C, let G(C) = (V,E) be the circuit graph of C, and IO denote the set of I/O nodes of G(C). If there exist subsets V_1, V_2, V_3 and V_4 of IO which satisfy the following conditions 1)-4).

1) $V_i \cap V_j = \phi$ $(1 \leq i < j \leq 4)$.
2) G(C) has a node-disjoint set P_1 of (V_1,V_3)-connecting paths.
3) G(C) has a node-disjoint set P_2 of (V_2,V_4)-connecting paths.

4) There exist four disjoint contiguous subboundaries B_1, B_2, B_3 and B_4 in clockwise order such that $V_i \subsetneq IO_i$ (i=1-4), where IO_i denotes the set of I/O nodes located on B_i.

Then, it follows that

$A(C) = \Omega(|P_1|\cdot|P_2|)$. □

Lemma 1 and 2 state a relationship between a circuit graph G(C) and the circuit area A(C). In order to obtain lower bounds on area of combinational circuits which compute a function f by using these lemmas, we examine some properties of the circuit graph for that function.

For a sequence $Z = (z_1,\ldots,z_k)$, a sequence $(z_{i_1},\ldots,z_{i_j})$, where $1 \leq i_1 <\ldots< i_j \leq k$, is called a subsequence of Z. For a sequence $Z = (z_1,\ldots,z_k)$, let $\bar{Z}$ denote the set $\{z_1,\ldots,z_k\}$.

Let Z_1 and Z_2 be subsequences of Z. If it holds that $\bar{Z}_1 \cap \bar{Z}_2 = \phi$ and $\bar{Z}_1 \cup \bar{Z}_2 = \bar{Z}$, then Z_2 is denoted by $Z-Z_1$.

Definition 2 Let $X = (x_1,\ldots,x_n)$ and $Y = (y_1,\ldots,y_m)$, and let $Y = f(X)$ be a function $\{0,1\}^n \to \{0,1\}^m$. Let $X_1 = (x_{i_1},\ldots,x_{i_h})$ denote a subsequence of X and let $X-X_1 = (x_{j_1},\ldots,x_{j_{n-h}})$. Let $Y_1 = (y_{k_1},\ldots,y_{k_\ell})$ be a subsequence of Y. Let $Q = (q_1,\ldots,q_{n-h}) \in \{0,1\}^{n-h}$.

$Y_1 = h(X_1)$ is a subfunction obtained from f, if it is obtained by assigning q_r to each input variable x_{j_r} of $X-X_1$ for f $(1 \leq r \leq n-h)$, and by restricting output variables to Y_1 of f. The subfunction h is denoted by $Y|Y_1 = f(X,Q)|X_1$.

A function $(y_1,\ldots,y_k) = g(x_1,\ldots,x_k)$ is a k-identity function, if it holds that $y_i = x_{p(i)}$ for each i $(1 \leq i \leq k)$, where $(p(1),\ldots,p(k))$ is a permutation of $(1,\ldots,k)$. □

Let C be a combinational circuit which computes a function f. In order to combine the function f with the circuit graph G(C), the following proposition and lemmas are needed. In what follows, a combinational circuit which computes f is denoted by C_f, unless otherwise stated.

Proposition 1 (Menger's theorem) [8]

Let $G = (V,E)$ be a directed graph. Let a be a node of G whose indegree is zero, and let b be a node of G whose outdegree is zero.

Let U be a subset of $V-\{a,b\}$ satisfying the following conditions:

1) Every directed path from a to b goes through a certain node in U,
2) U is minimal; i.e., for every $U' \subsetneq U$, U' does not satisfy 1).

Then, the maximum number of node-disjoint $(\{a\},\{b\})$-connecting paths is equal to the number of node in U. □

In order to use Proposition 1, the following directed graph is constructed from a circuit graph and special nodes a and b.

For a circuit graph $G(C) = (V,E)$, define the directed graph $\hat{G}(C)$ obtained from $G(C)$ to be

$\hat{G}(C) = (V \cup \{a,b\} , E \cup \{\langle a,p\rangle \mid p \in I\} \cup \{\langle q,b\rangle \mid q \in O\})$,

where $a, b \notin V$, and I and O denote the set of input nodes of $G(C)$ and the set of output nodes of $G(C)$ respectively.

Lemma 3 Let $(y_1,\ldots,y_k) = g(x_1,\ldots,x_k)$ be a k-identity function. For the directed graph $\hat{G}(C_g)$ obtained from $G(C_g) = (V,E)$, let U be a subset of V such that every path from a to b goes through a certain node in U. Then,

$|U| \geq k$.

(proof) Let $U = \{v_1,\ldots,v_j\}$ and assume that $j \leq k-1$. Since C_g is a combinational circuit, the output value of each logic gate corresponing to v_i ($\in U$) is uniquely determined by the input values, and the output values of g are also uniquely determined by the output values of the logic gates corresponding to $v_1,\ldots,v_j$. By the assumption that each logic gate can output only one bit amount of information, the number of possible values of the logic gates corresponding to $v_1,\ldots v_j$ is at most 2^j ($\leq 2^{k-1}$). On the other hand, since g is a k-identity function, the number of possible output values of g must be equal to 2^k. This implies that C_g cannot compute g correctly. Thus, $|U| \geq k$. □

Let $(y_1,\ldots,y_m) = f(x_1,\ldots,x_n)$ be a function, and let $G(C_f)$ be the circuit graph. Let IO denote the set of I/O nodes of $G(C_f)$. For $G(C_f)$ the I/O node mapping

$P : \{x_1,\ldots,x_n\} \cup \{y_1,\ldots,y_m\} \rightarrow IO$

is defined as the bijective mapping that indicates which I/O node an input or output variable corresponds to. For $W \subseteq (\{x_1,\ldots,x_n\} \cup \{y_1,\ldots,y_m\})$, let $P(W)$ denote $\{P(w) \mid w \in W\}$.

The following lemma combines a function f with the circuit graph $G(C_f)$.

Lemma 4 Let f be a function whose subfunction contains a k-identity function $Y = g(X)$. Then, the circuit graph $G(C_f)$ has k node-disjoint (V_1,V_2)-connecting paths, where $V_1 = P(\overline{X})$ and $V_2 = P(\overline{Y})$.

(proof) Let $\hat{G}(C_f)$ be the directed graph obtained from $G(C_f)$. Since C_f computes a k-identity function, Lemma 3 implies that

$|U| \geq k$

for $\hat{G}(C_f)$. Therefore, by Proposition 1 the maximum number of node-disjoint $(\{a\},\{b\})$-connecting paths in $\hat{G}(C_f)$ is at least k. From the construction of $\hat{G}(C_f)$, $G(C_f)$ must have k node-disjoint $(P(\overline{X}),P(\overline{Y}))$-

connecting paths. □

By Lemma 1 and 4, the following theorem holds.

Theorem 1 Let $Y = f(X)$ be a function. Assume that there exist subsequences X_1 and X_2 of X, and a subsequence Y_1 of Y which satisfy the following conditions 1)-3).

1) $\overline{X}_1 \cap \overline{X}_2 = \phi$.

2) There exist assignments Q_1 and Q_2 to $X-X_1$ and $X-X_2$ respectively, such that subfunctions $Y|Y_1 = f(X,Q_1)|X_1$, and $Y|Y_1 = f(X,Q_2)|X_2$ are $|\overline{Y}_1|$-identity functions.

3) There exist two disjoint contiguous subboundaries B_1 and B_2 of the boundary of C_f such that

i) $P(\overline{X}_1) \subseteq IO_1$ and $(P(\overline{X}_2) \cup P(\overline{Y}_1)) \cap IO_1 = \phi$,

ii) $P(\overline{Y}_1) \subseteq IO_2$ and $(P(\overline{X}_1) \cup P(\overline{X}_2)) \cap IO_2 = \phi$,

where IO_i (i=1,2) denotes the set of all I/O nodes located on B_i.

Then, it holds that

$$A(C_f) = \Omega(|\overline{Y}_1|^2).$$

(proof) By the condition 2) and Lemma 4, the circuit graph $G(C_f)$ has $|\overline{Y}_1|$ node-disdoint $(P(\overline{X}_1),P(\overline{Y}_1))$-connecting paths, and $|\overline{Y}_1|$ node-disjoint $(P(\overline{X}_2),P(\overline{Y}_1))$-connecting paths. And by the condition 1), it holds that

$$P(\overline{X}_1) \cap P(\overline{X}_2) = \phi.$$

Thus, the condition 1)-4) of Lemma 1 are satisfied. Since the condition 3) satisfies the condition 5) of Lemma 1, we have

$$A(C_f) = \Omega(|\overline{Y}_1|^2). \quad \square$$

Remark: If the relationship between input variables and output variables of f is exchanged, similar result holds. This is shown by the next theorem.

Theorem 2 Let $Y = f(X)$ be a function. Assume that there exist a subsequence X_1 of X, and subsequences Y_1 and Y_2 of Y which satisfy the following conditions 1)-3).

1) $\overline{Y}_1 \cap \overline{Y}_2 = \phi$.

2) There exist assignments Q_1 and Q_2 to $X-X_1$ such that $Y|Y_1 = f(X,Q_1)|X_1$ and $Y|Y_2 = f(X,Q_2)|X_1$ are $|\overline{X}_1|$-identity functions.

3) There exist two disjoint contiguous subboundaries B_1 and B_2 of the boundary of C_f such that

i) $P(\overline{Y}_1) \subseteq IO_1$ and $(P(\overline{X}_1) \cup P(\overline{Y}_2)) \cap IO_1 = \phi$,

ii) $P(\overline{Y}_2) \subseteq IO_2$ and $(P(\overline{X}_1) \cup P(\overline{Y}_1)) \cap IO_2 = \phi$,

where IO_i (i=1,2) denotes the set of all I/O nodes located on B_i.

Then, it holds that

$$A(C_f) = \Omega(|\overline{X}_1|^2). \quad \square$$

From Theorem 1, it can be concluded that combinational circuits which compute the addition, the multiplication, the maximum operation or the minimum operation require $\Omega(n^2)$ area, if the circuits have the separated I/O port locations, where n is the bit-size of the operand. Generally the following corollary can be obtained from Theorem 1.

A binary algebra $[S,\beta]$ is a set S with a binary operation $\beta: S\times S \rightarrow S$. It is assumed that the binary operation β is expressed as

$$(y_1,\ldots,y_m) = \beta(x_1,\ldots,x_n,w_1,\ldots,w_k),$$

where the operands $(x_1,\ldots,x_n)$, $(w_1,\ldots,w_k)$ and the result $(y_1,\ldots,y_m)$ are represented in the same coding system.

Let $n = k$, and $m \geq n$ for the binary operation β. An element $(s_1,\ldots,s_n)$ in S is called an identity of β, if it holds that

$$(a_1,\ldots,a_n,0,\ldots,0) = \beta(s_1,\ldots,s_n,a_1,\ldots,a_n) \text{ and}$$
$$(a_1,\ldots,a_n,0,\ldots,0) = \beta(a_1,\ldots,a_n,s_1,\ldots,s_n)$$

for any element $(a_1,\ldots,a_n)$ in S.

Corollary 1 Let $(y_1,\ldots,y_m) = f_b(x_1,\ldots,x_n, w_1,\ldots,w_n)$ be a binary operation which has an identity, where $m \geq n$. If the input ports corresponding to two operands and the output ports corresponding to the result are separated one another, then it follows that

$$A(C_{f_b}) = \Omega(n^2). \quad \square$$

A combinational circuit to compute the addition of two n-bit integers requires $\Omega(n^2)$ area if the input ports of the addend and the augend and the output ports are separated one another. However, there exists a construction of the n-bit addition with $O(n)$ by locating the input ports of the addend and the augend alternatively on the boundary. For the multiplication of two n-bit integers, Theorem 2 implies that even if the input ports of the multiplier and multiplicand are located alternatively on the boundary, the circuit requires $\Omega(n^2)$ area by locating the output ports corresponding to the result while preserving the bit order. Then, does there exist a combinational circuit to compute the multiplication with smaller area, if some I/O port locations are properly specified? It will be shown in the following section that it is impossible to construct these circuits. That is, if combinational multiplication circuits satisfy the boundary layout assumption, the circuits would require $\Omega(n^2)$ area independent of the I/O port locations. It is also shown that similar results hold for the division and the sorting.

4. A Lower Bound on Area of Combinational Circuits

4.1 Multiplication and Division

Consider the following N-bit shift function with selectors $s_0,\dots,s_{N-1}$

$$(y_1,\dots,y_N) = f_s(x_1,\dots,x_N,s_0,\dots,s_{N-1});$$

i) one and only one s_i is set to 1 among the selectors $s_0,\dots,s_{N-1}$,
ii) the i-th selector s_i is equal to 1 if and only if

$y_{j+i} = x_j$ for $1 \le j \le N-i$, and
y_j is undefined for $j < i$.

Since the multiplication and the division contain the shift function as a subfunction, obtaining lower bounds for the multiplication and the division is reduced to deriving a lower bound for the shift function. In the following, a lower bound on area of combinational circuits to compute the shift function is considered. In order to derive the lower bound, some definitions are needed.

Definition 3 Let $[k, k'] = \{i \in \mathbf{Z} \mid k \le i \le k'\}$, where $\mathbf{Z}$ is the set of integers. For an integer r and nonnegative integers a and b, let $L_r(a, b)$ denote the class of all subsets of exactly b elements of $[r+1, r+a]$.

For a subset $K = \{\ell_1,\dots,\ell_b\}$ of $\mathbf{Z}$, define the i-shift of K, $s_i(K)$ as $s_i(K) = \{\ell_1+i,\dots,\ell_b+i\}$. For two subsets $K_1 = \{\ell_1,\dots,\ell_b\}$ and $K_2 = \{m_1,\dots,m_b\}$, define the number of meets $m(K_1, K_2)$ between K_1 and K_2 to be $m(K_1, K_2) = |K_1 \cap K_2|$, where $|K_1 \cap K_2|$ denotes the number of elements in $K_1 \cap K_2$. □

For two sets K_1 and K_2 in the class $L_r(a, b)$, the following property holds. This property plays an important role for deriving the lower bound of the shift function.

Lemma 5 For any K_1, K_2 in $L_r(a, b)$, there exists an integer i $(-a \le i \le a)$ such that

$$m(K_1, s_i(K_2)) \ge \lfloor b^2/2a \rfloor .$$

(proof) For an element ℓ in K_1 and an element m in K_2, it holds that

$$r + 1 \le \ell \le r + a, \text{ and}$$
$$r + 1 \le m \le r + a.$$

Since the following inequality is satisfied

$$-(a - 1) \le \ell - m \le (a - 1),$$

there exists exactly one integer i $(-(a - 1) \le i \le a - 1)$ such that $\ell = m + i$ for every pair (ℓ, m). Thus,

$$\sum_{i=-(a-1)}^{a-1} m(K_1, s_i(K_2)) = \sum_{i=-(a-1)}^{a-1} |K_1 \cap s_i(K_2)|$$
$$= \sum_{i=-(a-1)}^{a-1} \sum_{\ell \in p} \sum_{m \in q} |\{\ell\} \cap \{m+i\}|$$
$$= \sum_{\ell \in p} \sum_{m \in q} \sum_{i=-(a-1)}^{a-1} |\{\ell\} \cap \{m+i\}|$$
$$= \sum_{\ell \in p} \sum_{m \in q} 1 = b^2$$

If Lemma 5 does not hold, for every i $(-a < i < a)$ it follows that

$$m(K_1, s_i(K_2)) < \lfloor b^2/2a \rfloor.$$

Therefore, we have

$$\sum_{i=-(a-1)}^{a-1} m(K_1, s_i(K_2)) < (2a-1) \cdot \lfloor b^2/2a \rfloor \leq b^2.$$

This is a contradiction. □

The following lemma enables us to use the result in preceding section.

Lemma 6 Let $(y_1,\dots,y_{3N}) = f_s(x_1,\dots,x_{3N}, s_0,\dots,s_{3N-1})$ be the 3N-bit shift function. Let X be an arbitrary subsequence of $(x_1,\dots,x_N)$ and Y be an arbitrary subsequence of $(y_{N+1},\dots,y_{2N})$ such that $|\overline{X}| = |\overline{Y}| = k \leq N$. Then, the shift function f_s contains an ℓ-identity function $Y_1 = f(X_1)$ as a subfunction which satisfies the following conditions.

1) $\overline{X}_1 \subseteq \overline{X}$ and $\overline{Y}_1 \subseteq \overline{Y}$, and
2) $\ell \geq \lfloor k^2/2N \rfloor$.

(proof) Let $K_1 = \{i+N \mid x_i \in \overline{X}\}$ and $K_2 = \{i \mid y_i \in \overline{Y}\}$. By definition, it holds that

$$K_1, K_2 \in L_N(N, k).$$

By Lemma 5, we have

$$m(K_1, s_i(K_2)) \geq \lfloor k^2/2N \rfloor,$$

for an integer i such that $-N \leq i \leq N$.
By letting $\overline{X}_1 = \{x_{j-N} \mid j \in K_1 \cap s_i(K_2)\}$ and $\overline{Y}_1 = \{y_j \mid j \in K_1 \cap s_i(K_2)\}$, we have $m(K_1, s_i(K_2))$-identity function

$$Y|Y_1 = f_s'(X, Q)|X_1,$$

where $Y = f_s'(X)$ is a subfunction of f_s, and where Q is the assignment of $(s_0,\dots,s_{3N-1})$ such that $s_{N+i} = 1$ and $s_h = 0$ $(h \neq N+i)$. □

The following theorem is a main result in this subsection and is obtained from Lemma 2 and 6.

Theorem 3 Let $(y_1,\dots,y_{3N}) = f_s(x_1,\dots,x_{3N}, s_0,\dots,s_{3N-1})$ be the 3N-bit shift function. Let C be a combinational circuit to compute a function which contains f_s as a subfunction. Then,

$A(C) = \Omega(N^2)$.

(proof) Consider the subset IO of I/O nodes corresponding to $x_1,\ldots,x_N$ and $y_{N+1},\ldots,y_{2N}$, that is,

$$IO = P(\{x_1,\ldots,x_N\} \cup \{y_{N+1},\ldots,y_{2N}\}).$$

Let $N = 4t + \delta$ $(0 \leq \delta \leq 3)$. Let B be the boundary of C. Since each node in IO is located on B, we can divide B into two contiguous subboundaries B_1 and B_2 such that each B_i $(i = 1, 2)$ contains at least 2t input nodes in IO. And either B_1 or B_2 contains at least 2t output in IO. Without loss of generality, it is assumed that subboundary B_2 contains at least 2t output nodes in IO.

The subboundary B_1 is divided into two contiguous subboundaries D_1 and D_2 such that both D_1 and D_2 contain at least t input nodes in IO, and the subboundary B_2 is also divided into two contiguous subboundaries F_1 and F_2 such that both F_1 and F_2 contain at least t output nodes in IO.

Consider the exactly t input nodes in IO located on D_1 and D_2 respectively, and let I_1 and I_2 denote the set of such nodes. And consider the exactly t output nodes in IO located on F_1 and F_2 respectively, and let O_1 and O_2 denote the set of such nodes.

By Lemma 4, 6, there exist ℓ_1 node-disjoint (I_1,O_1)-connecting paths, and ℓ_2 node-disjoint (I_2,O_2)-connecting paths, where $\ell_1, \ell_2 \geq \lfloor t^2/2N \rfloor$ and $t = \lfloor N/4 \rfloor$. Therefore, Lemma 2 implies that, for constant $c > 0$,

$$\begin{aligned} A(C) &\geq c \cdot \ell_1 \cdot \ell_2 \\ &\geq c \cdot (\lfloor 1/2N(\lfloor N/4 \rfloor)\rfloor)^2)^2 \\ &= \Omega(N^2). \quad \square \end{aligned}$$

Remark: The shift function f_s considered here is slightly different from usual one. A usual shift function has an encoded selector, that is, shift by i-bit $(0 \leq i \leq N-1)$ is specified by a binary number $a_{\log N} \ldots a_1$. however, Theorem 3 holds for shift functions with selectors of any form.

The multiplication of two n-bit integers and the division of 2n-bit integer by n-bit integer contain the shift function f_s as a subfunction[9]. Thus the following corollaries are directly obtained from Theorem 3.

Corollary 2 Let C be a combinational circuit to compute the multiplication of two n-bit integers. Then,

$A(C) = \Omega(n^2)$. $\square$

Corollary 3 Let C be a combinational circuit to compute the division of 2n-bit integer by n-bit integer. Then,

$$A(C) = \Omega(n^2). \quad \square$$

Remark 1: The lower bounds obtained in Corollary 2 and 3 are best possible in the sense that the multiplication and the division are both constructed with $O(n^2)$ area respectively [4].

Remark 2: In the derivation of the lower bound on area for the shift function, the convexity of a circuit region is not assumed. If the convexity is assumed, the same lower bound on area for the shift function (thus, the multiplication and the division) can be proved without the boundary layout assumption. The next theorem is shown by using Lemma 6 and the relationship between the area of convex region and the length of a chord perpendicular to the diameter [1].

Theorem 4 Let $(y_1,\ldots,y_{3N}) = f_s(x_1,\ldots,x_{3N}, s_0,\ldots,s_{3N-1})$ be the 3N-bit shift function. Let C be a combinational circuit to compute a function which contains f_s as a subfunction. Assume that C is embedded on a convex region. Then,

$$A(C) = \Omega(N^2).$$

(proof) Let R be a convex region on which C is embedded. Let D be a diameter of R, and L be a chord perpendicular to D.

Consider the input nodes corresponding to $x_1,\ldots,x_N$, and let I denote the set of such input nodes ($I = P(\{x_1,\ldots,x_N\})$). The chord L divides R into two parts R_1 and R_2 such that R_1 contains i input nodes in I, and R_2 contains N-i input nodes in I. We can assume that the input nodes in I are shrunk to infinitesimal size and that L does not intersect any input nodes in I, because the area of the input ports is not used in the proof. By sliding the intersection of L and D along D, we can arrange that both R_1 and R_2 contain at least $\lfloor N/2 \rfloor$ input nodes in I.

Since either R_1 or R_2 contains at least $\lfloor N/2 \rfloor$ output nodes in $P(\{y_{N+1},\ldots,y_{2N}\})$ (denoted by O), without loss of generality, R_2 contains at least $\lfloor N/2 \rfloor$ output nodes in O.

Consider the exactly $\lfloor N/2 \rfloor$ input nodes in I located on R_1, and let I_1 denote such input nodes. Similarly, consider the exactly $\lfloor N/2 \rfloor$ output nodes in O located on R_2, and let O_2 denote such output nodes. By Lemma 4 and 6, there exists ℓ node-disjoint (I_1, O_2)-connecting paths, where $\ell \geq \lfloor \lfloor N/2 \rfloor^2/2N \rfloor$. Then, since ℓ edges cross the chord L, it follows that

$$L \geq \ell \geq \lfloor \lfloor N/2 \rfloor^2/2N \rfloor.$$

By the relationship between A(C) and L [1], it holds that

$A(C) \geq L^2/2 \geq (\lfloor \lfloor n/2 \rfloor^2/2N \rfloor)^2/2 = \Omega(N^2)$. □

4.2 Sorting

A sorting of a list of n k-bit integers is a function

$$(y_{11},\ldots,y_{1k},\ldots,y_{n1},\ldots,y_{nk}) = f(x_{11},\ldots,x_{1k},\ldots,x_{n1},\ldots,x_{nk}) \quad (\{0,1\}^{nk} \rightarrow \{0,1\}^{nk})$$

such that $y_{ik}\ldots y_{i1}$ ($1 \leq i \leq n$) is the i-th integer in increasing (or decreasing) order among n integers $x_{1k}\ldots x_{11},\ldots,x_{nk}\ldots x_{n1}$. When the sorting is computed by a combinational circuit, a lower bound on area can be also shown by using the result in the preceding section.

Definition 4 [13] A boolean function $(y_1,\ldots,y_N) = f(x_1,\ldots,x_N, s_1,\ldots,s_b)$ computes a permutation group G, if for each permutation $g \in G$, there exist values for $s_1,\ldots,s_b$ such that $y_i = x_{g(i)}$ $(1 \leq i \leq N)$, where $(g(1),\ldots,g(N))$ denotes the permutation of $(1,\ldots,N)$ by g. □

It has been known that a function to sort a list of n k-bit integers ($k \geq \log_2 n$) contains a boolean function which computes the symmetric group $S_{\lfloor n/2 \rfloor}$ as a subfunction [3]. Whereas, a lower bound on area for the boolean function which computes the symmetric group S_N is considered more generally. The lower bound for the sorting is obtained from the result.

Theorem 5 Let $(y_1,\ldots,y_N) = f(x_1,\ldots,x_N,s_1,\ldots,s_b)$ be a boolean function which computes the symmetric group S_N. Then it follows that

$$A(C_f) = \Omega(N^2).$$

(proof) Let I and O denote the input nodes and the output nodes corresponding to $x_1,\ldots,x_N$ and $y_1,\ldots,y_N$ respectively, that is,

$I = P(\{x_1,\ldots,x_N\})$ and $O = P(\{y_1,\ldots,y_N\})$.

Let B be the boundary of C_f. Since each node in $I \cup O$ is located on B, we can divide B into three contiguous subboundaries B_1, B_2 and B_3 such that each B_i (i = 1, 2, 3) contains at least $\lfloor N/3 \rfloor$ output nodes in O. Then, there exists contiguous subboundary among B_1, B_2 and B_3, on which at least $\lfloor N/3 \rfloor$ input nodes of I are located. Without loss of generality, it is assumed that the subboundary B_1 contains at least $\lfloor N/3 \rfloor$ input nodes in I.

Consider the exactly $\lfloor N/3 \rfloor$ input nodes in I located on B_1, and the exactly $\lfloor N/3 \rfloor$ output nodes in O located on B_2 and B_3, respectively. Let I_1, O_2 and O_3 denote the sets of such nodes, i.e.,

$I_1 = P(\{x_{i_1},\ldots,x_{i_k}\})$,

$O_2 = P(\{y_{j_1},\ldots,y_{j_k}\})$, and

$O_3 = P(\{y_{h_1},\ldots,y_{h_k}\})$,

where $k = \lfloor N/3 \rfloor$ and $I_1 \subseteq I$, $O_2, O_3 \subseteq O$ and $O_2 \cap O_3 = \phi$.

Since the function f computes the symmetric group S_N, there exist permutations $g_1, g_2 \in S_N$ such that $y_{j_p} = x_{g_1(i_p)}$ $(1 \le p \le k)$ and $y_{h_q} = x_{g_2(i_q)}$ $(1 \le q \le k)$. By setting $X_1 = (x_{i_1},\ldots,x_{i_k})$, $Y_1 = (y_{j_1},\ldots,y_{j_k})$ and $Y_2 = (y_{h_1},\ldots,y_{h_k})$, the conditions 1)-3) in Theorem 2 are satisfied and $|\overline{X}_1| = k$ $(= \lfloor N/3 \rfloor)$. Thus, it follows that

$$A(C_f) = \Omega(N^2). \quad \square$$

Corollary 4 Let C be a combinational circuit to sort a list of n k-bit integers $(k \ge \log_2 n)$. Then

$$A(C) = \Omega(n^2). \quad \square$$

Remark: Sorting a list of n $\log_2 n$-bit integers is constructed by a combinational circuit with $O(n^2 \cdot \log_2 n)$ area [6], so the lower bound shown here is optimal within a logarithmic factor.

5. Conclusion

It is important to discuss the area complexity or the area-time complexity on the model more suitable for the current VLSI technology. In this paper, it has been shown that the practical restrictions such as the boundary layout assumption, and the restricted I/O port location assumption, possibly requires larger area than the functional complexity.

A lower bound on area of combinational circuits to compute the multiplication has been shown by Lipton-Tarjan using Planar Separator Theorem [7]. However, from the results of this paper, if the combinational circuit is embedded on a convex region, or it satisfies the boundary layout assumption, then it has been shown that the same lower bounds for the multiplication and the division hold. Furthermore, new lower bounds on area for the sorting, the decoder and the encoder can be obtained by using the techniques in this paper [17].

References

[1] R.P.Brent and H.T.Kung, "The Chip Complexity of Binary Arithmetic," Proc. 12th Annu. ACM Symp. on Theory of Comput., ACM, pp.190-200, April 1980.
[2] R.P.Brent and H.T.Kung, "On the Area of Binary Tree Layouts," Information Processing Letters, Vol.11, No.1, pp.46-48, Aug. 1980.
[3] B.Chazelle and L.Monier, "A Model of Computation for VLSI with Related Complexity Results," Dept. of Comput. Sci., Carnegie-Mellon Univ., Pittsburgh, Pa., Tech. Rep. CMU-CS-81-107, Feb. 1981.
[4] I.Deegan, "Concise Cellular Array for Multiplication and Division," Electronics Letters, Vol.7, No.23, Nov. 1971.
[5] R.B.Johnson Jr., "The Complexity of a VLSI Adder," Information Processing Letters, Vol.11, No.2, pp.92-93, Oct. 1980.
[6] D.E.Knuth, The Art of Computer Programming, Vol.3: Sorting and Searching, Addison-Wesley, Reading, Massachusetts, 1973.
[7] R.J.Lipton and R.E.Tarjan, "Applications of a Planar Separator Theorem," SIAM J., Vol.9, No.3, pp.615-627, Aug. 1980.
[8] K.Menger, "Zur Allgemeinen Kurventheorie," Fund. Math., Vol.10, pp.96-115, 1927.
[9] J.E.Savage, The Complexity of Computing, Wiley-Interscience, New York, N.Y., 1976.
[10] J.E.Savage, "Area-Time Tradeoffs for Matrix Multiplication and Related Problems in VLSI Models," Dept. of Comput. Sci., Brown Univ., Providence, R.I., Tech. Rep. CS-50, Aug. 1979.
[11] J.E.Savage, "Planar Circuit Complexity and the Performance of VLSI Algorithms," INRIA Rapports de Recherche, No.77, April 1981.
[12] C.D.Thompson, "A Complexity Theory for VLSI," Dept. of Comput. Sci., Carnegie-Mellon Univ., Pittsburgh, Pa., Tech. Rep. CMU-CS-80-140, Aug. 1980.
[13] J.Vuillemin, "A Combinational Limit to the Computing Power of V.L.S.I. Circuits," IEEE 21st Annu. Symp. on FOCS, pp.294-300, Oct. 1980.
[14] K.Wada, K.Hagihara and N.Tokura, "The Area-Time Complexity of n Variables Logical Functions," Trans. IECE Japan, Vol.J64-D, No.8, pp.676-681, Aug. 1981 (in Japanese).
[15] K.Wada, K.Hagihara and N.Tokura, "The Area Complexity on a VLSI Model," Trans. IECE Japan, Vol.J65-D, No.4, pp.478-485, April 1982 (in Japanese).
[16] K.Wada, K.Hagihara and N.Tokura, "Area and Time Complexities of VLSI Computations," Proc. of the 7th IBM Symp. on Math. Foundations of Comput. Sci., Math. Theory of Computations, IBM Japan, June 1982.
[17] K.Wada, K.Hagihara and N.Tokura, "The Area Lower Bounds of Combinational Circuits on a VLSI Model," Papers of Tech. Group on Automat. and Lang., AL82-30, IECE Japan, Sept. 1982 (in Japanese).
[18] H.Yasuura and S.Yajima, "On Embedding Problems of Logic Circuits in a VLSI Model," Papers of Tech. Group on Automat. and. Lang., AL81-49, IECE Japan, Sept. 1981 (in Japanese).

Hardware Algorithms and Logic Design Automation

--- An Overview and Progress Report ---

Shuzo YAJIMA and Hiroto YASUURA

Department of Information Science
Faculty of Engineering
Kyoto University
Kyoto, 606 JAPAN

1. Introduction

Advances in the fabrication technology of VLSI circuits will soon make it feasible to implement highly parallel computing systems consisting of hundreds or of thousands of computing elements. These highly parallel systems will operate cooperatively with software and achieve tremendous speed improvements of digital computing systems. Many researches have been carried out to establish effective design methods for either general or special purpose highly parallel systems[1].

Design of efficient algorithm for parallel computation is the key problem of design of highly parallel hardware systems as well as software. In this article, we will discuss the design problem of parallel algorithms, called hardware algorithms.

Various hardware algorithms have been proposed for several practically important problems such as sorting, arithmetic operations, matrix arithmetics and pattern matching[2]-[6]. To analyze these algorithms, several theoretical discussions have been going on and a new complexity measure of parallel computation suitable for VLSI, called area, has been proposed

[7][8] VLSI and hardware algorithms make a new area of theoretical research on computational complexity. A theory of design and analysis of hardware algorithms for VLSI systems will be established in several years ahead.

Logic design automation systems will be indispensable tools for design of large highly parallel systems. A silicon compilation system will be developed which generates mask patterns for VLSI fabrication from a high level hardware algorithm description[9]. Design verification tools are also important components of design automation systems, since highly parallel computation is inherently too complicated for designers to think without any tools such as simulator and verifier.

In this article, we will briefly survey topics of researches on hardware algorithms and design automation systems. Section 2 and 3 are an overview and progress report of the research on hardware algorithms. Section 4 is the progress report of development of an interactive logic design and verification support system ISS.

In the next section, hardware algorithms of integer multiplication and sorting are presented as examples of hardware algorithms. A general consideration of hardware algorithms for VLSI is also discussed. In section 3, some results of theoretical researches on the complexity of hardware algorithms are presented. Design automation systems and computer aided design will be discussed in section 4. As an example of design verification tools, ISS developed by our group is presented.

2. Design of Hardware Algorithms

2.1 Integer Multiplication

Integer multiplication is widely used as a basic operation in general purpose computers, in process controllers and signal processors. Many high speed algorithms for integer multiplication in software and hardware have been proposed and used practically. In table 1, several hardware algorithms for integer multiplication are compared.

In applications not required so much high speed computation, add-and-shift multiplication is generally used. The speed of

algorithm	size	area	speed of computation
add-and-shift multiplication (ripple carry adder)	n	n	n^2
add-and-shift multiplication (carry look-ahead adder)	n	n(log n)	n(log n)
serial multiplication	n	n	n
array multiplication	n^2	n^2	n
matrix generation-reduction multiplication	n^2	n^2(log n)	log n
Brent-Kung's algorithm	-	n(log n)	$\sqrt{n}$(log n)
Shonhage-Strassen's algorithm	n(log n)(log log n)	-	log n

n: the length of operands

Table 1. Hardware Algorithms for Integer Multiplication

n	8	16	32	64
matrix generation-reduction	22/672	24/2516	30/9064	34/35207
array	29/528	61/2336	125/9792	253/40064

(depth/size) using 4-input NOR/OR gates

Table 2. Depth and Size of Multipliers

computation is improved, when a carry look-ahead adder is adopted[10]. Serial multiplication is implemented in signal processing in which operands are input serially[11]. These above algorithms are implemented by sequential circuits.

For high-speed multiplication by combinational circuits, array multiplication and matrix generation-reduction scheme are developed and implemented for practical use. Array multiplication is attractive for their compactness and regularity of its iterative array structure using one basic circuit type, but their speed of operation increases linearly with the operand length and thus slow for large words[10]. Matrix generation-reduction scheme is much faster for large operands since their speed of operation increases with the logarithm of the operand length [12]. The basic idea of this algorithm was proposed by Karatsuba and Ofman, and the most popular circuit based on it is known as Wallace's tree[13][14]. Several papers discussed about multiplier based on this algorithm with higher performance[10][12].

In table 2, an evaluation of the size (the number of gates) and the depth (the speed of computation) of circuits designed according to these two algorithms is shown. Gates used in the design are 4-input NOR/OR gates. The circuit based on matrix generation-reduction algorithms is realized as a combination of Booth's algorithm, Wallece's tree and a carry look-ahead adder. The depth of the circuit is extremely smaller than the array multiplier for large n. This example shows that the design of a good hardware algorithm results in tremendous improvement on efficiency.

Although the size of circuits of these two algorithms are the same order, namely $O(n^2)$, the upper bounds of the area on VLSI have quite different order, $O(n^2)$ for array multiplication and $O(n^2 \log n)$ for matrix generation-reduction one. This difference is caused by the difference of complexity of interconnection in the circuits. In the next section, we will discuss a new circuit complexity analysis technique using the measure, called area.

Brent-Kung's algorithm in table 1 achieves the best upper bound of the area-time product[8]. Shonhage-Strassen's one is the best upper bound of the number of Boolean operations required to n-bit integer multiplication[15].

2.2 Sorting

Sorting is one of the most important operations in data processing. Many sequential and parallel sorting algorithms have been developed and practically used. In table 3, several sorting algorithms are compared.

Algorithm	the number of processing elements	the speed of computation
Bubble sort	O(1)	$O(n^2)$
Heap sort	O(1)	O(n log n)
Bitonic sort	O(n)	$O(\log^2 n)$
Rebound sort	O(n)	O(n)
Parallel enumeration sort	O(n)	O(n)
Parallel merge sort	O(log n)	O(n)
Parallel heap sort	O(log n)	O(n)
Sorting on mesh	O(n)	$O(n^{1/2})$
Parallel distributive sort	O(n)	O(log n)
Sorting by combinational circuit	$O(n^2)$	O(log n)

n: the number of sorted element

Table 3. Algorithms for Sorting

Bubble sort and Heap sort are software algorithms and have time complexity $O(n^2)$ and $O(n \log n)$, respectively[15]. Heap sort is one of the fastest algorithms of software, because it is easy to show that the lower bound of time complexity of software algorithms is $O(n \log n)$.

Many hardware algorithms for sorting have been proposed and some of them are implemented. Muller and Preparata showed that sorting of n elements can be performed by a combinational circuit with depth $O(\log n)$ and size $O(n^2)$[16]. Several hardware algorithms on multiprocessor systems have been proposed such as algorithms on mesh connected processors by Thompson-Kung[17] and by Nassimi-Sahni[18]. Parallel distributive sort by Maekawa[19] and Wilsow-Chow[20] are also this kind.

When we consider a sorting circuit which is attached to conventional computer systems, we will assume that data is transmitted one by one between the sorting circuit and memory devices. Rebound sort by Chen et. al.[21], Parallel enumeration sort by Yasuura and Takagi[22], Parallel merge sort by Todd[23], and Parallel heap sort by Tanaka[24] are all developed under the following assumptions; (1) a sorting circuit is separated from memory devices and (2) data transmission between the circuit and memory devices is serial. Under these assumptions, processing for sorting cannot be faster than data transmission. Since the time required for sorting in these algorithms is linearly proportional to the number of sorted elements, these algorithms achieve optimum order on time. In these algorithms, processes of sorting are efficiently overlapped with the input/output time.

Rebound sort and Parallel enumeration sort require $O(n)$ processing elements, but each element has constant size and realized by a very simple circuit. Circuits for these algorithms have linear array structure and simple communication structure. Rebound sort is suitable for implementation by magnetic bubble circuits. Parallel enumeration sorting is implemented on the Bus Connected Cellular Array structure, the detail of which is discussed in the next subsection.

Parallel merge sort and Parallel heap sort require only $O(\log n)$ processing elements. However, each processing element should possess $O(n)$ memory. Since memory can be integrated in higher density than logic circuits, several circuits have been implemented based on these algorithms using commercial LSI's.

2.3 Hardware Algorithms for VLSI

Kung and his group proposed the systolic algorithms which is suitable for VLSI implementation[4]-[6]. Systolic algorithms have the following properties[6].

(1) The algorithm can be implemented by only a few different types of simple cells.

(2) The algorithm's data and control flow is simple and regular, so that cells can be connected by a network with local and regular interconnections.

(3) The algorithm uses extensive pipelining and multiprocessing. Typically, several data streams move at constant velocity over fixed paths in the network, interacting at cells where they meet. In this way a large number of cells are active at one time so that the computation speed can keep up with the data rate.

The first property reduses the cost of design and test, since a designer only designs and tests a few different, simple cells. Regular interconnection in the second property implies that the design can be made modular and extensible. This also means that the area for wiring on a chip will be reduced and propagation delay caused by these wires will decrease. By pipelining and multiprocessing, one can meet the performance requirement of a circuit. Pipelining makes it possible to overlap processing and input/output effectively.

Kung and his coauthors developed many systolic algorithms on one dimensional cellular arrays, two dimensional square meshes and hexagonal meshes.

We developed hardware algorithms which are realized on Bus Connected Cellular Array (BCA). Algorithms on BCA possess the following properties added to (1)-(3) of systolic algorithms [22][25].

(4) The algorithm uses grobal communications through buses. The communication control of the grobal communication is simple and distributed.

(5) Input and output scheme is simple and it is easy to attach to other circuits in a system. One can extend the circuit only connecting chips which include smaller circuits without changing

input/output scheme.

(6) The restriction on performance of the algorithm shold be relaxed as much as possible. Processing time should depend on only the size of problem not on the size of the circuit.

The grobal communication using buses improves the speed of algorithm drastically and reduces the complexity of communication. Algorithms must be designed under realistic assumption of input/output protocols. Highly parallel input and output increases infeasiblly the complexity of communication of the outside of the algorithm, though the algorithm seems to achieve high performance. Hardware algorithms are inherently restricted their ability of processing by the size of circuits. However, algorithms should process problems smaller than their ability in time proportional to the size of problems. On BCA algorithms, these properties can be easily realized using grobal bus communication.

We proposed a sorting algorithm on BCA, called Parallel enumeration sort[22]. This algorithm can be introduced to conventional computer systems without changing their architecture. The processing time is linearly proportional to the number of data for sorting. The sorting circuit consists of a linear array of one type of simple cells each of which includes two registers, a comparator and a counter. These cells are connected by two buses. Since the circuit is extensible only by connecting the same circuit, we can implement a large circuit connecting chips including the circuits.

We have also developed BCA algorithms for pattern matching, matrix multiplication, very long integer multiplications and join opration in relational databases.

3. Analysis of Hardware Algorithms

3.1 Time Complexity

Combinational logic circuits are the most fundamental circuits in digital systems, which can realize the most highly parallel computation. Many researches have been carried out on depth and size of circuit required to realize Boolean functions [2][25][26]. Results of these researches can be applied to analysis of hardware algorithms.

The delay complexity of Boolean function is the smallest depth of circuits which realize f. Boolean functions which depend essentially on n variables have delay complexity $\Theta(\log n)$. Functions in several important classes such as linear functions, symmetric functions and threshold functions have delay complexity proportional to log n. It is also well known that there are functions required O(n) depth of circuits for their realization [26].

Integer addition and multiplication can be performed by circuits with depth O(log n), where n is the length of operands. For division and square rooting, algorithms with depth $O(\log^2 n)$ were developed. But it is not known whether these operations can be computed by circuits with depth O(log n) or not.

Unger proposed a method to construct a circuit with the logarithmic depth for a class of functions which can be computed by sequential circuits in linear time[27]. This class of functions includes many practical functions such as binary addition, comparison, counting and so on. We showed a method to reduce the depth of a circuit as a generalization of Unger's method[28].

Moreover, we proved an interesting general result on relation between complexity of software algorithms and of hardware algorithms as follows.

Theorem[29] A function which can be computed by a T(n)-time bounded and S(n)-tape bounded deterministic Turing machine can be computed by a combinational circuit whose depth is proportional to S(n)log T(n).

In the proof of this theorem, we obtained a construction method of the combinational circuit. Thus we can construct a combinational circuit with depth $O(\log^{k+1} n)$ for a function which is computed in $\log^k n$ space by a polynomial time software algorithm. This result shows an upper bound of speed up ratio of hardware algorithm and software one.

3.2 Area Complexity

In VLSI circuits, it is well known that the area of a circuit depends on not only the number of logic elements included in the circuit but also the area for the wiring and input/output terminals[4]. Thompson[7] and Brent-Kung[8] proposed mathematical models of VLSI circuits and provided several techniques to evaluate the area of circuits theoretically. On these mathematical models of VLSI, many works have been carried out to estimate performance of VLSI circuits by a measure, area-time product[3]. Leiserson[30] and Valiant[31] discussed the area required for layouts of tree and planer graph circuits into the VLSI model.

A VLSI model is defined as follows:

(1) A circuit is embedded into a convex planar region R.

(2) Wires have minimal width λ (a positive constant).

(3) At most ν ($\nu \geqq 2$, a positive constant) wires can overlap at any point of R.

(4) Logic elements each contain a $\lambda \times \lambda$ square and their shapes and area are given for each sort of elements.

(5) No logic element overlaps other logic elements and wires in R.

(6) Input and output of the circuit is performed through wires on the boundary of R.

The last condition of the definition of VLSI model was introduced by us and called the boundary conditions[32]. The boundary conditions is very realistic assumption of actual VLSI circuits. We can show that the area required for embedding of a binary tree circuit with n nodes is $\Theta(n \log n)$ under the boundary condition, though it is just $O(n)$ on the model except the boundary condition.

It is very important to estimate the area of a circuit at the stage of logic design before layout of the circuit. Several results have been obtained on evaluation methods of the area from feature of the circuit such as depth, width, size and structure which are easily measured from the logic design.

Trade-off between time and area is the most interesting subject of theoretical researches. Many results have been presented on area-time products of practical functions such as integer multiplication, integer addition, comparison, sorting, FFT, matrix multiplication and so on[3]. On combinational circuits, trade-off between depth d and area A for an n-variable function is shown as follows:

$A \log d = \Omega(n \log n)$[32].

4. Logic Design Automation

4.1 Computer Aided Design and Design Automation

Computer Aided Design (CAD) or Design Automation (DA) technology is almost indispensable in large, complicated logic design. A goal of researches on CAD/DA systems is to develop a silicon compilation system[9]. In the silicon compilation system, a designer describes a design of a hardware algorithm in a high-level hardware algorithms description language which is free from many constraints caused by physical implementation. The designer also specifies several conditions on realization of the algorithm by silicon devices such as speed, area, technology used for implementation and so on. The silicon compilation system generates a logic design which realizes the algorithm and satisfies the specified conditions. Moreover, the system translates the design into several information to control fabrication processes of VLSI. One can make his own VLSI chips only writing a specification of hardware algorithm in a high-level language as well as software in high-level languages. There are still many problems that should be resolved for implementation of the silicon compilation system.

Many kinds of CAD techniques and systems for logic design have been developed and are in use: hardware description language, logic diagram editor, logic function minimization programs, optimum circuit generation programs, logic simulator, logic design verifier, test pattern generator, etc. These systems are developed to relieve logic designers from troublesome, laborious work in logic design and verification. However, they are not utilized to a full extent because designers can not use them in an integrated way. Since the logic design stage is a repetition of design and verification, a logic design system should be an integrated one including a design language, editing system, design-aid programs, a design verification tool, etc. on a standardized data management. A CAD system for logic design usually needs some interaction by man for processing. An interactive sophisticated user interface is keenly needed for CAD systems.

We have developed Interactive Simulation System (ISS) --- an interactive logic design and verification support system for structured logic design [33]. One can carry out logic design in a structured way by utilizing functions of ISS all interactively. ISS has the following features.

(1) Interactive and Integrated Logic Design System

EDITOR, TRANSLATOR, LINKER, Interactive Simulator (IS) and other peripheral programs can be utilized easily by terminal commands (See Fig.1). A designer performs logic design and its verification alternately without bothering about management of many kinds of design data.

(2) Interactive Simulator(IS)

IS is used for design verification in ISS. A designer can control the simulation steps interactively to find out design errors at an early stage of design cycle. IS can simulate a design described at multi-levels (gate, functional, and register transfer) in a structured logic design.

(3) Structured Hardware Design Language (SHDL)

SHDL is used to describe logic designs in ISS. SHDL can describe a design as a hierarchically constructed set of modules. Each module is described with its structure or behavior. There are three kinds of behavioral description which enable multi-level description of a design.

(4) Design Data File

The file system storing description data or simulation data is configurated "module oriented". ISS has the file system as a common interface among programs and is integrated around it.

4.2 Design Verification in ISS

A logic simulator is used for design verification in order to find out design errors based on simulation results which do not satisfy the design specification. Verification by simulation shows that "the design contains errors", not that "the design is correct". The usefulness of a simulator as a design verification tool depends on how early a user can find out design errors with it. The turn around time to feed back the verification results to the design step becomes shorter, the earlier an error can be found.

Interactive Simulator IS in ISS is an interactive simulator which has following functions.

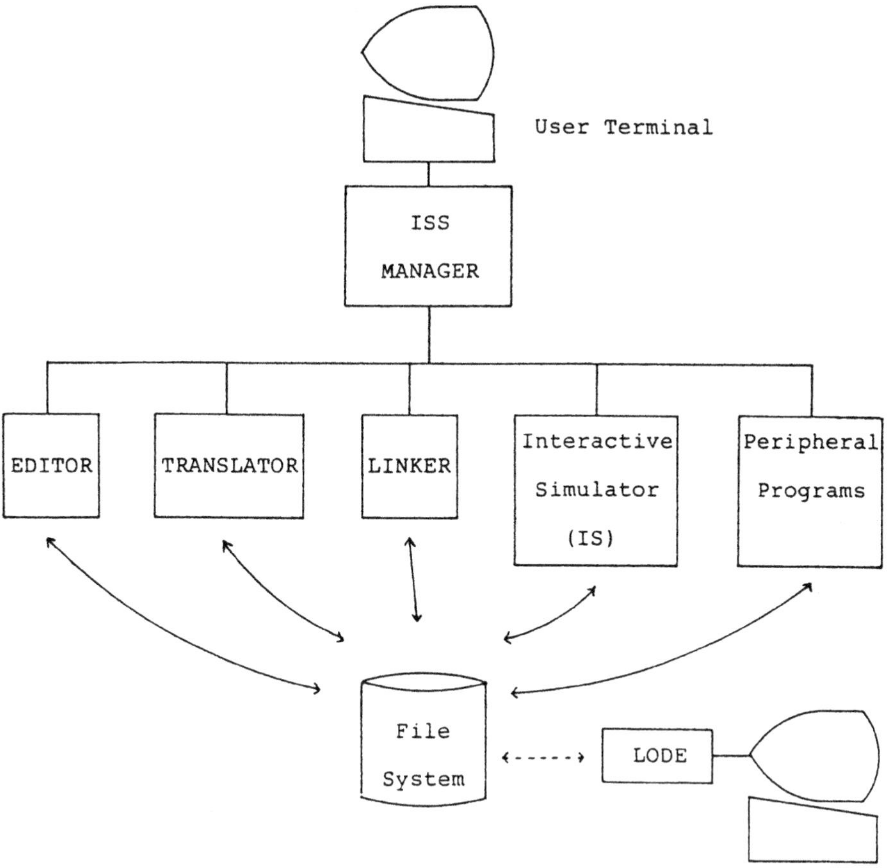

Fig. 1 Configuration of ISS

(1) Interrupting function

A user can set breakpoints in simulation steps using subcommand AT, STEP, WHEN and ON. When the following conditions occur, the simulation process is suspended and falls into the suspended state.

AT condition occurs when the simulation time reaches the specified one. AT subcommand is used to set this condition.

STEP condition occurs when the simulation for the specified time interval ends. STEP subcommand is used to set this condition.

WHEN condition occurs when the simulated module behaves in the specified manner or falls into the specified state. WHEN subcommand is used to set this condition.

INPUT CONSTRAINTS ERROR condition occurs when the simulated module or submodules (blocks) receive input patterns which do not satisfy the input constraints of the module or submodules. ON subcommand is used to make this condition in effect. In SHDL the design specification of a module can be partially described as constraints on its input patterns. In general the module is designed to have the specified function only of the input patterns which satisfy the input constraints. A designer can use this useful information in design verification with a simulator. When an input pattern to a module violates input constraints, the design containing the module must have some errors and the module behaves incorrectly. The errors are likely to be in the design of modules which supply the input pattern or in the input constraints itself. A designer can tell of what kind the error is from the type of the input constraints not satisfied. Moreover, a designer can easily prepare input patterns for a module according to input constraints description. Waste of computation time on unnecessary simulation for wrong input patterns. IS has a function to check input patterns during execution whether they satisfy input constraints of modules to which they drive. INPUT CONSTRAINTS ERROR condition occurs when these input constraints are violated.

While the simulation process stays in the suspended state, a user can examine the status of the module precisely and resume the simulation with modified input patterns specified adaptively.

(2) Display function of simulation results

A user can examine the values of signal lines and contents of memory elements at the specified simulation time immediately. LIST subcommand is used to display simulation results.

(3) Modification function of input patterns

A user can modify input patterns dynamically according to the simulation results. EDITWAVE subcommand is used.

(4) Simulation resuming function

A user can resume the simulation from the past as well as the present simulation time. When the input patterns are changed, he can go back to the time at which the changes are effetive and resume the simulation. This technique reduces much computation time. GO and RUN subcommand are available for this function.

Using these functions of IS, a designer can easily find design errors in the early stage of design cycle. It shortens the whole time spent for logic design. In Fig.2, an example of interactive simulation is shown. ISS is implemented on FACOM M-200 in Data Processing Center of Kyoto University. Programs are mostly developed in PL/I and about 18 thousands steps in total. Users can use ISS from their TSS terminal interactively.

5. Conclusion

In this paper, we discussed several problems in the design of hardware algorithms and logic design automation. The theory of complexty of logic circuits and parallel computation will form the foundation of design of hardware algorithms which will become more important for larger VLSI systems. Especially, the relation between the complexities of software and hardware is very important for practical system design, because systems are combination of software and hardware.

Design automation is one of the most highlighted fields in computer science. There are still many hard problems to solve in developing much efficient design automation system. Researches on high-level hardware design languages, automatic logic design from descriptions of these languages and design verification techniques for large systems have been increasing. Several techniques developed in the software engineering will be applied to these area.

Ackowledgement

The authors would like to express their sincere appreciation to Mr.Y.Ooi and Mr.N.Takagi for their many helpful co-operation in preparing the manuscript. The authors would like to thank Prof.Y.Kambayashi, Mr.H.Hiraishi and the members of the Yajima Lab. of Kyoto University for their discussions.

References

[1] "Highly Parallel Computing" Edited by L.S.Hayens, IEEE Computer, vol.15, no.1, pp.7-96, Jan. 1982.

[2] S.Yajima, H.Yasuura and Y.kambayashi, "Design of Hardware Algorithms and Related Problems", IECE Technical Rep. AL81-86, Dec. 1981 (in Japanese).

[3] N.Tokura, "VLSI Algorithms and Area-Time Complexity", Joho-Shori vol.23, no.3, pp.176-186, March 1982 (in Japanese).

[4] C.A.Mead and L.A.Conway, "Introduction to VLSI Systems", Addison-Wesley, Reading, Mass., 1980.

[5] H.T.Kung, "The Structure of Parallel Algorithms", Advanced in Computers, vol.19, Academic Press, 1980.

[6] M.Foster and H.T.Kung, "The Design of Special-Purpose VLSI Chips", IEEE Computer, vol.13, no.1, Jan. 1980.

[7] C.D.Thompson, "Area-Time Complexity for VLSI", Proc. 11th Symposium on the Theory of Computing, pp.81-88, May 1979.

[8] R.P.Brent and H.T.Kung, "The Area-Time Complexity of Binary Multiplication", JACM, vol.28, no.3, pp.521-534, July 1981.

[9] J.P.Gray, "Introduction to Silicon Compilation", Proc. 16th DA Conference, pp.305-306, June 1979.

[10] K.Hwang, "Computer Arithmetic:Principle, Architecture and Design", John-Wiley & Sons, Reading, Mass., 1979.

[11] L.B.Jackson, S.F.Kaiser and H.S.McDonald, "An Approach to the Implementation of Digital Filters," IEEE Trans. Audio Electro., AU-16, Sept. 1968.

[12] W.J.Stenzel, W.J.Kubitz and G.H.Garcia, "A Compact High-Speed Parallel Multiplication Scheme," IEEE Trans. on Comput., vol.C-26, no.10, pp.948-957, Oct. 1977.

[13] A.Karatsuba and Y.Ofman, "Multiplication of Multidigit Numbers with Computers", Dokl. Akad. Nauk. SSSR, no.145, Feb. 1962.

[14] C.S.Wallace, "A Suggestion for a Fast Multiplier", IEEE Trans. on Electro. Comput., vol EC-13, no.1, pp.14-17, Feb. 1964.

[15] A.V.Aho, J.E.Hopcroft and J.D.Ullman, "Design and Analysis of Computer Algorithms", Addison-Wesley, Reading, Mass., 1974.

[16] D.E.Muller and F.P.Preparata, "Bounds to Complexities of Networks for Sorting and Switching", JACM, vol.22, no.2, pp.195-201, Apr. 1975.

[17] C.D.Thompson and H.T.Kung, "Sorting on a Mesh-Connected Parallel Computer", CACM, vol.20, no.4, Apr.1977.

[18] D.Nassimi and S.Sahni, "Bitonic Sort on a Mesh-Connected Parallel Computer", IEEE Trans. Comput., vol.C-28, no.1, Jan. 1979.

[19] M.Maekawa, "Parallel Sort and Join for High Speed Database Machine Operations", AFIPS Conf. Proc., vol.50, June 1981.

[20] L.E.Winslow and Y.C.Chow, "Parallel Sorting Machines :Their Speed and Efficieny", AFIPS Conf. Proc., vol.50, June 1981.

[21] T.C.Chen, V.Y.Lum and C.Tung, "The Rebound Sorter:An Efficient Sort Engine for Large Files", Proc. 4th VLDB, pp.312-318, Sept. 1978.

[22] H.Yasuura and N.Takagi, "A High-Speed Sorting Circuit Using Parallel Enumeration Sort", Trans. IECE, vol.J65-D, no.2, pp.179-186, Feb.1982 (in Japanese).

[23] S.Todd, "Algorithm and Hardware for a Merge Sort Using Multiple Processors", IBM Journal of R. & D., vol.22, no.5, Sept. 1978.

[24] Y.Tanaka, Y.Nozawa and A.Masuyama, "Pipeline Searching and Sorting Modules as Components of a Data Flow Database Computer", Proc. IFIP80, pp.427-432, Oct. 1980.

[25] H.Yasuura, "Hardware Algorithms for VLSI", Proc. Joint

Conf. of 4 Institutes Related on Electric Engineering, 34-4, Oct. 1981 (in Japanese).

[26] J.E.Savage, "The Complexity of Computing", Wiley-Interscience, Reading, Mass., 1976.

[27] S.H.Unger, "Tree Realizations of Iterative Circuits", IEEE Trans. Comput., vol.c-26, no.4, pp.365-383, Apr. 1977.

[28] H.Yasuura, Y.Ooi and S.Yajima, "On Macroscopic Depth Reduction for Combinational Logic Circuits", IECE Technical Rep. EC81-1, Apr. 1981 (in Japanese).

[29] H.Yasuura, "Width and Depth of Combinational Logic Circuits", Information Processing Letters, vol.13, no.4, 5, End, pp.191-194, 1981.

[30] C.E.Leiserson, "Area-Efficiency Graph Layout (for VLSI)", Proc. 21st FOCS, Oct. 1980.

[31] L.G.Valiant, "Universality Considerations in VLSI Circuits", IEEE Trans. on Comput., vol.C-30, no.2, pp.153-157, Feb.1981.

[32] H.Yasuura and S.Yajima, "On Area of Logic Circuits in VLSI", Trans. IECE, vol.J.65-D, pp.1080-1087, Aug. 1982.

[33] T.Sakai, Y.Tsuchida, H.Yasuura, Y.Ooi, Y.Ono, H.Kano, S.Kimura and S.Yajima, "An Interactive Simulation System for Structured Logic Design -- ISS", Proc. 19th DA Conf., pp.747-754, June 1982.

List Processing with A Data Flow Machine

Makoto AMAMIYA, Ryuzo HASEGAWA,
and Hirohide MIKAMI

Musashino Electrical Communication Laboratory, N.T.T.

1. Introduction

A data flow machine, whose basic idea was offered by J. B. Dennis [1] and for which several researches are pursued at several places throughout the world [2,3,4,5,6], is a very attractive concept for computer architecture from the following viewpoints:

(1) A data flow machine exploits the parallelism inherent in problems, and executes it in a highly concurrent manner.

(2) Recent advances in VLSI technology have been noteworthy. One of the major problems of computer architecture is how systems, which utilize a large amount of VLSI devices, should be constructed. A data flow machine enables the implementation of a distributed control mechanism, which is a key technique when using VLSI devices.

(3) How to increase a program productivity, verifiability, testability and maintainability, are serious problem in software engineering. One solution for this problem is to write side-effect free programs based on a functional programming concept. A data flow machine can effectively execute side-effect free programs written in a functional language such as pure Lisp, due to inherent parallelism.

(4) Non-deterministic execution will become an important mechanism in computer systems, when the problem solving concepts obtained from AI research are applied to real world problems. A data flow machine is expected to execute non-deterministic programs effectively, due to its parallel processing capabilities.

However, many problems remain to be solved before a data flow machine can be of practical use in a real environment. Looking especially at Items (3) and (4), it is necessary to clarify the applicability of the data flow machine to non-numerical problems. This implies the necessity of solving the problem of structure memory construction.

List processing is typical of non-numerical data processing. This paper discusses list processing with a data flow machine, with the Lisp data structure and operations in mind. The main reasons for why Lisp is considered are that Lisp has simple and transparent program and data structure, and that it contains basic problems in structure data manipulation even in the simplified structure.

In conventional Lisp implementations, it is inevitable to introduce such side-effect facilities as a prog-feature in order to achieve high speed execution under a sequential execution control environment and efficient memory usage under a centralized memory management on von-Neumann type computers. However, if machines can obtain highly parallel execution in a data flow control environment, and if VLSI technology can offer high level functional memory devices for computer architecture, side-effect free and pure functional list processing will make sense in practical use.

This paper shows that a data flow control is effective for list processing. First, the parallelism in list processing is discussed, and it is pointed out that a high degree of parallelism can be achieved by parallel evaluation of function arguments, and partial execution of the function body. Then it is shown that the parallelism increases dramatically when a "lenient cons" concept is introduced into the data flow execution control. That is, the lenient cons makes possible a pipelined execution among activated functions for the data flow, because each activated function executes each data item (in the list) as soon as it is partially constructed. The effect of a lenient cons on list processing is shown through analyses of several programs.

Finally a garbage collection algorithm based on the reference count method is discussed. Garbage collection is also an important problem in list processing because of the need to utilize memory cells effectively. The algorithm is essentially parallel in the sense that cells are reclaimed independently of the foreground list manipulation whenever they become useless.

All programs throughout this paper are described in the functional language Valid [7], which has been designed as a high level programming language for the data flow machine presented in this paper.

2. List processing in a data flow control environment

The remarkable effects of data flow execution control are as follows.

(1) It exploits, to the maximum degree, the parallelism inherent in a given program, both on a low level (primitive operation level) and on a high level (function activation level).

(2) It effectively executes programs constructed on the basis of functional programming, the concept for which contains no notion of program variables and side effects (i.e., re-writing of the global variables).

Parallelism at the primitive operation level can be obtained from the data driven control principle. Each operation is initiated without any attention being paid to other operations when all operands have arrived. The parallelism at the function activation level is obtained from the partial evaluation mechanism. This mechanism has the following characteristics.

(a) Each argument of a function is evaluated in parallel.

(b) The execution of a function is initiated when one of the arguments of the function is evaluated, and the caller function resumes its execution when one of the return values is obtained during the execution of invoked function.

In a conventional cons mechanism, there exists a case in which the partial execution of a function body is not effective, because an execution, which uses list data generated by another function, has to wait until its construction is completed. However, if a "lenient cons" concept is introduced, the consumer function, which partially uses list elements, can start its execution as soon as the producer function generates a partial fragment of the list, before the whole list is constructed.

In following sub-sections, the parallel execution mechanism will be examined through several examples, the programs of which are written in Valid (some Valid features are described in Appendix A).

2.1 Parallel evaluation of arguments

Programs written in Valid are equivalently transformed to pure functional representation, i.e., the form of prefix notation, and equally translated into data flow graphs. For instance, Program1 which reverses a given list in each level is translated by the Valid compiler into the data flow graph shown in Fig. 2.1. In this program, three value definitions and one return expression within Block2, which is enclosed with clause and end, are evaluated according to their data dependency. Therefore, Block2 is equivalently represented in the prefix notation

fulrev(cdr(x), cons(fulrev(car(x), nil), y)) .

In this expression, the arguments cdr(x) and cons(...) of the function fulrev are evaluated in parallel. Before evaluating the argument cons(fulrev(...),y), its two arguments fulrev(...) and y are evaluated in parallel, and so on. Thus, in general, the evaluation of a function proceeds from the inner to the outer (i.e., innermost evaluation), which results in the highly parallel evaluation of the innermost arguments. This means that each evaluation is independent of the other evaluations under the condition that the evaluation is initiated only when all values of the arguments are obtained, (which is called data driven control).

```
-- Program1
-- Mirror image of tree

fulrev: function (x,y) return (list)
= case
    null(x) -> y;
    atom(x) -> x;
    others  -> clause
               ( u = cdr(x);
    block2    {  v = fulrev(car(x),nil);
               { w = cons(v,y);
               ( return fulrev(u,w)
               end
  end

-- Top level invocation is
-- fulrev(a,nil)
```

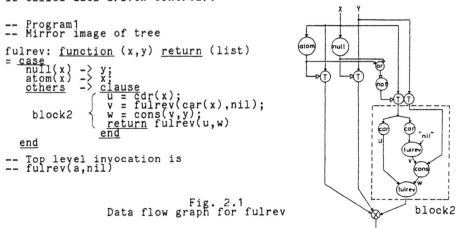

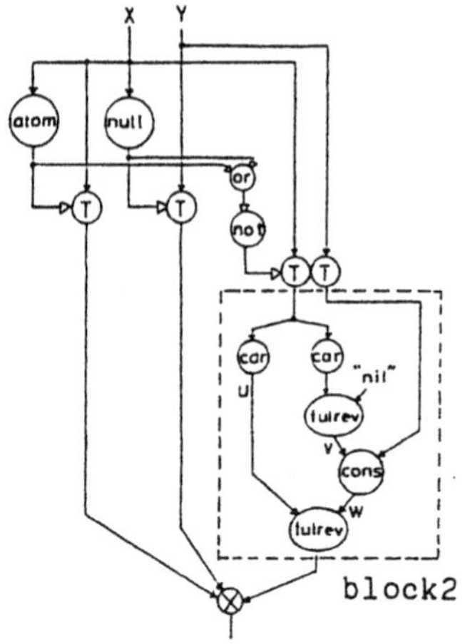

Fig. 2.1
Data flow graph for fulrev

Another parallelism, suggested in Lisp 1.5 interpreter [8] is also achieved by implementing the interpreter on the basis of the data flow control concept. An implementation is represented in Appendix B. In the implementation, the parallel execution is exploited in the evcon and evlis recursion. In the course of argument evaluations, evlis forks eval functions for each argument, and each eval-ed argument value is ultimately cons-ed, as shown in Appendix B. The parallel activation of eval in evlis is effective, since the Lisp interpreter also has the ability to evaluate each argument of the Lisp function in parallel. The following program written in m-expression, which uses a divide and conquer algorithm for factorial calculation, can be executed in a highly parallel manner (here, N!=fact[0;N]).

```
fact[m;n]
= [equal[m;n] -> n;
   t -> times[fact[m;quotient[plus[m;n];2]];
              fact[add1[quotient[plus[m;n];2]];n]]]
```

2.2 Partial execution of function body

The parallelism, based on the parallel evaluation of arguments of each function, is limited because the nesting of arguments is limited in the source text. This restriction on parallelism, however, can be overcome by partially executing the function body.

Partial evaluation has the following advantage over data driven control. If the data driven control principle is applied to the function activation, as in the case of primitive operations, every function is activated only after all its arguments are evaluated. In this case, time is wasted unnecessarily in each function activation because it is necessary to wait for the completion of all its argument evaluations. However, if each value of the arguments is passed into the function body immediately after it is evaluated, and the execution of the body proceeds partially every time the value is passed in, efficient execution can be obtained. This is because unnecessary waiting time is eliminated at the function activation. In the same way, each of the return values can be passed back to the calling function as soon as it is generated, and the calling function can resume and proceed with its partial execution every time the return value is passed back from the called function. (Here, each function is permitted to return multiple values, i.e., a tuple of values, under the data flow control environment.)

These function activation and parameter passing mechanism can be easily realized in the data flow control environment.

The function activation and argument passing mechanism for partial execution of function is implemented as shown in Fig. 2.2. The data flow graph in Fig. 2.2(c) represents the activation control mechanism for the function

[y1,y2,...yn] = f(x1,x2,...xm) .

The copy node, which creates a new instance for the activated function (or logically makes a new copy of the body), is initiated by or-gating nodes, when one of the tokens (values) has arrived. The orgate implementation uses a T/F switch as shown in Fig. 2.2(b).

When the new instance is created and the body is ready to run, the token "in" (instantiation name of the activated function) is sent to link nodes and rlink nodes. Each link node passes on each argument value x1, x2, ... xm to the body of the activated function every time each value has arrived. Each rlink node passes on information concerning the place where the return value is sent. This information y1', y2',... yn', each of which is determined at compilation phase corresponding to y1, y2, ... yn, are attached to each return value in order to identify its destination.

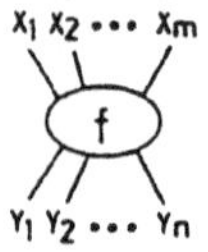

(a) Function invocation node

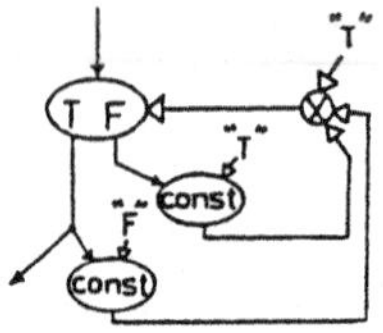

(b) Orgate implementation

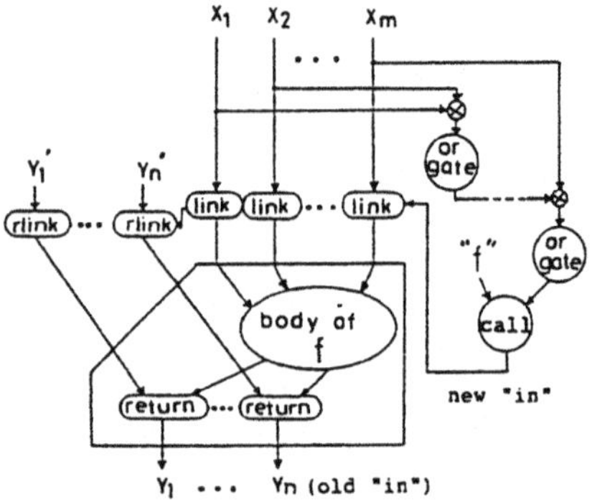

Fig. 2.2 Fuction activation mechanism

This mechanism has the effect that the parallel evaluation of arguments goes beyond the barrier of function invocation and restrictions mentioned above.

For instance, in Program2, which sorts list x with Hoare's quicksort algorithm, the function partition is activated when the argument cdr(x) or y is evaluated in the function sort. In this case, because the evaluation of y in sort means the evaluation of the expression list(car(x)), the value of cdr(x) is generated just before y is generated. In this way, when the second argument is passed into the body of function partition, the then-part or the else-part is ready to run. The function partition returns three values. In the else-part of the partition body, the values w1, w2 and w3 are generated by the function partition and are used in the return expression. Thus, if each of w1, w2 and w3 is returned immediately when it is generated, each of these values is used and passed back to the calling function; in case where x1=y1, for example, w1 and w3 are passed back to the calling function directly, while w2 is appended to list(x) to generate a new list which is passed back. In the innermost recursion phase, the partition independently returns each of three values nil, y, and nil.

```
-- Program2
-- Quicksort

sort: function (x) return (list)
= if null(x) then x
           else clause
                  y = list(car(x));
                  [y1,y2,y3] = partition(cdr(x),y);
                  return append(sort(y1),append(y2,sort(y3)))
                end;

partition: function (x,y) return (list,list,list)
= if null(x) then (nil,y,nil)
           else clause
                  [w1,w2,w3] = partition(cdr(x),y);
                  x1 = car(x); y1 = car(y);
                  return
                   case
                    x1=y1 -> (w1,append(list(x1),w2),w3);
                    x1<y1 -> (append(list(x1),w1),w2,w3);
                    x1>y1 -> (w1,w2,append(list(x1),w3))
                   end
                end
```

In this way, many partition functions are activated and they execute their bodies partially. As the same may be said for the function sort, Program2 can sort the input list due to the highly parallel execution.

The partial execution of the function body also has an effect on the Lisp 1.5 interpreter. This is because the interpreter can execute Lisp programs in a highly parallel manner due to the partial execution of each body of the function eval, evlis, and evcon.

Note, if function evcon is implemented on the basis of the recursion concept, eval functions, which evaluate each predicate part, would be activated in parallel as shown in Appendix B. This implementation would achieve much higher parallelism than the one which uses a loop control mechanism as shown below, so as to keep the Lisp 1.5 semantics where the predicate parts of cond expression are evaluated in sequential order.

```
Evcon: macro (c,a) return (list,list)
= for (x,y):(c,a)
  do if Eval(caar(x),y) then return(cadar(x),y)
                        else recur(cdr(x),y)
```

In order to make the parallel control structure of evcon more effective and to assure that every condition is mutually exclusive, the conventional Lisp program should be modified. For instance, the conventional equal function should be modified as follows:

```
equal[x;y]
= [and[atom[x];atom[y]] -> eq[x;y];
   not[or[atom[x];atom[y]]] -> and[equal[car[x];car[y]];
                                   equal[cdr[x];cdr[y]]];
   t -> f]
```

Although parallel execution of evcon has a potential non-deterministic control [9], its incarnation and integration in Valid semantics remain to be solved.

2.3 Lenient cons and parallelism by pipelined processing

Although the partial execution of a function yields higher parallelism, it is not sufficient for maximally exploiting the parallelism inherent in the given program. For instance, Program2 is expected to be executed in a highly parallel fashion among activated functions of sort and partition. This parallelism, however, is not effective in reducing the execution time in the order, because the time spent to sort the list of length n is proportional to the square of n in the worst case. (It is proportional to n in the best case.) The reason is that; since each of the value y1, y2 and y3 is not returned until the append operation is completed in the partition body, the execution of function sort, which uses those values, must wait until they are returned, and the waiting time is proportional to the length of the list made by the append operation.

The Lisp interpreter is another example which shows that the parallelism is not maximal. Because the operation of the apply to each eval-ed value which is returned from the function evlis must wait until all of the eval-ed values are constructed to a list by the cons operation which resides in the last position in the evlis body, the execution of the Lisp function body can not partially proceed.

If the former parts of the list which are partially generated are returned in advance during the period when the latter parts are appended, the execution which uses the former parts of the list can proceed. Thus the producer and consumer executions overlap each other. As the append is the repeated applications of cons as shown by Program3, this problem can be solved by introducing leniency into the cons operation.

```
-- Program3
append: function (x y) return (list)
= if null(x) then y
               else cons(car(x),append(cdr(x),y))
```

Lenient cons, which is slightly different from the concept of "suspended cons" [10], means the following: For the operation of cons(x,y), the cons operator creates a new cell and returns its address as a value in advance before its operand x or y arrives. Then the value x and y are written in the car field and the cdr field of the cell, respectively, when each arrives at the cons node. In the implementation, the cons operator is decomposed into three primitive operators, getcell, writecar, and writecdr, as shown in Fig. 2.3. The getcell node is initiated on the arrival of a signal token, which is delivered when the new environment surrounding the cons operation is initiated. The getcell operator creates a new cell, and sends its address to the writecar node, the writecdr node, and all nodes which wait for that cons value.

Each memory cell has, in addition to the garbage tag, a car-ready tag and a cdr-ready tag, each of which controls read accesses to the car field and the cdr field, respectively (Fig. 2.3 (b)). The getcell operator resets both ready tags to inhibit read accesses. The writecar (or writecdr) operator writes the value x (or y) to the car field (or cdr field), and sets the ready tag to allow read accesses to the field.

The lenient cons has a great effect on list processing. It naturally implements the stream processing feature in which each list item is processed as a stream [4,11] even for programs which are normally written according to the list processing concept, not having the explicit stream notion.

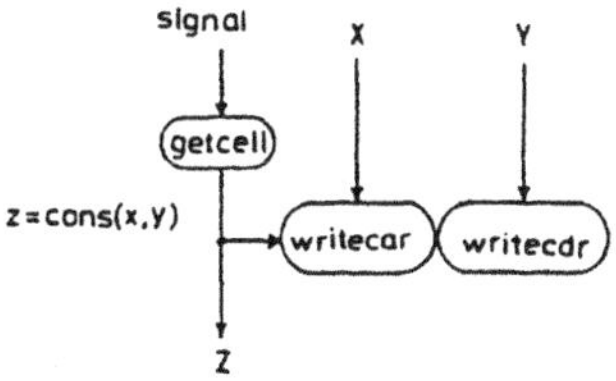

(a) Cons mechanism

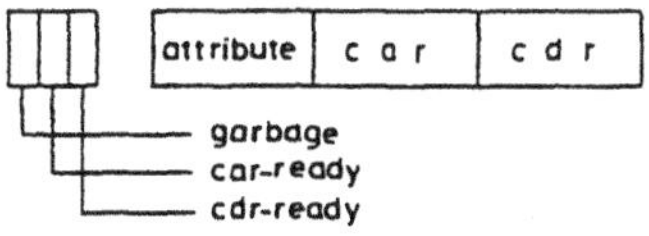

(b) Data cell structure

Fig. 2.3 Lenient cons implementation

3. Estimation of the lenient cons effect

The parallelism enhanced by pipelined processing, which is made possible with lenient cons, is analyzed so as to estimate the lenient cons effect. The sieve algorithm for finding prime numbers and the quicksort algorithm described in chapter 2 are taken as examples.

The following are assumed in the evaluation of a parallel algorithm:

(1) There exist an infinite number of resources, namely, the operation units (processors) required for calculations are available at any time, and the time for resource allocation and network delay is ignored.

(2) Any primitive operation can be completed in a unit time.

(3) The time required for function linkage is ignored.

3.1 Sieve program for finding prime numbers

Program4 finds prime numbers using the Eratosthenes's sieve method. Each cons operation appearing in this program is implemented with lenient cons. In the lenient cons implementation, a signal token is used to initiate the cons operations when a block which contains the cons operations is initiated. Such signal tokens are automatically created with the Valid compiler. The example of signal token "s" is shown in Fig. 3.1.

Program4 -- Sieve program

```
primenumber: function (n) return (list)
= sieve(intseq(2,n));

sieve: function (n) return (list)
= if null(n) then nil
             else cons(car(n),
                       sieve(delete(car(n),cdr(n))));

delete: function (x,n) return (list)
= if null(n) then nil
             else if remainder(car(n),x)=0
                  then delete(x,cdr(n))
                  else cons(car(n),delete(x,cdr(n)));

intseq: function (m,n) return (list)
= if m>n then nil
         else cons(m,intseq(m+1,n))
```

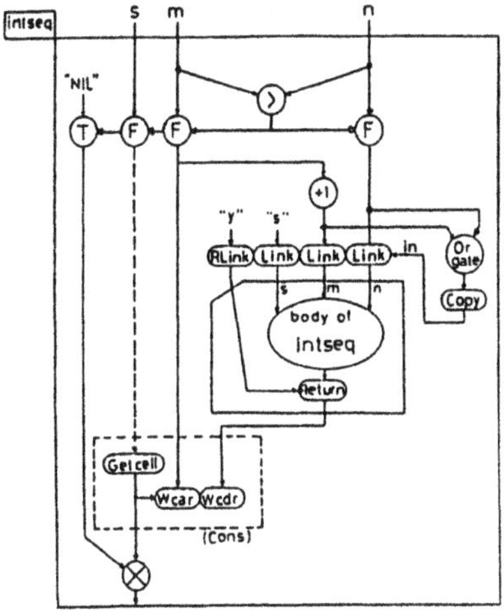

Fig. 3.1 Data flow graph for intseq

Table 1 Execution of primenumber(2,n)

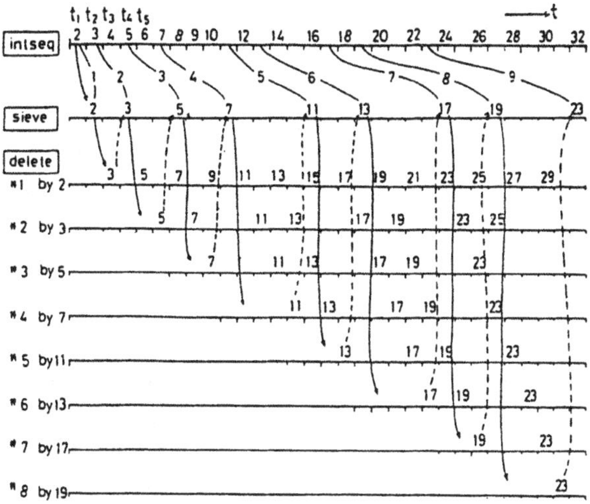

Primenumber(n) obtains the sequence of prime numbers by sieving the sequence (2 3 ... n) generated by intseq(2,n). The lenient cons enables intseq to send out the sequence of numbers (2 3 ... n) one after another. To illustrate this more clearly, a data flow graph of intseq(m,n) is shown in Fig. 3.1., in which s is a signal token name which does not appear in the source text but is created by the compiler.

When the function intseq(m,n) is initiated, if m is not greater than n the getcell node is fired and the new cell address is returned immediately. Then, the value m and the value of intseq(m+1,n) are written into the car and cdr part of the new cell respectively, after each of them is obtained. Consequently, intseq sends out new numbers one by one every time it is initiated. Sieve(n) holds the first element (car(n)) of the sequence sent from intseq, and invokes the function delete to remove the numbers in the rest (cdr(n)), which are divisible by car(n). The function delete sends out the elements indivisible by car(n) one by one, which are, in turn, passed on to the sieve function recursively. In this way, the intseq, sieve and delete executions overlap.

Table 1 traces the invocations of intseq, sieve and delete, and depicts how the algorithm works. Each row gives the sequence of numbers generated in each invocation. The intseq generates the sequence of numbers (2 3 ... n) one by one in a unit time. The top level sieve is initiated immediately when the first element of the sequence is returned from the intseq. After the activation, the sieve immediately invokes the $delete_1$, in which the suffix represents the invocation sequence number.

The $delete_1$ takes the sequence (2 3 ... n) from the intseq, and deletes the numbers which are divisible by 2. The $delete_i$ which is initiated in the i-th level sieve takes the output sequence of the $delete_{i-1}$, and deletes the numbers which are divisible by the i-th prime number. The $delete_1$ returns the values one time unit after the intseq, as seen in the table.

As the time needed for invocation of the delete in the sieve is constant and the $delete_i$ returns values one time unit after the $delete_{i-1}$ returns the value, the sieve returns the ith prime number i time units after the intseq returns that number.

Since there are n/log n primes asymptotically in the natural numbers set {2,3,...,n}, the computation time needed in this program is n+n/log n, i.e. the order of n.

3.2 Quicksort program

This section analyzes the computation time of the quick sort program described in Section 2.3. To simplify the analysis, the worst case behavior of the algorithm, in which input data are in the reverse order, is studied. It is assumed that the time required for the append operation is proportional to the length of a list to which another list is appended. Table2 traces the function invocation sequence to illustrate the action of sort((3 2 1)).

Let y be (x1 x2 ... xn), where $x_i > x_{i+1}$ for i = 1, 2, ... n-1. The function sort(y) acts as follows: The activated function $partition_1$ partitions a list into three lists y11, y12, and y13 each of which contains the elements less than, equal to, and greater than x1, respectively. For the input data of the worst case, y11 is (x2 x3 ... xn), y12 is x1, and y13 is nil. The result of sort(y), the top level activation of the sort, is obtained by appending the result of append(y12, sort(y13)) to the result of sort(y11). Due to the lenient cons effect, the function sort(y11) is invoked immediately after the first element of y11, i.e. x2, is obtained. The values y12 and y13 are returned from the $partition_1$ 2*N(y11) time units after the first value of y11 is returned from that function. This is because it takes N(y11) time units for the first element of y (=x1) to be passed into the innermost recursion phase of the $partition_1$ where y12 (=(x1)) and y13 (=nil) are generated. Further it takes the same time for those values to be returned (here, N(yi) means the number of elements in yi).

Next, in the computation of sort(y11), the second level activation of the sort, the function $partition_2$ generates three values y21, y22, and y23. [y21=(x3 x4 ... xn), y22=x2, y23=nil.] Due to the lenient cons effect, $partition_2$ returns the first element of y21, i.e. x3, one time unit after $partition_1$ returns element x3. As it takes C (a constant, which equals 2 in this case) time unit for each activated partition to generate the second element after generating the first element, the first element of y21 is returned C1 (=C+1) time unit after the first element of y11 is returned. Since y12 and y13 are returned at time 2*N(y)=2n, and y22 and y23 are returned at time C1+2*N(y11)=2n+C-1, y22 and y23 are returned C-1 time unit later than y12 and y13. In the same way, $partition_3$ which is invoked in the computation of sort(y21) returns the result y31 (=(x4 x5 ... xn)) C1 time units after $partition_2$ returns result y21. And it returns results y32 (=x3) and y33 (=nil) C-1 units time later than results y22 and y23, and so on.

Since n partition functions are activated for sorting n elements, it takes n*C1 time units until the last partition is completed. Each result of the partitions has to be constructed to a list by the append operation. Since it takes n time units for the first element to be returned from the top level append, and it takes n-1 time units for the append operation to be completed, the total computation time for sorting n elements is n*C1+2n-1, i.e. in the order of n.

In the best case, the total computation time is in the order of log n. This is because the partitioning operation cuts the length of each list in half, and the invocation tree of the sort function is balanced with the height of log n; that is, the execution achieves the maximal effects of the divide and conquer strategy.

Table 2 Example of the worst case ; Z ← sort(3 2 1)

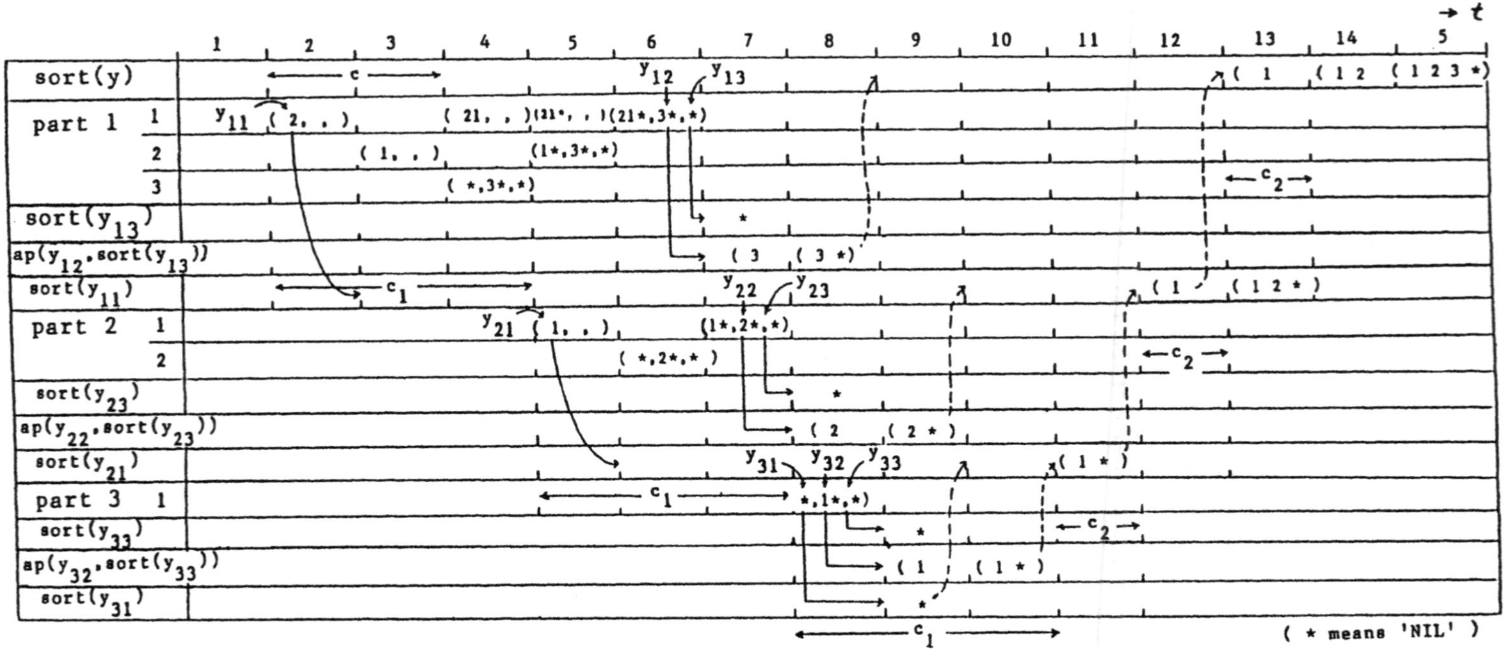

4. Garbage collection

As a number of data are copied, used, and thrown away very often in the course of side-effect free data manipulation, it is very important to resolve the problem of how to utilize structure memory cells effectively. That is, it is important to come up with an effective garbage collection method under a parallel processing environment.

Although mark-scan methods are used generally as a garbage collection method in conventional machines, the reference count method is adopted here. The reason is that: (a) since the data tokens, which are pointers to list data entries in the structure memory, are scattered in such various parts of the machine as instruction memory units, communication networks and operation units [12,13], it is very difficult to extract the active cell without suspending execution; and (b) as the list manipulations have no side-effect, and every list is made only by the cons operation, no circular lists are created.

A basic method will be described at first, though it is not efficient, in order that the fundamental idea behind it be made clear. Then a revised method, which is much more efficient, will be described.

In the method presented here, each data cell has a reference count field which is updated every time the data cell is accessed. The reference count handling algorithm in the fundamental method is described below by means of Algol-like language for each primitive list operation. Here, r(x), z and d respectively denote the reference count field for cell x, the value of each operation, and the number of operation nodes which are waiting for the value z.

```
procedure Car(x,d);
  begin
     z := car(x);
     Red(x);
     r(z) := r(z)+d
  end;

procedure Cdr(x,d);
  begin
     same as Car(x,d)
  end;

procedure Atom(x);
  begin
     z := atom(x);
     Red(x)
  end;

procedure Eq(x,y);
  begin
     z := eq(x,y);
     Red(x);
     Red(y)
  end;
```

```
procedure Cons(x,y,d);
  begin
     z := cons(x,y);
     r(z) := d
  end;
procedure Red(x);
  begin
     r(x) := r(x)-1;
     if r(x)=0 then begin
                       Red(car(x));
                       Red(cdr(x))
                    end
  end
```

One problem involving completeness with the reference count method appears in the execution of a conditional expression. In the case of ' if p then f(x) else g(y) ' , for example, since g(y) is never executed when p is true, the reference count for cell y is left un-decremented. As a result, cell y is never reclaimed though it is garbage in virtual. In order to avoid this, a special operator erase, which executes only procedure Red, has been prepared. (Fig. 4.1)

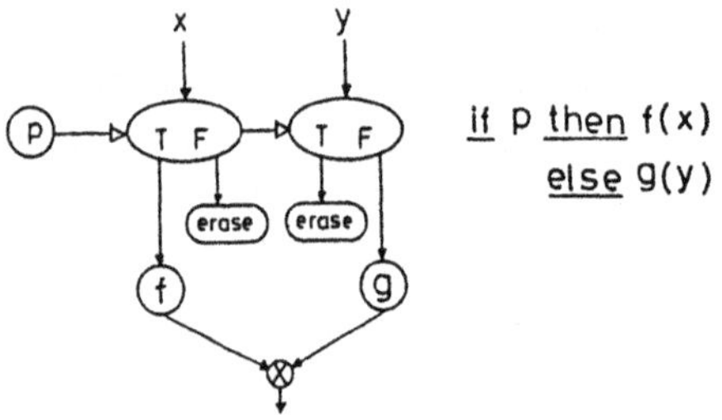

Fig. 4.1 Conditional expression and erase operation

The reference count handling overhead is serious, because reference count updating is needed in all operations not only in the five primitive operations but also in the switch and gate operations, as can be seen in the conditional expression case. However, by making use of the features of high level language, i.e., the block structure and scope rule, redundant reference count updating operations can be eliminated.

Instead of implicitly updating the reference count in every operation, the revised method updates it explicitly. It does so by using the increment and decrement operators which the Valid compiler generates through use of several Valid features which are suitable for eliminating the redundant reference count updating operations. The reference count of a cell indicates the number of value names which denote the cell in the program text. Here, the value names are explicitly defined by the value definition or implicitly defined as cons values or function values. The reference count of a cell which is newly denoted in a block is incremented when the block is initiated, and decremented when it is terminated. Since, due to the features of Valid, all value names are defined uniquely in a block and are local to the block defining them, each value name refers to only one cell, and the cell is not referred to by any value name outside the block. Therefore the reference count is incremented only once, even if a number of operations refer to the value within the block.

A Valid source program fragment, such as

```
[x,y] = clause
          x = E1; y = E2; z = E3;
          return(E(x,y),x)
        end,
```

is compiled in the data flow graph shown in Fig. 4.2. Here, E1, E2, E3, and E represent expressions. The return expression generates two values, which are implicitly denoted by ret1 and ret2, and are explicitly denoted by u and v in the environment outside the block. The reference count for local values x, y and z should never be decremented before the reference count of return value ret1 and ret2 are incremented, so as to prevent the cells, which are pointed to by the return values, from being reclaimed during the transient time of return value passing. The and-gate node and gate nodes keep the increment and decrement operations order safe for garbage collection.

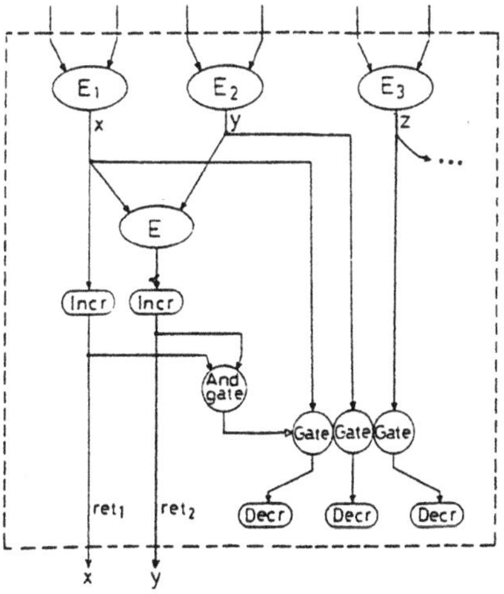

Fig. 4.2

Reference count management of local value name

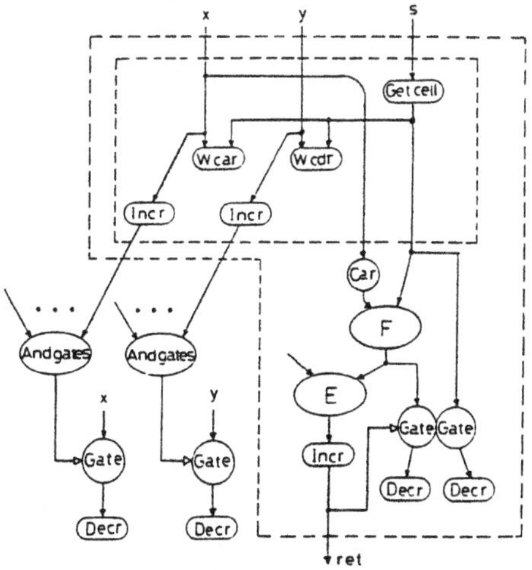

Fig. 4.3

Reference count management in the expression with cons operation

Even the values of cons operations or function applications, which are not denoted explicitly but are obscured in expressions, have to be treated as if they are denoted, in order to prevent the cells created by cons operations or function applications from being reclaimed. Therefore, an expression, which contains cons operations or function applications as sub-expressions in it, is treated as if it is composed of the pseudo blocks.

For example, the expression E(..., F(cons(x,y), car(x)), ...)

is interpreted as

```
clause
  u' = clause z'=cons(x,y); return z' end;
  v' = F(u',car(x));
  return E(...,v',...)
end
```

and compiled in the graph shown in Fig. 4.3. Here, F represents a function application. The readers should note that the getcell operator initializes the reference count of the created cell to one, and the function body, when it returns the values, increments the reference count of the cell pointed to by the returned value.

Another problem affecting safety arises with the case of lenient cons. As in the described example above, if the value (token) x or y is pointed to by a cell which is newly created in this block, its reference count must not be decremented before it is confined that the reference count for each cell, x, y, is incremented. This is because through write-car or write-cdr operation, each of the cells is to be pointed to by the cell which is created by the getcell operation.

The general rule for guaranteeing safety in garbage collection with cons operation is as follows.

In cons operation,

$$\text{cons}(u,v) \qquad u \equiv E_x(x1,x2,\ldots xn)$$
$$v \equiv E_y(y1,y2,\ldots ym)\ ,$$

E_x and E_y are expressions composed of the value names x1,... xn, and y1,... ym, each of which is a name denoting a specific value either explicitly or implicitly (namely, the cons operation or function application is replaced by xi or yi). The decrement operation for the reference count of each cell pointed to by x1, ..., xn, y1, ..., and ym, is postponed until the increment operations to the reference count for each cell, pointed to by u or v, are completed.

5. Conclusions

This paper discussed some issues in list processing within the context of a data flow control environment, and from the viewpoint of parallelism. The basic philosophy behind the data flow machine architecture presented in this paper is that highly parallel execution can be achieved by data flow control concept, both at the primitive operation level and at the function activation level.

The mechanism of partial execution for each function was shown through some examples including an Lisp interpreter implementation, to be effective for exploitation of the parallelism in list processing. The lenient cons mechanism was shown through analysis of two programs, to be effective for maximally exploiting parallelism. These two were parallel sieve program to find prime numbers and Hoare's quicksort program, which proved that each execution was made in the order of linear time.

A garbage collection mechanism based on the reference count method was also described. The algorithm works well as a parallel garbage collection algorithm in the sense that the garbage is reclaimed at all times, in parallel with the foreground list operations.

A description of the data flow machine architecture, which is on the basis of associative memory and logic-in-memory concepts, was omitted in this paper. However, the characteristics of the machines's are that it is composed of a number of modules, say 1000 modules, so as to realize logical exploitation of the parallelism described in this paper. The details of the architecture are described in [12,13].

Many problems remain to be solved before the machine will be available for practical use. Several studies are in progress for the purpose of examining the effectiveness of the theory described in this paper. These studies involve the construction of a software simulator, the design of an experimental hardware system, and the implementation of a Valid compiler.

The simulator is now running and collecting the statistical data for evaluation of: the effect of the lenient cons; the effect of cons strategy and memory partition; the garbage collection overhead; etc. Design of an experimental hardware system will provide data for estimation of the cost performance ratio. The Valid compiler generates the object code for the simulator and the experimental hardware system. The Valid language is a tool for establishing a functional programming environment for the data flow machine.

References

[1] Dennis,J.B., "A Preliminary Architecture for a Basic Data Flow Processor", The Second Annual Symposium on Computer Architecture, Jan., 1975, pp.126-132.

[2] Plas,A., "LAU System Architecture: A Parallel Data-Driven Processor Based on Single Assignment", Proceedings of the International Conference on Parallel Processing, 1976, pp.293-302.

[3] Watson,I. and J.Gurd, "A Prototype Data Flow Computer with Token Labelling", AFIPS Conference Proceedings 48, 1979, pp.623-628.

[4] Arvind, K.P.Gostelow and W.Plouffe, "An Asynchronous Programming Language and Computing Machine", Report TR 114a, Department of Information and Computer Science, University of California, Irvine, California, December, 1978.

[5] Davis,A.L., "The Architecture and System Method of DDM1: A Recursively Structured Data Driven Machine", Proceedings of the Fifth Annual Symposium of Computer Architecture, Apr., 1978, pp.210-215.

[6] Keller,R.M., G.Lindstrom and S.Patil, "An Architecture for a Loosely-Coupled Parallel Processor", UUCS-78-105, University of Utah, Salt Lake City, Utah, 1978.

[7] Amamiya,M., "Design Philosophy behind the High Level Language VALID for a Data Flow Machine", Proceedings of IECEJ Annual Conference, 1981, NO.1486, in Japanese.

[8] McCarthy,J., et.al., "Lisp 1.5 Programmer's Manual", MIT Press, Cambridge, Massachusetts, 1960.

[9] Dijkstra,E.W., "Guarded Commands, Non-Determinacy, and Formal Derivation of Programs", Comm. ACM, 18, 8, 1975, pp.453-457.

[10] Friedman,D.P. and D.S.Wise, "CONS should not evaluate its arguments", Automata, Language and Programming, S.Michaelson and R.Milner, Eds, Edinburgh Univ. Press, 1976.

[11] Dennis,J.B. and K.S.Weng, "An Abstract Implementation for Concurrent Computation with Streams", Proceedings of International Conference on Parallel Processing, 1979, pp.35-45.

[12] Amamiya,M., R.Hasegawa, O.Nakamura and H.Mikami, "A List-Processing-Oriented Data Flow Machine Architecture", Proceedings of National Computer Conference, 1982, pp.143-151

[13] Amamiya,M., R.Hasegawa and H.Mikami, "A List Processing Oriented Data Flow Machine and Its Software Simulator", Proceedings of Meeting on Computer Architecture, IPSJ, 40-8, January 1981, in Japanese.

Appendix A. Brief description of Valid language features

Valid is a high level language designed for programming on a data flow machine. "Valid" is an abbreviation for VALue IDentification language. The basic design philosophy is that Valid semantics supports programming based on functional concept, while its syntax offers conventions for writing of programs in Algol style. This appendix describes certain Valid features which should help readers to understand the sample programs used in this paper.

The characteristics of Valid are as follow.

(1) The basic structures of Valid are expressions and definitions, i.e. Valid contains no concept of program variables and assignments. Instead, values are denoted by value names (variables), if necessary. This concept is different from a single assignment concept in the sense that variables only mean those names which are assigned to each value.

(2) A Valid program consists of function definitions, macro definitions and value definitions. A function (or macro) definition defines a function name and its body, which returns multiple values. The body is an expression. Functions are invoked during execution time by their names, while macros replace their names with their bodies at the compilation phase. Value definitions also define multiple values. For example, the value definition

```
[x1,x2,...xn] = expression
```

defines n values generated by expression, and denotes them by value names x1, x2, ... xn.

The function definition

f: <u>function</u> (a1,a2, ... an) <u>return</u> (b1,b2, ... bm)
  = expression

defines the function f which has formal arguments (value names) a1, a2, ..., an, and returns values of type b1, b2, ..., bm.

(3) Valid has a block structure. A block is a set of definitions. A block is an expression which is enclosed with a clause and end. In a block, all value definitions are sequence-free. That is, each value definition is evaluated concurrently within the block according to the data dependency. In Valid, the delimiter semicolon (;) which delimits value definitions, is used in a different way from the semicolon in Algol. The semicolon (;) in Valid separates two constructs in parallel, i.e. A;B is identical

to B;A. The delimiter comma (,), on the other hand, separates two constructs keeping the sequence in the program text, i.e. A,B is not the same as B,A (the reader should note that sequence does not mean the sequence of execution in this context).

Functions and macros may be defined in any block locally.

All function names, macro names and value names can be referred to from other positions in the program text under scope rule constraints, as in Algol. That is, every identifier denotes a local value which may be referred to within a block surrounding its definition. A block becomes an expression, and values of the block are generated by a return expression.

(4) Conditional execution is described in two ways, by conditional expression and by case expression.

The conditional expression is written as follows.

```
if predicate-1 then expression-1
elsif predicate-2 then expression-2
  . . .
elsif predicate-n then expression-n
else expression
```

In this expression, all predicates are evaluated in parallel, and their values are examined in the order from predicate-1 to predicate-n. Thus, if predicate-i is the first predicate which is evaluated to be true, then the values of this conditional expression are the values of expression-i.

The case expression is written in the following form.

```
case
   predicate-1 -> expression-1;
   . . .
   predicate-n -> expression-n;
   others -> expression
end
```

In the case expression, all predicates are evaluated and examined in parallel (the current version of Valid permits a case expression in which at most one predicate is evaluated to be true).

(5) All iterations are described based on a recursion concept. A conventional loop is described using a recurrence expression of the following form.

```
for (<iteration value names>) : (<initial values>)
do <block including recur expression>
```

An example of a recurrence expression is shown below. It is a program for computing remainder and quotient of y to x.

```
[remainder,quotient]
 = for (r,q) init (y,0) do
    if r<x then (r,q) else recur(r-x,q+1)
```

Other features of Valid as a general purpose high level language were omitted because they are irrelevant to the discussion of this paper. Omitted are as include items (6) and (7) below.

(6) Valid permits several data types. These include Boolean, integer, real, array, record, list, signal and function types. Each value is specified with its type name (if necessary). Since this paper deals with only the list type, type specification is omitted.

(7) Parallel expression, which is based on a fork-join concept, is powerful for describing blocks executable in parallel, such as with vector operation and the mapcar function in Lisp.

Appendix B. Data flow implementation of Lisp 1.5 interpreter

Though each function in the Lisp 1.5 interpreter is defined as recursive, it is necessary to reconstruct them so as to realize efficient execution control, such as in the case for conventional von Neumann machines. Lisp interpreter functions implemented on the data flow machine are designed from this viewpoint.

The implementation has been made on the following principle:

(1) The tail-recursive structure, which has no explicit parallelism, is implemented by the loop control mechanism in such a manner as to reduce the function invocation overhead.

(2) Explicit parallelism, such as parallel argument evaluation, is exploited only in a non-tail-recurvive structure.

Thus, the apply and eval functions are implemented by using the loop control mechanism, where the apply1 and eval1 programs are executed alternatively, as is shown in Fig. B(a,b). On the other hand, the evlis and evcon functions, which have non-tail-recursive structures, are implemented by using recursive functions. These recursive functions are executed in parallel, since eval functions are forked by recursive execution of the evlis or evcon body, as is shown in Fig. B(c,d).

```
-- Lisp 1.5 interpreter
Apply: function (fn,args,alist) return (list)
= for (x,y,z):(fn,args,alist)
   do clause
       [t,xx,yy] = Apply1(x,y,z);
       xxx = if t then xx
             else clause
                   [t1,x1,y1,z1] = Eval1(xx,yy);
                    x2 = if t1 then x1
                         else recur(x1,y1,z1);
                   return(x2)
                  end;
       return(xxx)
       end;
Apply1: macro (fn,args,alist) return (list,list,list)
=  for (x,y,z):(fn,args,alist)
    do case
         atom(x) -> case
                     x='car' -> return('t',caar(y),nil);
                     x='cdr' -> return('t',cdar(y),nil);
                     x='cons' -> return('t',cons(car(y),cadr(y)),nil);
                     x='eq' -> return('t',eq(car(y),cadr(y)),nil);
                     others -> recur(Eval(x,y),y,z)
                     end;
         car(x)='lambda' -> clause
                              x1 = caddr(x);
                              y1 = Pairlis(x,y,z);
                              return(nil,x1,y1)
                             end;
         car(x)='label' -> clause
                             x1 = caddr(x);
                             z1 = cons(cons(cadr(x),caddr(y)),z);
                             recur(x1,y,z1)
                            end
       end;

Eval: function (e,a) return (list)
= for (x,y):(e,a)
   do clause
       [t,z1,z2,z3] = Eval1(x,y);
       xx = if t then z1
            else clause
                  [t1,y1,y2] = Apply1(z1,z2,z3);
                  return(if t1 then y1 else recur(y1,y2))
                 end;
       return(xx)
       end;
```

```
Eval1: macro (e,a) return (list,list,list,list)
= for (x,y):(e,a)
   do
     case
       atom(x) -> return('t',cdr(Assoc(x,y)),nil,nil);
       car(x)='quote' -> return('t',cadr(x),nil,nil);
       car(x)='cond' -> clause
                          [x1,y1] = Evcon('f',cdr(x),y);
                          recur(x1,y1)
                        end;
       others -> clause
                   y1 = car(x);
                   y2 = Evlis(cdr(x),y);
                   y3 = y;
                   return(nil,y1,y2,y3)
                 end
     end;
Evcon: function(b,c,a) return(list)
= if null(c) then nil
             else clause
                    y = Eval(caar(c),a);
                    x = Evcon(b y,cdr(c),a);
                    return( if b then nil
                            elsif y then Eval(cdar(c),a)
                            else x )
                  end;

Evlis: function (m,a) return (list)
= if null(m) then nil
             else clause
                    x = Eval(car(m),a);
                    y = Evlis(cdr(m),a);
                    return(cons(x,y))
                  end;

Assoc: function (x,a) return (list)
= for (y1,y2):(x,a)
   do case
        null(y2) -> return(nil);
        equal(y1,caar(y2)) -> return(car(y2));
        others -> recur(y1,cdr(y2))
      end;

Pairlis: function (v,e,a) return (list)
= for (x,y,z):(v,e,a)
   do if null(x) then return(z)
                 else clause
                        x1 = cdr(x);
                        y1 = cdr(y);
                        z1 = cons(cons(car(x),car(y)),z);
                        recur(x1,y1,z1)
                      end;
```

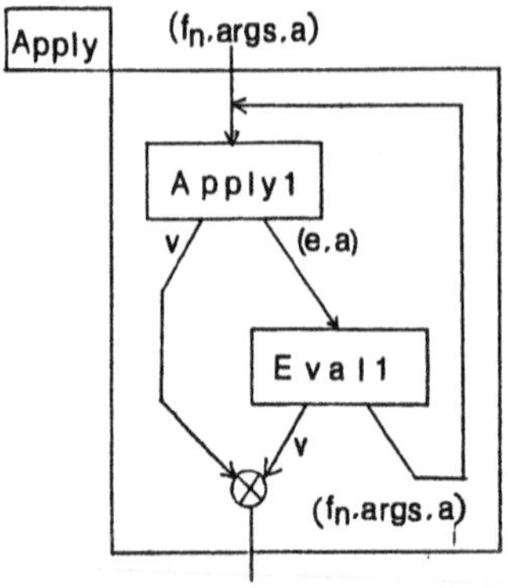

(a) Apply

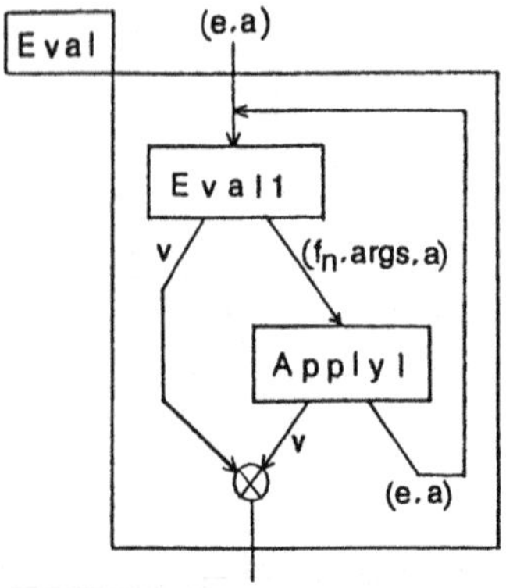

(b) Eval

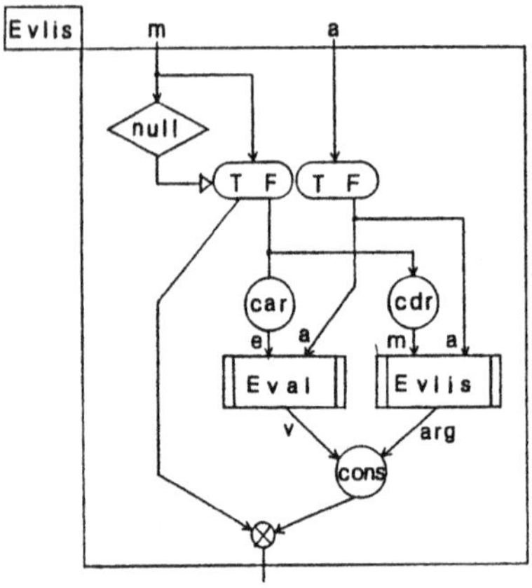

(c) Evlis

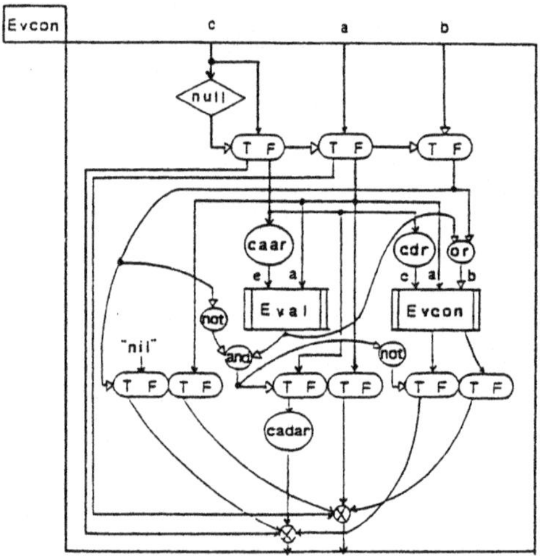

(d) Evcon

Fig. B. Data Flow Representation for Lisp 1.5 Interpreter

Relational Algebra Machine

GRACE

Masaru Kitsuregawa, Hidehiko Tanaka, and Tohru Moto-oka

University of Tokyo
Faculty of Engineering, Department of Information Engineering
Bunkyo-ku, Tokyo, 113 Japan

Abstract

Most of the data base machines proposed so far which adopts a filter processor as their basic unit show poor performance for the heavy load operation such as join and projection etc., while they can process the light load operations such as selection and update for which a full scan of a file suffices. These machines which incorporates n processors takes it $O(N*M/n)$ time to execute join of two relations whose cardinalities are N and M respectively.

GRACE adopts a novel relational algebra processing algorithm based on hash and sort, and can join in $O((N+M)/n)$ time. GRACE exhibits high performance even in join dominant environment. In this paper, hash based relational algebra processing technique, its implementation on parallel machine, and architecture of GRACE are presented.

1. Introduction

1.1 History of Data Base Machine

The relational model proposed by Codd has saddled a software implemented data base management system with a big performance burden in return for its high data independence, simplicity, uniformity in the operator set and its firm logical foundation. This is one of the largest motivation that stimulates the research on the data base machine.

The origin of Data Base Machine(DBM) is considered to be 'Logic per Track' concept by Slotnick[1]. Many of the machines proposed so far adopted this idea as basis with some enhancement. As is shown in RAP.1[2] or CASSM[5], each head of a disk is equipped with some simple logic and it can efficiently perform selection of records which satisfy a certain condition. Since this logic can filter out the unnecessary records, it is called a filter processor. The filter processing can reduce the amount of data that must be transferred between the secondary storage device and the main memory of a host computer. Later the storage media are changed from disks to electronic devices such as CCD and magnetic bubble memories. And the machine of this type, namely the one which is constructed by many identical cells, where a cell is composed of a pair of a processor and a memory bank, comes to be known as 'the cellular logic type data base machine'[7]. In the support of a relational data base, this outperforms the conventional one by orders of magnitude in the execution of relatively light load operations such as selection and update. Because there is no auxiliary data structure such as indexes in many of the machines, the heavy overhead caused by the consistent update of the indexes disappears. Elimination of the large software for the complex access method is supposed to be another merit of these machines. In heavy load operations such as join and projection including duplicate elimination, however we can not expect large scale of performance improvement but only slight one[3]. In RAP.1 many revolutions of the disk are required to implement an implicit join. The brute force application to join of the filter processing approach pioneered by Slotnick, which is very powerful to the operation for which one scan of the file is sufficient, has gradually revealed its limitation. That is, most of the machines are regarded as basically filter processors and it is difficult to judge that they hold a full efficiency in join and duplicate elimination which are essential operations for a relational data base management. There seems to be no machine which provides sufficient performance for join.

1.2 Join Processing Techniques

Here we examine the join processing method on several machines. There might be lots of another interesting features peculiar to the individual machine but we ignore them and concentrate on join.

In the cellular logic type DBM, the processing load of join is proportional to the product of each relation's cardinality, and it is processed in parallel on each cell. That is, tuples of the source relation is broadcast to the cells comprising the target relation, and then all target cells simultaneously compare the obtained tuples with its own tuples. Therefore the processing time is proportional to $N*M/n*k$ *where M and N are the cardinalities of both relations, n is the number of cells and k is the number of comparators per cell. RAP[2,3,4] and EDC[8] etc. belong to this category. As was shown in performance evaluation of RAP[3], DBM of this type is not always suited for join operation and the performance gain against a conventional machine is not so definite.*

In RELACS[9,10] which is characterized by its extensive use of associative processors, the processing load itself is same but the parameter k is relatively large, while in cellular approach the processing power of one cell is small and simplified because its cost increases linerly as the volume of the data base becomes enlarged. Anyway it also employs exhaustive matching algorithm, so it is difficult to attain high performance. The separation of the high performance processing unit and the memory banks necessitates the data transfer path having high bandwidth between them.

In DIRECT[11,12] which actualizes the page level control, the load of join is $O(n*m)$ *where n and m are the number of pages occupied by two relations respectively, and it is processed by using* $max(m,n)$ *processors in* $O(min(m,n))$ *time. Join is implemented in the manner that many processors activated by the controller are assigned one page of the source relation and scan all the pages of the target relation. This machine does not adopt a special processing unit such as a filter processor but employs a general purpose* μ *processor and the page itself is processed in the conventional manner; the page processing consists of 3 phases, page loading, page processing based on sort, and page storing.*

Systolic array based DBM[13] relies on technological advances in VLSI circuitry. The special purpose VLSI chip activates many comparators in pipeline fashion and can join two relations by only feeding them in counter direction. But this dose not generate a joined relation but only a joinability matrix. Highly Concurrent Tree Machine[14] also assumes the VLSI implementation and can join in about $2*(N+$ *no. of result tuples) where N is the cardinality of both relations. While these two machines are very fast when the related data can be accommodated in the machine, there remains a partitioning problem for cases where the problem size exceeds its capacity. The fact that both require at least* $O(N)$ *comparators should be noticed.*

Join processor in DBC[15,16] also relies on the development of VLSI technology and takes $O(N/n + M)$ *time in join. The join algorithm is basically the same as the cellular logic type DBM, namely exhaustive matching of both relations.*

Data stream data flow data base computer[17] is composed of two main functional modules, sort engine and search engine, which realize $O(N)$ *sort and* $O(\log N)$ *search respectively. On join processing, the source relation is fed into a sort engine and its sorted data is then led to a search engine. When loading of a search engine completes, the target relation is put into it, and a join operation is performed in pipeline fashion. In this manner, two pages of both relations are joined in* $O(N)$ *time where N is the page size. The control scheme at page level is not clarified.*

DPNET based DBM[18] which takes similar approach to ours, completes join in $O(N*M/n^2*k)$ *time. The data streams from each head of a disk are partitioned dynamically and sent to the same number of processors. It assumes to use an associative memory in the processing unit in order to follow the data transfer rate of the disk. Therefore its capacity is limited and it needs several revolutions like RAP when the source relation cannot be fitted into.*

CAFS[19] employs an unique algorithm based on the hashed bit array for join operation which is examined in detail in the next section. It can be regarded as joinability filter; it can filter out many of tuples which cannot be joined. This leads to a large load reduction but actual join must be done at a host machine. The same approach can be found in LEECH[20] and CASSM[6].

It is obvious from the above survey that there is no machine that can perform efficient execution of the relational algebra operation, especially join. In this paper, we propose a novel relational algebra machine based on hash and sort algorithm[21,22], which can join in $O((N+M)/n)$ *time where N and M are the cardinalities of two relations and n is the number of memory banks.*

2. Hash Based Relational Algebra Processing

2.1 Utilization of Hash

Hash is one of the efficient search strategies, where necessary records can be obtained with almost one access provided that load factor remains low, and this is one of the fastest access method. It is not such a conventional static usage as an access method but a dynamic usage of hash technique that we utilize in our machine. The clustering feature of hash operation can reduce the load of join.

The simple join algorithm takes time proportional to the product of two relations' cardinalities. However, if two relations are clustered on the join attribute, that is, the tuples are grouped into disjoint buckets based on the hashed value of the join attribute, there is no joining between tuples from buckets of different id(hashed value of the join attribute). Tuples of i-th bucket in one relation cannot be joined with those of j-th bucket($j \neq i$) in the other relation but with only i-th bucket. So the total load of join operation is reduced into join between buckets of the same id. Let

$$N = \sum_{i=1}^{s} n_i \;, \quad M = \sum_{i=1}^{s} m_i$$

where N and M are the cardinalities of two relations, n_i and m_i are the sizes of the i-th bucket in each clustered relation , and s is a number of buckets. The total processing time T can be expressed as follows

$$T \propto \sum_{i=1}^{s} n_i * m_i$$

*This load reduction effect is depicted in fig.1, where two axes denote two relations which are divided into s intervals and the cross section reveals the join load of n*m. The shaded areas correspond to the processing load. Therefore, this clustered approach can dramatically diminish the load in comparison with nonclustered ordinary approach.*

Duplicate elimination task in projection also used to be a big burden in relational data base systems. The above approach can be applied to projection quite as well as join. Through hashing extracted fields of the tuples, the given relation is broken up into many disjunctive buckets. All the same tuples fall into the same bucket. Therefore duplicate elimination can be done in each bucket independently. No need for inter-bucket comparisons. A database machine utilizing this clustering approach would attain a very rapid relational algebra execution.

Another technique utilizing hash is 'Joinability Filter' which can be found in CAFS and LEECH. In this approach tuples from one relation are hashed on the join attribute and its corresponding bit store is set. Then the other relation is read and hashed also on the join attribute, and the proper bit in the hashed bit array is checked. If the bit is set, it is assumed that this tuple has possibility to be joined. This hashed bit array as a joinability filter makes many tuples that cannot be joined sieved out and thereafter the cardinalities of both relations fall into small sizes, which results in large load reduction. While this method is very powerful as preprocessing, remaining tasks such as elimination of spurious tuples and tuple concatenation in explicit join must be done on the host machine. As was shown clear, this method is quite different from the previous clustering approach.

Two approaches discussed above are mutually independent, so both joinability filter processing which decreases the candidate tuples and clustered processing can be integrated together

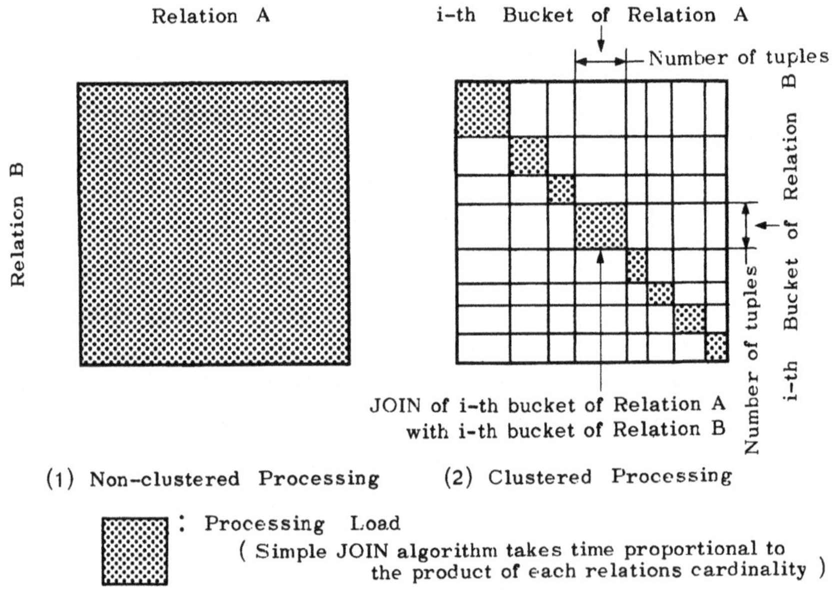

Fig. 1. Clustering Effects By Hashing In JOIN Operation

2.2 Implementation on Parallel Processing Machine

The great reduction of processing load of join and projection was shown to be actualized through the hash based relational algebra processing method. Now, we will consider how to materialize this method for a database machine.

The buckets generated by hash are independent each other. Therefore, rather than processing them serially using only one processor, the relational algebra operation can be executed much faster by processing each bucket in parallel using many processors. Note that there is no inter-processor communication during bucket processing. Some problems can be identified in designing a relational algebra machine which realizes this bucket parallel processing.

First problem is how to exploit the bank parallelism. All the tuples of the relation could be processed serially. But the stream may become very long for large data bases, and it is desirable to divide a long stream by distributing the relation over several memory banks and to process segmented streams in parallel. By exploiting bank parallelism, if the relation could be staged in the working page space which consists of multiple memory banks, processing time can be independent of the cardinality of the relation and is determined by the memory bank capacity which is constant.

Bucket processing is second problem. By hash, processing load can be reduced from $O(s^2)$ *to* $O(s)$ *at bucket level, where s is the number of buckets. Processing of a bucket itself should be fast in order not to disturb the data stream. Namely* $O(n)$ *processing of a bucket rather than* $O(n^2)$ *is preferable where n is a size of a bucket. Integration of both* $O(s)$ *processing at bucket level and* $O(n)$ *processing within bucket makes it possible to organize a relational algebra machine of high performance.*

In order to attain high degree of parallelism, efficient bucket allocation to processors must be achieved, which is the third problem. Buckets can be processed in parallel by allocating them to many processors. In this bucket allocation, it is necessary that each processor can gather the data of the allocated bucket efficiently. Generally relations might be very big and the number of buckets generated by hash is larger than that of available processing units. So buckets beyond the number of available processors must be processed serially, where bucket serial data stream needs to be generated efficiently by memory banks.

The size of buckets might vary so much. The mechanism is required to handle the fluctuation of processed objects. Handling of nonuniformity in bucket size is the forth problem. The capacities of buckets are not necessarily uniform but may differ each other so much. There might be a bucket overflow, which here stands for the case that the bucket size is larger than the capacity of a processor. This phenomenon of nonuniformity .caused by hash function is inevitable. On the other hand, for the machine architecture the fluctuation of the amount of each object to be processed generally leads to degradation of resource utilization and system performance. We have to manage these problems inherent to hash and to provide means to handle such an exceptional phenomena as bucket overflow.

To sum up, we can enumerate four major problems to be considered in the design of a data base machine which realizes parallel execution of our hash based processing method. These are utilization of bank parallelism, processing of a bucket, bucket allocation to processors, and nonuniformity of buckets. We will discuss these problems in more detail at next section.

3. Design Consideration

We will explain our treatment of problems described at the previous section.

1) Bank Parallelism

As is known from the brief survey of join processing in section 1, Systolic Array, Tree Machine and SOE-SEE method can execute join in $O(N)$ time, but it is difficult to do in $O(N/n)$ time with n modules. We say that 'Bank Parallelism' is fully exploited in case $O(N/n)$ time is achieved with n banks. For example, DBC Join processor cannot execute join in $O((M+N)/n)$ time but in $O(M/n+N)$ time. In DIRECT, if as much processors as the product of the number of pages occupying the operand relations were to be activated and each processor could process a page in liner time, $O((M+N)/n)$ join would be possible. However it is too expensive. Our aim is to seek a machine which can reflect bank parallelism at reasonable cost.

We proposed the hash based join method in section 2, where the relations are hashed into several disjunctive buckets and thereafter each of them is processed serially. Incorporating bank parallelism in this method, we can identify two approaches.

a) Bucket converging method

b) Bucket spreading method

Suppose the relation stored over multiple source memory banks are to be hashed and transferred to the same number of destination banks. If the correspondence between the source and the destination is fixed, the tuples composing a certain bucket would be spread over banks. On bucket processing, a processor has to gather the tuples of its allocated bucket from all the banks. In order to avoid this situation, the bucket converging method can be derived naturally, where tuples of a bucket over source banks are converged into a single destination bank. There are two major problems in this approach. One is a bank overflow problem which is a conventional one caused by nonuniformity of the hash function. Since the data distribution can not be uniform and a bank capacity is limited, a situation would occur where some buckets have to accept the tuples beyond its capacity. At least we have to prepare a larger space than the actual capacity of the relation. The determination of load factor is very hard, which is also related to the efficiency of the storage utilization. This difficulty in memory management is crucial in the bucket converging method. DPNET which adopts this method provides no solution to this problem, In DPNET, the distribution terminates when one of banks overflows in source loading. Another facet of this nonuniformity is discussed in the next paragraph. The other problem is data confliction during transfer. The conflict occurs when a number of tuples are sent to the same bank at the same time. We have to manage this problem by an appropriate method such as introducing some buffer. This problem is due to the fact that multiple data streams are hashed simultaneously.

In the bucket spreading method, tuples of a bucket have to be gathered from banks since they spread over them. But the gathering process itself can be pipelined, that is, a processor visits memory banks serially and a bank outputs the tuples of the bucket allocated to that processor (bucket allocation mechanism is discussed at 2) of this section.), so a processor runs through the pipe composed of banks. Thus we can activate the same number of processors as the banks and can expect fair performance improvement exploiting bank parallelism. Moreover there is no memory management difficulty of the overflow as is found in the bucket converging method. This approach also accompanies some problems. Pipeline processing works well when each segment time of it is equal. If some segment time is very large, it'll become a bottle neck of the pipeline. In the case where the correspondence between the source and the destination is fixed as described before, the number of tuples which belong to a certain bucket differs much among banks. This means that the segment time of the pipeline varies dynamically. It is anticipated to cause performance degradation. Provided that the hash function is random, it'll not so large. In order to resolve this segment time fluctuation in pipeline, we have to make the tuples of a bucket almost equally spread over the banks and to make all the buckets themselves almost equal in size. We do 'bucket flat distribution' to satisfy the first condition where a tuple emitted by a source bank is controlled to fall into the destination bank which has the bucket of that tuple least in number. We do 'bucket size tuning' to guarantee the second condition, which is discussed in the later paragraph. With the adoption of this bucket spreading approach with some enhancement, we can attain high performance by activating banks in parallel.

2) O(n) processing of a bucket

The processor is required to process a bucket in O(n) time not to disturb the data stream. And it is evident that the relational algebra operation can be completed in O(n) time when the relations are sorted on the attribute participating in that operation. So we decided to attach the processor with the O(n) hardware sorter. (which is discussed in section 5.1) Of course an associative processor or some functional unit with the capability of parallel comparison also may be possible, which in fact many of the proposed machines such as RAP, RELACS, and DPNET, etc. employ. It can process the data stream very rapidly and keep up with the stream completely. Its capacity, however, is generally limited in current technology and this imposes the condition that at least the cardinality of one relation should be very small. On the other hand, in our approach of using a sorter there is not so severe capacity limitation and operand relations can be treated equivalently. So either of operand relation need not be so small. The sorter can not complete sorting until the last data item arrives, so it takes longer to process a bucket than the associative processor. But this hardly affects the performance because buckets are processed in a pipelined fashion.

3) Bucket Allocation

Bucket serial processing is necessary under the environment of finite resources of processors. In order to complete an operation in O(n) time, efficient generation of bucket serial data stream must be realized. It is required to output tuples of a bucket continuously, which are not necessarily placed adjacently each other inside the memory devices. Of course it is possible to do some preparatory processing when a memory bank inputs the data, but it also needs to be performed not disturbing the input stream. It is almost impossible to do with magnetic disk. On the contrary, it is clearly possible to achieve it by RAM Recently the development of semiconductor technology is extraordinary and it is found in a disk cash even at present time, so it will be feasible to use RAM as the staging buffer. However the size of the relation is very large even after the filter processing by an associative disk, so it is meaningful to seek a memory device lying between RAM and disk on memory hierarchy. We can find a magnetic bubble memory satisfying our conditions. The chip capacity of bubble is at present four or more times greater than that of semiconductor RAM and much more development is expected by contiguous disk technology. Transfer rate is relatively low but it can be improved by activating multiple chips in parallel. And dual conductor technology also seems to make a big contribution on it. Recent rapid progress promises well for the future. Moreover bubble has another advantages such as nonvolatility and flexible start/stop mechanism. Considering above features, we think magnetic bubble memory is also a good candidate for a staging buffer medium. We put some modification to a conventional major/minor bubble chip and make it suitable for efficient bucket serial generation.

4) Nonuiformity of Buckets

In a Direct Access Method the bucket overflow often degrades its performance largely because the storage medium is disk. In our case we are going to use magnetic bubble memory or large capacity and low speed RAM for storage media of staging buffers where a bucket is constructed using linked list in the former and mark bits in the latter, so there is no conventional overflow problem such as degradation in access time and memory efficiency. But the bucket size fluctuation itself arises another type of problems discussed before. In pipeline processing of bucket gathering, it is desirable that the size of each bucket is uniform. And the size itself had better be close to the processor capacity from the point of the processor utilization efficiency. If it takes $O(n^2)$ time to process a bucket, the smaller the bucket is, the faster a bucket can be processed. On the other hand, in the environment of our case where a bucket can be processed in O(n) time, we do not have to have the size of a bucket so small. Excessive bucket generation with the purpose of bucket size reduction *incurs extra overhead rather than the load reduction because the bucket allocation cost must be paid. Therefore it is desirable that the size is close to the processor's capacity. However, it is difficult to find a hash function dynamically which generates buckets with the size of processor's capacity. So we do 'bucket size tuning'. Namely we at first partition the relation into more buckets and then integrate some of them into a larger bucket with the volume less than the capacity of a processor. Through this preprocessing, we can have buckets of near uniformity. This 'bucket size tuning' process is well known as Bin Packing Problem which is NP-complete, and the pseudo optimal solution is already obtained. The overhead brought about by bucket integration is not so large. This can be overlapped with the data stream generation of the bucket. Once the first integrated bucket is obtained, then the data stream generation of the bucket can be overlapped with the bucket integration. Even by this bucket size tuning, it is impossible to make the size of each bucket completely uniform. It might still occur bucket overflow, in which case a large bucket is processed by concatenating a number of processors dynamically through one channel of a ring bus with a few overhead time.*

4. Query Execution on Abstract Architecture of GRACE

4.1 Abstract Architecture

As shown in fig.2-1, the abstract architecture of GRACE is composed of three major components; DSP(Data Stream Processor), DSG(Data Stream Generator), and SDM(Secondary Data Manager).

DSP processes an allocated bucket sent from DSGs and produces result tuples on another DSGs. DSP consists of hardware sorter, filter processor, tuple manipulation unit, and hash unit etc. Each bucket is at first sorted with O(N) hardware sorter and relational algebra processing is applied to the sorted data stream in the tuple manipulation unit in DSP.

DSG provides the working space for the temporary relation to the subsequent processing and for the result relation. The relation filtered and hashed on SDM, as well as the result relation generated in DSP, is transferred into DSGs. DSG generates the bucket serial data stream to DSPs. Semiconductor RAM or magnetic bubble memory seems to be the most appropriate for its medium at present.

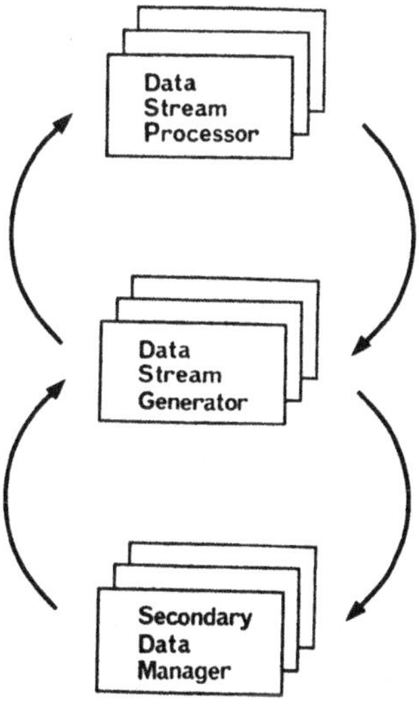

Fig.2.1 Abstract Architecture Of GRACE

SDM stores the relations of the data base on disk. The disk can transfer data from all heads of a cylinder in parallel, and it is equipped with filter processing function containing selection, restriction, bit map manipulation, and simple projection(attribute selection). On-the-fly processing limits the allowable complexity of the predicate in selection. The remainder of the predicate which cannot be evaluated in the filter processor of SDM is rendered to DSP. The hashing unit hashes the tuples on specified attribute and generates bucket ids.

4.2 Query Execution on GRACE

Here we consider how the query is executed on GRACE. The query is assumed to be complex and has many joins and projections, etc. The query processing consists of two major phases: staging phase and processing phase.

At the staging phase, relations necessary for the first join operation are staged from SDMs into DSGs. The data stream from disk is led to the filter processor in SDM, where selection and simple projection are performed on the fly. And then the filtered stream is hashed on the attribute which participates in that operation and hashed id is attached with each tuple. These hashed data streams are transferred from SDMs to DSGs over network between them. Once the SDMs begin to output their data streams, DSGs receive the tuples and maintain buckets corresponding to the hashed value. The relations are clustered overlapped with data transfer during the staging phase. When DSG completes data stream input, the clustered relation depicted in fig.1 is conceptually produced in DSG.

At the processing phase, actual processing is performed between DSGs and DSPs. After the staging phase DSGs literally generate data stream bucket-serially to DSPs. As a bucket is equally spread over DSGs, the DSP has to attach the appropriate data stream and gather the tuples which belong to that bucket. This proceeds in pipeline fashion and the data streams generated by DSG are not disturbed so much. The data gathering process itself is overlapped with the sorting process. A hardware sorter in DSP sorts the input tuples keeping up with the stream. When all the tuples of a bucket are taken in a DSP, it begins to operate on the sorted data stream from the sorter. Most of the operations necessary for relational data base support can be performed efficiently on the sorted stream. After DSP processes a bucket, it then proceeds to next bucket. Buckets are processed in parallel using multiple DSPs. One relational algebra operation terminates when all the buckets are consumed.

4.3 Operator Level Pipeline

Each operation in a query tree is executed as described above. As is shown in fig.2-2, one operation corresponds to one data flow cycle: a data stream is generated at first in DSG and then passes through DSP and at last is returned back to DSG. In a cycle, all the tuples in the source DSGs is transferred to the target DSGs. A complex query comprising multiple operations is implemented by repeating such cycles. As we can see in fig.2-2, once a data flow cycle terminates, new cycle of the next operation begins. Here we should notice that we don't have to interleave the hashing cycle of a result relation for the next operation. When a DSP processes a bucket, it hashes the result tuple on the subsequent operation and outputs a result data stream to DSG. Namely clustering operation of a result relation for the next operation is overlapped with the actual processing of the present operation. We named this 'Operator Level Pipeline'. By this processing schema, vanishes the overhead which we are afraid to be caused by clustering as preprocessing. The first clustering processing is overlapped with the staging phase. We don't have to execute operators one at a time for the cases where sufficiently large space in DSG is available. More than one operation could be performed simultaneously. In that case, as many cycles as the height of a relational algebra tree for the query would be required to get the result. For example, a relational algebra tree for the query in fig.2-3 is executed in a way shown in fig.2-4. The query includes join of four relations. For simplicity selections for the base relations and for the derived relations are not included explicitly. The operations which can be processed on the fly is executed implicitly overlapped with the join operation.

As mentioned above, our machine GRACE can execute a complex query very efficiently with repetitive data flow cycles. Accordingly, we can expect that GRACE can execute join sequence much faster than any DBM proposed so far.

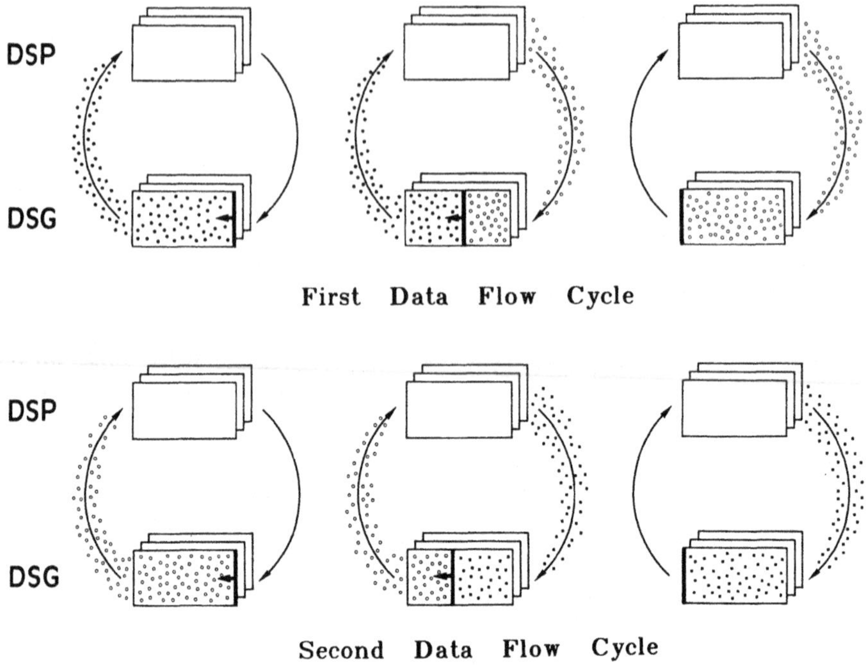

Fig. 2.2 Execution Overview

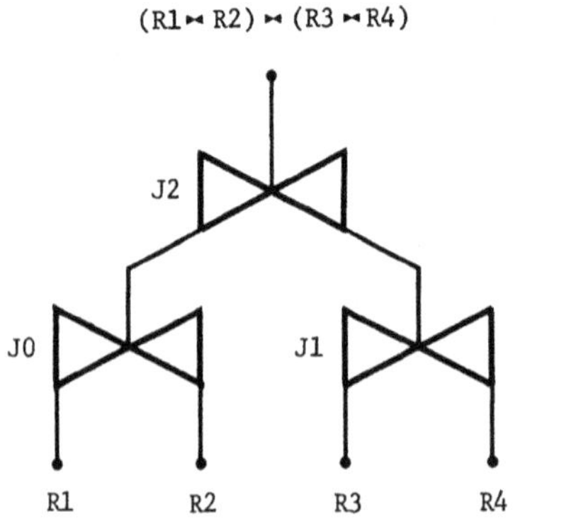

Fig.2-3. An Example of Relational Algebra Tree with 4 Joins.

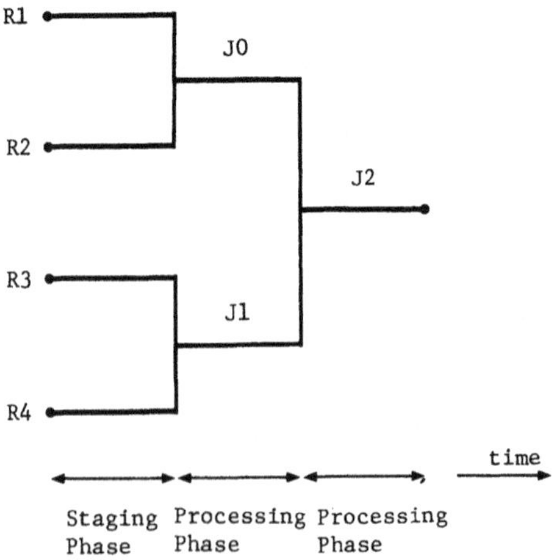

Fig.2-4. Execution Overview for The Query in Fig.2-3.

5. Hardware Architecture of GRACE

Here we proceed the hardware architecture of GRACE. As shown in fig.3, GRACE consists of four kinds of fundamental modules: processing module, memory module, disk module and control module. The first three correspond to DSP, DSG, and SDM in the abstract architecture of fig.2-1 respectively. The modules are connected with two ring buses. The processing modules and memory modules are connected by a processing ring and memory modules and disk modules are connected by a staging ring. The relations stored in disk modules are staged into memory modules with staging ring and then the data streams generated in memory modules are processed in processing modules through the processing ring. The time division multi-channel buses make many modules run simultaneously. Here we look at the organization of the individual modules.

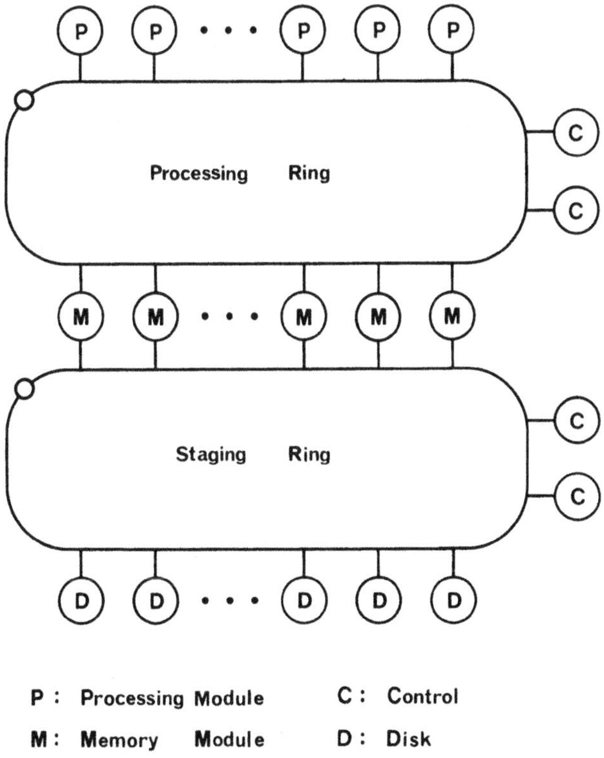

Fig.3 Global Architecture Of Data Manipulation Subsystem In GRACE

5.1 Processing Module

The role of processing module is to process a bucket efficiently. As was discussed in the previous section we determined to equip it with O(n) hardware sorter (it completes the sort in O(n) time) in order to realize data stream oriented processing. The tuples are generated serially by memory module, so it is required that the sorter can sort the stream of tuples efficiently. The sorters which assume that all the data to sort must be in its memory beforehand are not adequate. Among hardware algorithms, we choose the pipeline merge sort[24][25] which satisfies our conditions. This can sort the data keeping up with the input data stream and when it inputs all the data, then the sorted data stream is available after only a few delay.(see fig.4) The hardware structure of this sorter is depicted in fig.5. In a k-way merge sorter, the i-th processor has memory with the capacity of $k^{i-1}*(k-1)$ *items and the total capacity of memory also sums up to O(n). We constructed this sorter of microprogrammed control. The processing rate is mainly depends on the access time of the memory and our experimental one attained the rate of 3 MB/sec. The detail design considerations about it is to be discussed in the future paper. Anyway the bucket-serial data stream generated in memory module is led into this sorter of processing module. The key fields, that is, the join or projection attributes are arranged to be at the top of a record and after the key the relation id is added so that the join of the relations with the attribute of the same domain can be performed at once. Once the bucket input completes, its sorted stream enters a tuple manipulation unit where join or projection are performed. A qualification predicates over the joined tuple is evaluated and the result tuple is formed from only necessary attributes. Another functional unit in processing module is hashing unit. The tuple in an appropriate format from the tuple manipulation unit is sent to hashing unit and is hashed over the attributes for the next operation. After a hashed value is added to a tuple, it is transmitted into the processing ring through a ring bus interface unit.*

5.2 Memory Module

The role of a memory module is to generate an efficient bucket-serial data stream to processing module. Since the processing module can process the data at the rate almost same as that of the input data stream, performance of GRACE mainly lies upon the effective transfer rate of the stream.

Memory module is required to provide a large space for temporary relations and result relations. Semiconductor RAM and magnetic bubble memory are candidates for the medium of the memory module. Since the usage of RAM is straightforward, here we discuss the bubble memory based organization of the module. The pseudo random access mechanism of major/minor magnetic bubble memory is quite different from that of disk. Namely the unnecessary record

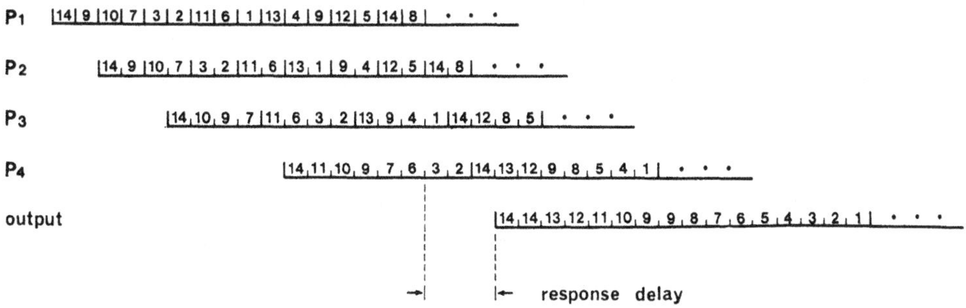

necessary processors ----- $\log_k N$

response delay ----- $\log_k N - 1$

total sort time ----- $2N + \log_k N - 1$

Fig. 4 Sorting Process Overview Of Stream Driven Sorter

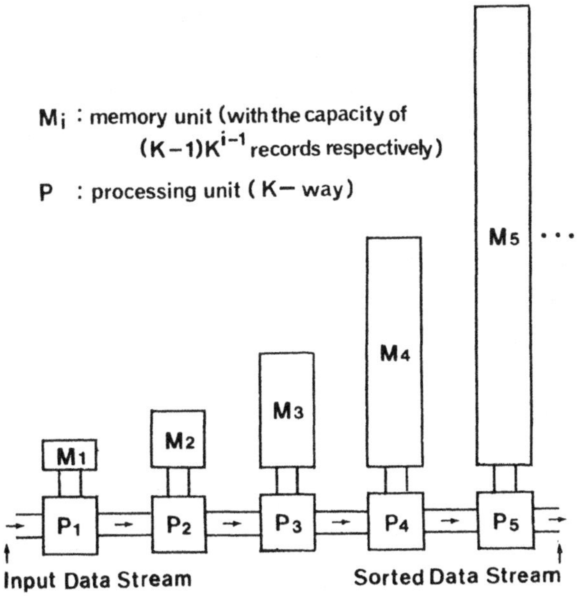

Fig. 5 Global Architecture Of Stream Driven Sorter

stored over minor loops can be skipped in a period of 1 bit field rotation time, while a head has to traverse all the fields of the record in disk whether the record is necessary or not. This quick skip mechanism improves the access time and the effective transfer rate of the magnetic bubble memory. Especially in data base processing, instead of individual record access by an address, set oriented access is more usual where a set of records satisfying some conditions are required and the retrieval sequence itself is not a matter. Therefore the effective transfer rate rather than the access time is much more influential.

We modified a conventional major/minor magnetic bubble memory to obtain heigher transfer rate. In our bubble chip organization, small buffer loops are added between major line and minor loops, in which records can be buffered while another record is output. Record output through major line and transfer of records from minor loops to buffer loops can be overlapped and this can improve effective transfer rate of the bubble for the set oriented access of records. Since the buffer loop length is short, tuples can be output almost continually. A block replication/transfer gate and swap gate are used between major line and buffer loops and between buffer loops and minor loops respectively. With swap gate, look ahead buffering of the bucket into buffer loops is possible and the bucket serial stream can be generated efficiently.

The gates of the bubble memory are controlled using mark bit RAM which synchronizes the magnetic field rotation of the bubble. The hashed value of a tuple is stored in this RAM when it enters memory module. The bubble control unit controls the gates at the appropriate time by referring the hashed value of the record in the RAM. To generate bucket-serial data stream, all the tuples with the same hashed value which belong to a bucket have to be processed before the next bucket processing. So the number of tuples in each bucket is maintained.

We described bubble memory based organization for the memory module. The modified magnetic bubble memory chip with buffer loops is available at present. But it's also possible to use semiconductor RAM. Future development of the these memory technology has to be observed.

5.3 Disk Module

GRACE adopts disk as medium for the secondary storage which makes the capacity of the data base very large. Relations are stored on the disk and are staged into memory modules through the staging ring. Disk module employs a filter processor which evaluates simple predicate on the fly where all the track can be read in parallel. This mechanism can be found in many of the machines proposed so far such as RAP and TIP in DBC. The irrelevant attributes in the tuples are also removed. This reduce the amount of the data which must be transferred between disk module and memory module. If the complexity of the predicate is beyond the filter processor's ability, it can be processed in processing module. The data stream which passed through the filter processor is led into the hashing unit which can be also found in processing module, where tuples are hashed over the attributes for the next operation such as join or projection. After a hashed value is added to the tuple, it is sent to memory modules through the staging ring.

As for the storage organization in disk, various kinds of mapping from the logical structure of relation into the physical storage media are possible. Many file organizations have been investigated so far. In most of data base machines such as RAP there is no auxiliary data structure and the tuples are arranged in the form like a sequential file. This eliminates the software burden caused by a complex access method in secondary storage mapping. However any retrieval requires to scan all the tuples of the relation. As the data base expands, full scan of the large file becomes prohibitive. On the other hand DBC employs structure memory(SM) where indexes are maintained and a partitioned content addressable memory is realized with disk and TIP. GRACE adopts multi-dimensional clustering technique in secondary storage mapping and reduces the number of pages to be retrieved. Disk module also manage the auxiliary data structure consistently.

6. Summary

According to the classification by Paula Hawthorn[23], some of the data base application provides the join dominant environment. For a simple query containing selection, the previous machines based on exhaustive matching with full scan of the file would suffice. For a complex query which has many joins and projections, however, it is difficult to attain a high performance with the ordinary approach of the filter processor. GRACE adopted a novel relational algebra processing algorithm based on hash and sort, and can execute not only simple query but also complex one comprising many joins or set operations rapidly. While the data streams are fixed in a secondary storage in the previous machines, the clustered data streams appropriate for the given operation are generated dynamically in GRACE and can be processed keeping up with the stream. This allows a complex operation also to complete with one data flow cycle, and moreover due to the operator level pipeline, time overhead caused by hashing is effectively canceled.

In this paper, we have described the processing algorithm and it has been shown that GRACE can execute a relational algebra complex very efficiently. The abstract architecture and execution overview on it are only briefly explained and the details about the actual implementations, data stream control mechanism, and performance evaluation are prepared in the future paper.

Acknowledgement

The authors would like to acknowledge S.Suzuki, H.Akamatsu, Y.Ogi, S.Fushimi, and S.Sakai for their valuable contribution to the data base machine project.

References

1. Slotnick,D.L., Logic per Track Devices, Advances in Computers, Vol.10, J.Tou,ed., Academic Press, New York, pp.291-296 (1970)

2. Ozkarahan,E.A., Schuster,S.A. and Smith,K.C., RAP-An Associative Processor for Data Base Management, Proc. AFIPS NCC, Vol.45, pp.379-387 (1975)

3. Ozkarahan,E.A., Shuster,S.A. and Sevcik,K.C., Performance Evaluation of a Relational Associative Processor, ACM Trans. Database Syst., Vol.2, No.2, pp.175-195 (1977)

4. Oflazer,K. and Ozkarahan,E.A., RAP.3-A multi-micro cell architecture for the RAP database machine, Proc of the Int. Workshop on High Level Language Computer Architecture, pp.108-119 (1980)

5. Copeland,G.P., Lipovski,G.J. and Su,S.Y.W., The Architecture of CASSM: a Cellular System for Non-numeric Processing, Proc. 1st Annu. Symp. Computer Architecture, pp.121-128 (1973)

6. Su,S.Y.W., Nguyen,L.H., et al., The Architectural Features and Implementation Techniques of the Multicell CASSM, IEEE Trans. Comput. Vol.C-28, No.6, pp.430-445 (1979)

7. Su.S.Y.W., On Logic-per-Track Devices: Concept and Applications, IEEE COMPUTER, Vol.12, No.3, pp.11-25 (1979)

8. Uemura,S., Yuba,T., Kokubu,A., et al., The Design and Implementation of a Magnetic-Bubble Database Machine, IFIP 80, pp.433-438 (1980)

9. Oliver,E.J. and Berra,P.B., RELACS A Relational Associative Computer System, Proc. of the Fifth Workshop on Computer Architecture for Non-Numeric Processing, pp.106-114 (1980)

10. Berra,P.B. and Oliver,E.J., The Role of Associative Array Processors in Data Base Machine Architecture, IEEE Computer, Vol.12, No.3, pp.53-61 (1979)

11. DeWitt,D.J., DIRECT-A Multiprocessor Organization for Supporting Relational Database Management Systems, IEEE Trans. Comput., Vol. C-28, No.6 (1979)

12. DeWitt,D.J., Query Execution in DIRECT, Proc. ACM-SIGMOD 1979, pp.13-22 (1979)

13. Kung,H.T. and Lehman,P.L., Systolic (VLSI) Arrays for Relational Database Operations, Proc. of ACM-SIGMOD pp.105-116 (1980)

14. Song,S.W., A Highly Concurrent Tree Machine for Database Applications, Proc. of the 1980 Int. Conf. on Parallel Processing, pp.259-268 (1980)

15. Banerjee,J., Hsiao,D.K. and Kannan,K, DBC-A Database Computer for Very Large Databases, IEEE Trans. Comput., Vol.C-28, No.6, pp.414-429 (1979)

16. Menon,M.J. and Hsiao,D.K., Design and Analysis of a Relational Join

Operation for VLSI, Proc. Int. Conf. on Very Large Data Bases, pp.44-55 (1981)

17. Tanaka,Y., Nozaka,Y., et al., Pipeline Searching and Sorting Modules as Components of a Data Flow Database Computer, IFIP 80, pp.427-432 (1980)

18. Oda,Y., Database Machine Architecture using Data Partitioning Network, IECEJ Technical Group Meeting, EC80-72 (1981) (in Japanese)

19. Babb,E, Implementing a Relational Database by Means of Specialized Hardware, ACM Trans. Database Syst., Vol.4, No.1, pp.1-29 (1979)

20. McGregor,D.R., Thomson,R.H. and Dawson,W.N., High Performance Hardware for Database Systems, Systems for Large Data Bases, North-Holland, pp.103-116 (1976)

21. Kitsuregawa,M., Suzuki,S., Tanaka,H., and Moto-oka,T., Application of Hash to a Data Base Machine, The 23rd Information Processing Society National Convention (1981) (in Japanese)

22. Kitsuregawa,M., et al., Relational Algebra Machine based on Hash and Sort, IECEJ Technical Group Meeting, EC81-35 (1981) (in Japanese)

23. Hawthorn,P., The Effect of Target Applications on the Design of Database Machines, Proc. of ACM-SIGMOD, pp.188-197 (1981)

24. Kitsuregawa,M., Fushimi,S., Kuwabara,k., Tanaka,H., and Moto-oka,T., Organization of Pipeline Merge Sorter, IECEJ Technical Group Meeting, EC82-32 (1982) (in Japanese)

25. Todd,S., Algorithm and Hardware for a Merge Sort Using Multiple Processors, IBM J.RES.DEVELOP., Vol.22, pp509-517 (1978)

Vocabulary Building for Database Queries

Yuzuru Tanaka
Faculty of Engineering
Hokkaido University
Sapporo, 060 JAPAN

ABSTRACT

The introduction of natural language features into database query languages seems to be classified into three stages. The first stage is characterized by the introduction of the syntactic flexibility of natural languages, while the second stage introduces the access flexibility. In the first stage languages, database users can afford syntactic structures similar to a natural language. However, he has to specify, in a procedural manner, how to retrieve information. The second stage languages offer facilities to define a variety of virtual access paths as well as actual access structures of databases. Their definition may be expressed as a semantic network or a set of logical rules. The third stage, that is the most sophisticated one, needs the introduction of vocabulary semantics. Sharing a common vocabulary, human beings can communicate with each other without referring to the conceptual structures of their knowledge. Vocabulary plays one of the most important roles in human communications. The purpose of this paper is the introduction of the similar facilities to the communication between a database and its users. We call it vocabulary semantics. The paper clarifies how to build vocabulary and how to define its formal semantics.

1. Introduction

The introduction of natural language features into query languages seems to be classified into three stages. The first stage is the introduction of the syntactic flexibility of natural languages. In this stage, users have to know a lot of things about the definite logical access structure defined by the database system. Queries have to describe a desired logical access path in a procedural manner, following certain syntactic rules similar to those of a natural language. The only semantics that users can afford is a one-to-one correspondence between the names of attributes and the actual attributes of the database. While such a system may be able to optimize queries, it can not work if the instruction about which access path is chosen is not given by the user.

The second stage is the introduction of the access flexibility. Such a system can define a variety of virtual access paths as well as an actual access structure. The definition of the virtual access paths may be given by a semantic network or a logical program written in such a logical language as PROLOG. Since they are mostly based on the first order logic, they are not flexible enough to define the semantics of a variety of words including those that modify the meanings of predicates.

In my personal view, the third stage, that has not been much studied untill now, is characterized by the introduction of flexible vocabularies. The roll of vocabularies will be well understood if we replace a database with a person who understands English and knows the same information stored in the database. Let us call him a DB person. Suppose that a database consists of two relations R(person, birthday, sex, address) and S(person, parent). What kind of queries will you put to a DB person when you want to know the birthday of someone's brother, or the address of someone's aunt? You will not tell him the detail procedure to retrieve the desired information from the two relations. Instead, you will say to him, for example, "tell me the birthday of John Smith's brother." And it will be sufficient because you and he have the common understanding about what brother and aunt mean. You need not know the access structure of information in the DB person, wheather it is actual or virtual. The communication among human beings deeply depends on the shared vocabulary. The logical structures of the stored information may vary from person to person. However, we can communicate with each other usig a shared vocabulary. Besides, the DB person does not have the semantic definitions respectively for "the birthday of the brother" and "the address of the brother". Instead, he has the intensional definition of "of the brother". It is considered as a function that specifies relationships among persons. This will become obvious if R is modified to have another attribute 'skill'. If a DB person must have the definition of "the skill of the brother", he can not answer a query about the skill of John Smith's brother before someone teaches him what "the skill of the brother" means in his database. However, this is not the case. He will be able to answer the query. This is because he can construct the meaning of "the skill of the brother" from the already known meanings of basic constituents "the skill" and "of the brother". The meaning of the phrase "of the brother" is intensionally defined independently from the contexts in which it is used. Some philosophers considered it possible to define the meanings of words and phrases independently from the contexts in which they are used. This principle is called Frege's principle, or the compositionality principle. According to this principle, the meaninig of a sentence can be constructed from the meanings of the basic components in that sentence. Therefore, the vocabulary and its semantics play the most essential roles in the semantics of natural languages.

The purpose of this paper is the introduction of the vocabulary semantics to a database query language. For this purpose, we have to clarify how to define formal semantics of words. In our natural languages, the semantics of words can be explained using a rather small set of basic words. You can build your vocabulary based on your present vocabulary. The same mechanism will be also introduced to our query language. For these goals, chapter 2 develops a formal theory of relations with null values and generalized relational operations. Based on these, chapter 3 defines a database access space with infinitely many access paths, while chapter 4 shows how the stepwise building of a vocabulary defines the formal semantics of a database.

2. Generalized Operations on Relations

2.1 A partial relation

Let Ω and D be two enumerable sets respectively called an attribute set and a value set, where we assume a symbol ϕ belongs to D while another special value "$\perp$" does not. By $(X \to Y)$ we denote a set of all the functions from a set X to another set Y. A subset of $(\Omega \to D)$ is called a relation over Ω, while a subset of $(\Omega \to D')$ for $D'=D \cup \{\perp\}$ is called a partial relation over Ω. An element of $(\Omega \to D)$ is called a tuple over Ω, while one of $(\Omega \to D')$ is a partial tuple over Ω. An attribute value ϕ means the nonexistence of this attribute, while $\perp$ means that the value is unknown. For the sake of convenience, we identify ϕ with an empty set. For a partial relation R over Ω, we define $\omega(R)$ as Ω. A special set $(\phi \to D')$ is considered as a singleton $e=\{\varepsilon\}$, i.e.,

$$e=(\phi \to D')=\{\varepsilon\}.$$

Let f be a function from X to Y, and Z be another set. A restriction of f within Z denoted by $f|_Z$ is a function defined as

$$f|_Z \in (Z \to Y \cup \{\perp\}),$$

$$\text{and} \quad A \in X \cap Z \quad f|_Z(A) = f(A),$$

$$A \in Z - X \quad f|_Z(A) = \perp.$$

For each $R \subset (\Omega \rightarrow D')$ and a set X, we define a projection of R onto X as

$$[X]R = \{\mu|_{X \cap \omega(R)} \mid \forall \mu \in R \text{ s.t. } \mu|_{X \cap \omega(R)} \in (X \cap \omega(R) \rightarrow D)\},$$

which is always a relation.

Lemma 2.1
A partial relation R is a relation if and only if $[\omega(R)]R=R$.

(proof) Obvious.

Different from the case of relations, it does not always hold for a partial relation R that $[X][Y]R=[X \cap Y]R$. The left hand side imposes the condition that the Y-values should be known, while the right hand side imposes only the certainty of $X \cap Y$-values.

The natural join of two relations r and s is defined as

$$r*s = \{\mu \mid \forall \mu \in (\omega(r) \cup \omega(s) \rightarrow D) \text{ s.t. } \mu|_{\omega(r)} \in r \wedge \mu|_{\omega(s)} \in s \wedge \mu|_{\omega(r) \cap \omega(s)} \in (\omega(r) \cap \omega(s) \rightarrow D-\{\phi\})\}.$$

This definition imposes the condition that the values of join attributes should be known and existent, i.e., they should be neither ϕ nor $\perp$.

2.2 Grouping of values and generalized projection

Here we formalize the so-called "GROUP-BY" operations that are fundamental to enrich our formal semantics of queries.

Let Ω_h be a set defined as

$$\Omega_h = \Omega \cup \{X/Y \mid X, Y \subset \Omega\}.$$

An attribute (X/Y) in an extended attribute set Ω_h is read as "X grouped by each Y value". An extended value set D_h is defined as

$$D_h = \bigcup_i P^i(D),$$
$$\text{where } P^0(D) = D,$$
$$P^{n+1}(D) = P^n(2^D).$$

For a subset X of Ω_h and a tuple μ of a relation r over Ω, we define a partial tuple $\mu^h_{r,X} \in (X \rightarrow D_h \cup \{\perp\})$ as follows;

$$A \in \Omega \cap X \quad \mu^h_{r,X}(A) = \mu(A),$$

and

$$\forall V/W \in X$$

$$\mu^h_{r,X}(V/W) = \begin{cases} \{\nu(V) \mid \forall \nu \in [VW]r \text{ s.t. } \\ \qquad \nu|_W = \mu|_W \text{ and } \nu|_V \in (V \to D-\{\phi\}) \\ \qquad \text{if } \mu|_W \in (Y \to D-\{\phi\}), \\ \perp \quad \text{otherwise.} \end{cases}$$

For a subset X of Ω_h, flat(X) denotes a subset of Ω defined as

$$\mathrm{flat}(X) = (X \cap \Omega) \cup (\bigcup_{V/W \in X} (V \cup W)).$$

A generalized projection of a relation r onto a subset X of Ω_h is defined as

$$[X]_h r = \{\mu^h_{r,X}|_{\omega(r)_h \cap X} \mid \forall \mu \in r \text{ s.t. } \mu^h_{r,X}|_{\omega(r)_h \cap X} \in (X \to D_h)\},$$

while a generalized projection of a partial relation R onto a subset X of Ω_h is defined as

$$[X]_h R = [X]_h[\mathrm{flat}(X)]R.$$

Now we consider a set of computable functions F defined as

$$F = \{g \mid g \text{ is computable, } \exists k \geq 0 \;\; g \in ((D_h)^k \to D_h \cup \{\perp\})\}.$$

For an attribute set Ω, let Ω_f be a set defined as

$$\Omega_f = \Omega \cup \{g(A_1, A_2, \ldots, A_k) \mid \forall i \;\; A_i \in \Omega, \; \forall g \in F\}.$$

For a subset X of Ω_f and a tuple μ of a relation r over a subset of Ω, we define a partial tuple $\mu^f_{r,X} \in (X \to D_h \cup \{\perp\})$ as follows;

$$\forall A \in \Omega \cap X \quad \mu^f_{r,X}(A) = \mu(A),$$

and

$$\forall g(A_1, A_2, \ldots, A_k) \in X$$
$$\mu^f_{r,X}(g(A_1, A_2, \ldots, A_k))$$
$$= \text{if } v = g(\mu^f_{r,X}(A_1), \mu^f_{r,X}(A_2), \ldots, \mu^f_{r,X}(A_k)) \text{ is defined then } v, \text{ else } \perp.$$

For a subset X of Ω_f, arg(X) denotes a subset of Ω defined as

$$\arg(X) = (X \cap \Omega) \cup (\bigcup_{g(Y) \in X} Y).$$

A generalized projection of a relation r onto a subset X of Ω_f is defined as

$$[X]_f r = \{\mu^f_{r,X}|_{\omega(r)_f \cap X} \mid \forall \mu \in r \text{ s.t. } \mu^f_{r,X}|_{\omega(r)_f \cap X} \in (X \to D_h)\},$$

while a generalized projection of a partial relation R onto a subset X of Ω_f is defined as

$$[X]_f R = [X]_f[\arg(X)]R.$$

A more general attribute set Ω_g is defined as follows;

(1) $\Omega \subset \Omega_g$,

(2) $\forall X, \forall Y \subset \Omega_g \quad X/Y \in \Omega_g$,

(3) $\forall A_1, \forall A_2, \ldots, \forall A_k \in \Omega_g \quad \forall g \in F$

$g(A_1, A_2, \ldots A_k) \in \Omega_g$,

(4) only those defined by a finite number of applications of the above rules are elements of Ω_g.

For a partial relation R over a subset of Ω, its projection onto a subset X of Ω_g is defined as

$$[X]_g R = \begin{cases} [X]R & \text{if } X \subset \Omega, \\ [X]_f[\arg(X)]_h[\text{flat}(\arg(X))]_g R & \text{otherwise.} \end{cases}$$

Some important functions are defined below, where S denotes a subset of D.

$$\text{sum}(S) = \begin{cases} \Sigma_{v \in S}\, v & \text{if S is a set of numbers,} \\ \perp & \text{otherwise.} \end{cases}$$

$$\max(S) = \begin{cases} \max_{v \in S}\, v & \text{if S has some definite order,} \\ \perp & \text{otherwise.} \end{cases}$$

$$\min(S) = \begin{cases} \min_{v \in S}\, v & \text{if S has some definite order,} \\ \perp & \text{otherwise.} \end{cases}$$

$$\text{count}(S) = \text{cardinarity of S.}$$

For any attributes A, B, and C in Ω_g, we define average(A; B/C), that is also an element of Ω_g, as

$$\text{average}(A;\ B/C) = \frac{\text{sum}(\text{sum}(A/BC)/C)}{\text{count}(B/C)}.$$

<u>example 2.1</u>

We show a computation process for a generalized projection [average(A; B/C), C]r of an example relation r over {A, B, C, D}.

r	A	B	C	D
	1	a	c	e
	1	a	c	f
	3	b	c	f
	3	a	c	e
	4	b	c	g
	2	a	d	h
	1	a	d	h
	1	b	d	$\perp$

[average(A; B/C), C]r

$$= [\frac{\text{sum(sum(A/BC)/C)}}{\text{count(B/C)}}, \text{C}]_f[\text{sum(sum(A/BC)/C), count(B/C), C}]_f$$

$[\text{sum(A/BC)/C, B/C, C}]_h[\text{sum(A/BC), B, C}]_f[\text{a/BC, B, C}]_h$

[A, B, C]r

r_1 = [A, B, C]r		
1	a	c
3	b	c
3	a	c
4	b	c
2	a	d
1	a	d
1	b	d

$r_2 = [\text{A/BC, B, C}]_h r_1$		
{1, 3}	a	c
{3, 4}	b	c
{2, 1}	a	d
{1}	b	d

$r_3 = [\text{sum(A/BC), B, C}]_f r_2$		
4	a	c
7	b	c
3	a	d
1	b	d

$r_4 = [\text{sum(A/BC)/C, B/C, C}]_h r_3$		
{4, 7}	{a, b}	c
{3, 1}	{a, b}	d

$r_5 = [\text{sum(sum(A/BC)/C),count(B/C),C}]_f r_4$		
11	2	c
4	2	d

$r_6 = [\frac{\text{sum(sum(A/BC)/C)}}{\text{count(B/C)}}, \text{C}]r_5$	
5.5	c
2	d

2.3 Generalized restrictions of relations

Let $P(\underline{x})$ be an n-place predicate, $\underline{A} = (A_1, A_2, \ldots, A_n)$ be an n-dimensional vector of attributes in Ω, and $\underset{\sim}{A}$ be the corresponding set $\{A_1, A_2, \ldots, A_n\}$. A predicate $P(\underline{A})(\mu)$ for $\mu \in (\Omega \to D)$ is defined as

$$P(\underline{A})(\mu) \quad \text{iff} \quad (\forall A \in \tilde{\underline{A}} \ \ \mu(A) \text{ is defined} \wedge \mu(A) \neq \bot) \wedge P(\mu(A_1), \mu(A_2), \ldots, \mu(A_n)).$$

$P(\underline{A})$ is called a predicate scheme over Ω.

A restriction of a partial relation R over some subset of Ω by a predicate scheme $P(\underline{A})$ over Ω is a relation defined as

$$[P(\underline{A})]R = \{\mu \mid \forall \mu \in R \ \text{ s.t. } \ P(\underline{A})(\mu)\}.$$

<u>Lemma 2.2</u>

$$\forall X \subset \Omega \quad [x][P(\underline{A})]R = [X][P(\underline{A})][X \cup \tilde{\underline{A}}]_g R.$$

(proof)
Obvious from the definition.

A generalized restriction of a partial relation R over some subset of Ω by a scheme $P(\underline{A})$ for $\tilde{\underline{A}} \subset \Omega_g$ is defined as

$$\forall X \subset \Omega_g \quad [X]_g[P(\underline{A})]_g R = [X]_g[P(\underline{A})][X \cup \tilde{\underline{A}}]_g R.$$

3. <u>Intensional Relations and an Information Space</u>

For a partial relation R over some subset of Ω, we define a function ${}^{int}R$ from Ω_g to a set of relations as

$${}^{int}R = \lambda x. \ [x]_g R.$$

This function is called a intension of R over Ω. A restriction of ${}^{int}R$ by a scheme $P(\underline{A})$ is defined as

$$[P(\underline{A})]^{int}R = \lambda x. \ [x]_g[P(\underline{A})]^{int}R(x \cup \tilde{\underline{A}}).$$

<u>Theorem 3.1</u>

$$\forall X \subset \Omega_g \quad ([P(\underline{A})]^{int}R)(X) = [X]_g[P(\underline{A})]_g R.$$

(proof)

$$\begin{aligned}[P(\underline{A})]^{int}R &= \lambda x. \ [x]_g[P(\underline{A})]^{int}R(x \cup \tilde{\underline{A}}) \\ &= \lambda x. \ [x]_g[P(\underline{A})][x \cup \tilde{\underline{A}}]_g R \\ &= \lambda x. \ [x]_g[P(\underline{A})]_g R \text{ (from the definition of } [P(\underline{A})]_g).\end{aligned}$$

A relation $([P(\underline{A})]^{int}R)(X)$ can be informally interpreted as all the information about X that can be obtained from R and satisfies the condition $P(\underline{A})$. We define a set of all the intensional relations over Ω as

$$IR_{\Omega} = (2^{\Omega_g} \rightarrow ER_{\Omega}),$$

where ER_{Ω} is a set of extensional relations over Ω defined as

$$ER_{\Omega} = \bigcup_{\Omega' \in \Omega_g} 2^{(\Omega' \rightarrow D_h)}.$$

Let L be an enumerable set called a set of labels. We define a set Ω^L as follows;

(1) $\Omega \subset \Omega^L$,

(2) $\forall A \in \Omega^L, \forall l \in L \quad lA$ is an element of Ω^L,

(3) $\forall X \subset \Omega^L, \forall Y \subset \Omega^L \quad (X/Y)$ is an element of Ω^L,

(4) $\forall A_1, \forall A_2, \ldots, \forall A_k \in \Omega^L, \forall g \in F$
$g(A_1, A_2, \ldots, A_k) \in \Omega^L$,

(5) only those defined by a finite number of applications of the above rules constitute Ω^L.

An elementary adjective for Ω is a triple $(P^{n,m}(\underline{x}; \underline{y}), \underline{A}, \underline{B})$, where $P^{n,m}(\underline{x}; \underline{y})$ is a (n+m)-place predicate and $\underline{A}$, $\underline{B}$ are n and m dimensional vectors of attributes in Ω_g. For an elementary adjective $\theta = (P(\underline{x}; \underline{y}), \underline{A}, \underline{B})$, its inverse θ^- is defined as

$$(P^*(\underline{y}; \underline{x}), \underline{B}, \underline{A}),$$

$$\text{where} \quad P^*(\underline{y}; \underline{x}) \text{ iff } \forall \underline{x}, \forall \underline{y} \quad P(\underline{x}; \underline{y}).$$

Let Θ be some set of elementary adjectives. Each $\theta \in \Theta$ is considered as a label for attributes and, for simplicity, $\theta = (P(\underline{x}; \underline{y}), \underline{A}, \underline{B})$ is denoted by $P(\underline{A}; \theta\underline{B})$, where $\theta\underline{B} = (\theta B_1, \theta B_2, \ldots, \theta B_m)$. Besides, $\underline{A}$ and $\underline{B}$ of $\theta = P(\underline{A}; \theta\underline{B})$ is denoted by $\underline{A}_\theta$ and $\underline{B}_\theta$.

For a pair of same dimensional vectors $\underline{A}$ and $\underline{B}$ of attributes in Ω^Θ, a renaming operator $\underline{A}\%\underline{B}$ renames attribute name B_i of a relation to a new name A_i, i.e.,

$$(\underline{A}\%\underline{B})r = \{\mu^* \mid \forall \mu \in r\},$$

$$\text{where} \quad \mu^* \in ((\omega(r)-\tilde{\underline{B}}) \cup \tilde{\underline{A}} \rightarrow D_h),$$

$$\text{and} \quad \forall A \in \omega(r)-\tilde{\underline{B}} \quad \mu^*(A) = \mu(A)$$

$$\forall i \text{ s.t. } B_i \in \tilde{\underline{B}} \cap \omega(r) \quad \mu^*(A_i) = \mu(B_i).$$

For simplicity, $(\underline{A}\%\underline{B})$ is denoted by $(\tilde{\underline{A}}\%\tilde{\underline{B}})$ if its meaning is clear.

Definition 3.1

Let $\underline{r}_0$ be an intensional relation over Ω, and Θ be a set of elementary adjectives for Ω. An information space for $(\Omega, \Theta, \underline{r}_0)$ is an intensional relation $\underline{r}$ over Ω^Θ satisfying

(1) $\forall X \subset \Omega_g \quad \underline{r}(X) = \underline{r}_0(X)$,

(2) $\forall \theta = P_\theta(\underline{A};\ \theta\underline{B}) \quad \forall X$ s.t. $X \cap \theta(\Omega^\Theta) = \phi$

$$\underline{r}(X\theta(Y))$$
$$= [X\theta(Y)]_g[P_\theta(\underline{A};\ \theta\underline{B})]_g(\underline{r}(X \cup \tilde{\underline{A}}) * \underline{r}(\theta(Y) \cup \theta(\tilde{\underline{B}}))),$$

(3) $\forall \theta \in \Theta \quad \underline{r}(\theta(Y)) \subset (\theta Y \% Y)\underline{r}(Y)$,

(4) for each subset X of Ω^Θ, $\underline{r}(X)$ is the maximal set satisfying the above conditions.

Theorem 3.2

The condition (2) (3) and (4) in Definition 3.1 may be replaced with the following condition.

$$\forall \theta = P(\underline{A};\ \theta\underline{B}) \quad \forall X \text{ s.t. } X \cap \theta(\Omega^\Theta) = \phi \quad \forall Y$$
$$\underline{r}(X\theta(Y))$$
$$= [X\theta(Y)]_g[P(\underline{A};\ \theta(\underline{B}))]_g(\underline{r}(X \cup \tilde{\underline{A}}) * \theta\underline{r}(Y \cup \tilde{\underline{B}})),$$

where $\forall r \in ER_\Omega \quad \theta r = (\theta(\omega(r))\%\omega(r))r$.

(proof)

Let $\underline{s}$ be an intensional relation satisfying the new condition, and $\underline{r}$ be an information space defined by Definition 3.1.

We first prove that

$$\forall X \quad \underline{s}(X) \supset \underline{r}(X)$$

by mathematical induction on rank(X), where rank(X) is defined as follows;

$$\forall X \subset \Omega_g \quad \text{rank}(X) = 0,$$
$$\text{rank}(XY) = \text{rank}(X)+\text{rank}(Y),$$
$$\text{rank}(\theta) = \text{rank}(\underline{A}_\theta)+\text{rank}(\underline{B}_\theta)+1 = 1 \ (\because\ \tilde{\underline{A}}_\theta,\ \tilde{\underline{B}}_\theta \subset \Omega_g),$$
$$\text{rank}(\theta(X)) = \text{rank}(X)+\text{rank}(\theta),$$
$$\text{rank}(X/Y) = \text{rank}(X)+\text{rank}(Y).$$

For each X satisfying rank(X)=0, i.e., X Ω_g, $\underline{s}(X)$ equals to $\underline{r}(X)$, and hence $\underline{s}(X) \supset \underline{r}(X)$ holds for rank(X)=0. Let us assume that $\underline{s}(Y) \supset \underline{r}(Y)$ holds for any Y whose rank is less than k, and let rank(Z) be k. We can assume without loss of generality that Z is $X\theta(Y)$ with $X \cap \theta(\Omega^\Theta)=\phi$. Then $s(X\theta(Y))$ is equal to

$$[X\theta(Y)]_g[P(\underline{A};\ \theta(\underline{B}))]_g(\underline{s}(X\cup\tilde{\underline{A}}) * \theta\underline{s}(Y\cup\tilde{\underline{B}})).$$

Since it holds that

$$k = \mathrm{rank}(X\theta(Y)) = \mathrm{rank}(X)+\mathrm{rank}(Y)+1,$$

$$\mathrm{rank}(X\cup\tilde{\underline{A}}) = \mathrm{rank}(X)+\mathrm{rank}(\tilde{\underline{A}}) = \mathrm{rank}(X) \lneq k,$$

$$\mathrm{rank}(Y\cup\tilde{\underline{B}}) = \mathrm{rank}(Y)+\mathrm{rank}(\tilde{\underline{B}}) = \mathrm{rank}(Y) \lneq k,$$

it can be assumed that

$$\underline{s}(X\cup\tilde{\underline{A}}) \supset \underline{r}(X\cup\tilde{\underline{A}}),$$

$$\underline{s}(Y\cup\tilde{\underline{B}}) \supset \underline{r}(Y\cup\tilde{\underline{B}}).$$

Hence it holds that

$$\underline{s}(X\theta(Y)) \supset [X\theta(Y)]_g[P(\underline{A};\ \theta(\underline{B}))]_g(\underline{r}(X\cup\tilde{\underline{A}}) * \theta r(Y\cup\tilde{\underline{B}})).$$

From the condition (4) of Definition 3.1, $\theta\underline{r}(Y\cup\tilde{\underline{B}})$ must be a super set of $\underline{r}(\theta(Y)\cup\theta(\tilde{\underline{B}}))$. Hence we can conclude that

$$\begin{aligned}\underline{s}(X\theta(Y)) &\supset [X\theta(Y)]_g[P(\underline{A};\ \theta(\underline{B}))]_g(\underline{r}(X\cup\tilde{\underline{A}}) * \theta\underline{r}(Y\cup\tilde{\underline{B}}))\\ &\supset [X\theta(Y)]_g[P(\underline{A};\ \theta(\underline{B}))]_g(\underline{r}(X\cup\tilde{\underline{A}}) * \underline{r}(\theta(Y)\cup\theta(\tilde{\underline{B}})))\\ &= \underline{r}(X\theta(Y)).\end{aligned}$$

Since we have proved that, for any X, $\underline{s}(X)$ includes $\underline{r}(X)$, $\underline{s}$ is an information space if it satisfies the conditions (1) (2) (3) of Definition 3.1. Obviously, $\underline{s}$ satisfies the condition (1). Let us check the condition (3). From the definition of $\underline{s}$, $\underline{s}(\theta(Y))$ is equal to

$$[\theta(Y)]_g[P(\underline{A};\ \theta(\underline{B}))]_g(\underline{s}(\tilde{\underline{A}}) * \theta\underline{s}(Y\cup\tilde{\underline{B}})),$$

which is included by

$$\begin{aligned}&[\theta(Y)]_g\theta\underline{s}(Y\cup\tilde{\underline{B}})\\ &\subset \theta\underline{s}(Y).\end{aligned}$$

This proves the condition (3). Since $\underline{s}(\theta(Y)\cup\theta(\tilde{\underline{B}}))$ is

$$[\theta(Y)\cup\theta(\tilde{\underline{B}})]_g[P(\underline{A};\ \theta(\underline{B}))]_g(\underline{s}(\underline{A}) * \theta\underline{s}(Y\cup\tilde{\underline{B}}))$$

from the definition of this theorem, the substitution of $\underline{r}(X\cup\tilde{\underline{A}})$ and $r(\theta(Y)\cup\theta(\tilde{\underline{B}}))$ in the condition (2) by $\underline{s}(X\cup\tilde{\underline{A}})$ and the expression given above gives

$$\begin{aligned}&[X\theta(Y)]_g[P(\underline{A};\ \theta(\underline{B}))]_g\\ &\qquad(\underline{s}(X\cup\tilde{\underline{A}}) * [\theta(Y)\cup\theta(\tilde{\underline{B}})]_g[P(\underline{A};\ \theta(\underline{B}))]_g(\underline{s}(\tilde{\underline{A}})*\theta\underline{s}(Y\cup\tilde{\underline{B}})))\\ &= [X\theta(Y)]_g[P(\underline{A};\ \theta(\underline{B}))]_g(\underline{s}(X\cup\tilde{\underline{A}})*\underline{s}(\tilde{\underline{A}})*\theta\underline{s}(Y\cup\tilde{\underline{B}}))\\ &= [X\theta(Y)]_g[P(\underline{A};\ \theta(\underline{B}))]_g(\underline{s}(X\cup\tilde{\underline{A}})*\theta\underline{s}(Y\cup\tilde{\underline{B}}))\\ &= \underline{s}(X\theta(Y)).\end{aligned}$$

Hence $\underline{s}$ satisfies Definition 3.1, and we have proved the theorem.

Theorem 3.3

An information space $\underline{r}$ does not depend on the way to choose θ in the condition (2) of Definition 3.1.

(proof)

We prove, for X satisfying $X \cap \sigma(\Omega^\Theta) = X \cap \tau(\Omega^\Theta) = \phi$, $\underline{r}(X\sigma(Y)\tau(Z))$ is uniquely evaluated independently from which adjective should be chosen first.

If σ is first chosen then

$$\underline{r}(X\sigma(Y)\tau(Z))$$
$$= [X\sigma(Y)\tau(Z)]_g[P(\underline{A};\ \sigma(\underline{B})]_g(\underline{r}(X \cup \tilde{\underline{A}} \cup \tau(Z)) * \underline{r}(\sigma(Y)\sigma(\tilde{\underline{B}}))).$$

Since $\tilde{\underline{A}} \subset \Omega_g$ holds, $(X \cup \tilde{\underline{A}})$ and $\tau(\Omega^\Theta)$ are disjoint. Hence $\underline{r}(X \cup \tilde{\underline{A}} \cup \tau(Z))$ is equal to

$$[X\tilde{\underline{A}}\tau(Z)]_g[Q(\underline{C};\ \tau(\underline{D}))]_g(\underline{r}(X \cup \tilde{\underline{A}} \cup \tilde{\underline{C}}) * \underline{r}(\tau(Z) \cup \tau(\tilde{\underline{D}}))).$$

Therefore, $\underline{r}(X\sigma(Y)\tau(Z))$ is

$$[X\sigma(Y)\tau(Z)]_g[P(\underline{A};\ \sigma(\underline{B}))]_g$$
$$(([X\tilde{\underline{A}}\tau(Z)]_g[Q(\underline{C};\ \tau(\underline{D}))]_g(\underline{r}(X \cup \tilde{\underline{A}} \cup \tilde{\underline{C}})*\underline{r}(\tau(Z) \cup \tau(\tilde{\underline{D}})))$$
$$*\underline{r}(\sigma(Y)\sigma(\tilde{\underline{B}}))),$$

which is equal to

$$[X\sigma(Y)\tau(Z)]_g[P(\underline{A};\ \sigma(\underline{B}))]_g[Q(\underline{C};\ \tau(\underline{D}))]_g$$
$$(\underline{r}(X \cup \tilde{\underline{A}} \cup \tilde{\underline{C}})*\underline{r}(\tau(Z) \cup \tau(\tilde{\underline{D}}))*\underline{r}(\sigma(Y)\sigma(\tilde{\underline{B}})))$$
$$= [X\sigma(Y)\tau(Z)]_g[P(\underline{A};\ \sigma(\underline{B}))\ \ Q(\underline{C};\ \tau(\underline{D}))]_g$$
$$(\underline{r}(X \cup \tilde{\underline{A}} \cup \tilde{\underline{C}})*\underline{r}(\tau(Z) \cup \tau(\tilde{\underline{D}}))*\underline{r}(\sigma(Y) \cup \sigma(\tilde{\underline{B}})))$$

from the definition of a generalized projection. The last expression is independent of the order between σ and τ.

Let Ω_θ, $\bar{\Omega}_\theta$ be

$$\Omega_\theta = \theta(\Omega^\Theta),$$
$$\bar{\Omega}_\theta = \Omega^\Theta - \Omega_\theta.$$

For each $\theta \in \Theta$, θ^* and $\bar{\theta}^*$ are defined as

$$\theta^* = \lambda x.\ y \quad \text{s.t. } \theta(y) = x \cap \Omega_\theta,$$
$$\bar{\theta}^* = \lambda x.\ x \cap \bar{\Omega}_\theta.$$

Besides, for each intensional relation $\underline{r}$ and $\theta \in \Theta$, $\underline{r}$ is defined as

$$\theta\underline{r} = \lambda x.\ \theta\underline{r}(x).$$

Theorem 3.4

An information space $\underline{r}$ for $(\Omega, \Theta, \underline{r}_0)$ satisfies the following;

$$\forall \theta = P(\underline{A};\ \theta(\underline{B}))\quad \forall X \text{ s.t. } X \cap \theta(\Omega^{\Theta}) \neq \phi$$

$$\underline{r}(X) = ([\theta]((\theta\underline{r}\theta^*) * (\underline{r}\bar{\theta}^*)))(X).$$

(proof)

Obvious.

For an information space $\underline{r}$ for $(\Omega, \Theta, \underline{r}_0)$, we define Θ^* as follows;

(1) $\Theta \subset \Theta^*$,

(2) $\forall \theta \in \Theta \quad \theta^- \in \Theta^*$,

(3) $\forall \sigma, \forall \tau \in \Theta^*$

$\sigma \circ \tau \in \Theta^* \quad (\sigma \circ \tau)^- = \sigma^- \circ \tau^-$,

$\sigma + \tau \in \Theta^* \quad (\sigma + \tau)^- = \sigma^- + \tau^-$,

$\sigma - \tau \in \Theta^* \quad (\sigma - \tau)^- = \sigma^- - \tau^-$,

$\sigma * \tau \in \Theta^* \quad (\sigma * \tau)^- = \sigma^- * \tau^-$,

(4) only those defined by a finite number of applications of the above rules constitute Θ^*.

Besides, we define Ω^* as follows;

(1) $\Omega^{\Theta} \subset \Omega^*$,

(2) $\forall X, \forall Y \subset \Omega^* \quad X/Y \in \Omega^*$,

(3) $\forall A_1, \forall A_2, \ldots, \forall A_n \in \Omega^* \ \forall g \in F$

$g(A_1, A_2, \ldots, A_n) \in \Omega^*$,

(4) $\forall A \in \Omega^* \ \forall \theta \in \Theta \quad \theta A \in \Omega^*$,

(5) only those defined by a finite number of applications of the above rules constitute Ω^*.

An intensional relation $\underline{r}^*$ over Ω^* is defined as

(1) $\forall X \subset \Omega^{\Theta} \quad \underline{r}^*(X) = \underline{r}(X)$,

(2) $\forall \theta \in \Theta \ \forall X \text{ s.t. } X \cap \theta(\Omega^*) \neq \phi$

$$\underline{r}^*(X) = ([\theta]((\theta\underline{r}^*\theta^*) * (\underline{r}^*\bar{\theta}^*)))(X),$$

where the domains of all the operations are assumed to be extended from Ω^{Θ} to Ω^*,

(3) $\forall (\sigma \circ \tau) \in \Theta^* \quad \underline{r}^*(X(\sigma \circ \tau)(Y))$

$$= (X(\sigma \circ \tau)(Y) \% X\sigma(\tau(Y)))\underline{r}^*(X\sigma(\tau(Y))),$$

$$(4)\ \forall(\sigma+\tau) \in \Theta^* \quad \underline{r}^*(X(\sigma+\tau)(Y)) = (X(\sigma+\tau)(Y)\%X\sigma(Y))\underline{r}^*(X\sigma(Y)) \cup (X(\sigma+\tau)(Y)\%X\tau(Y))\underline{r}^*(X\tau(Y)),$$

$$(5)\ \forall(\sigma-\tau) \in \Theta^* \quad \underline{r}^*(X(\sigma-\tau)(Y)) = (X(\sigma-\tau)(Y)\%X\sigma(Y))\underline{r}^*(X\sigma(Y)) - (X(\sigma-\tau)(Y)\%X\tau(Y))\underline{r}^*(X\tau(Y)),$$

$$(6)\ \forall(\sigma*\tau) \in \Theta^* \quad \underline{r}^*(X(\sigma*\tau)(Y)) = (X(\sigma*\tau)(Y)\%X\sigma(Y))\underline{r}^*(X\sigma(Y)) \cap (X(\sigma*\tau)(Y)\%X\tau(Y))\underline{r}^*(X\tau(Y)).$$

4. Stepwise Vocabulary Building for Database Queries

Definition 4.1

A stepwise vocabulary building process for a database with a partial relation Δ over a finite attribute set Ω_0 is a sequence of triples $(\Omega_i, \Theta_i, \underline{r}_i)$ defined as follows;

(1) $\Omega_0 \subset \Omega_1 \subset \dots \subset \Omega_n = \Omega^\infty$,

$\phi = \Theta_0, \Theta_1, \dots, \Theta_n, \qquad \Theta^\infty = \bigcup_i \Theta_i$,

$\Delta = \underline{r}_0, \underline{r}_1, \dots, \underline{r}_n = \underline{r}^\infty$,

(2) Θ_i is a set of elementary adjective of Ω_{i-1},

(3) Ω_i is a set $\Omega_{i-1}{}^*$ for $(\Omega_{i-1}, \Theta_{i-1}{}^*, \underline{r}_{i-1})$,

(4) $\underline{r}_i$ is an information space $\underline{r}^*$ of $(\Omega_{i-1}, \Theta_{i-1}{}^*, \underline{r}_{i-1})$.

Then $\underline{r}^\infty$ is called a universal information space of $(\Omega_0, \{\Theta_i\}, \Delta)$ and $\{\Theta_i\}$ is called the stepwise basic vocabularies of adjectives, and Ω^∞ is called the lexicon.

Theorem 4.1

For each $X \subset \Omega^\infty$, $\underline{r}^\infty(X)$ is computable if X and Δ are finite.

(proof)

This can be proved by mathematical induction on rank(X) that is a modified version of the previous definition and is defined as follows;

$\forall\ X \subset \Omega_{0g} \quad \mathrm{rank}(X) = 0,$

$\mathrm{rank}(XY) = \mathrm{rank}(X)+\mathrm{rank}(Y),$

$\forall\ i,\ \forall\theta \in \Theta_i \quad \mathrm{rank}(\theta) = \mathrm{rank}(\underline{A}_\theta)+\mathrm{rank}(\underline{B}_\theta)+1,$

$$\text{rank}(\theta(X)) = \text{rank}(X)+\text{rank}(\theta),$$

$$\text{rank}(X/Y) = \text{rank}(X)+\text{rank}(Y)+1,$$

$$\text{rank}(g(X)) = \text{rank}(X)+1,$$

$$\text{rank}((\sigma^\circ\tau)(X)) = \text{rank}(\sigma(\tau(X)))+1,$$

$$\text{rank}((\sigma\cdot\tau)(X)) = \text{rank}(\sigma(X))+\text{rank}(\tau(X))+1,$$

where $\cdot$ is one of $+$, $-$, $*$,

$$\text{rank}(\theta^-) = \text{rank}(\theta)+1.$$

Here we will list up several important forms of adjectives.

Definition 4.2

$\forall A \in \Omega^\infty \quad \hat{A} = ((x=y)(x;\ y),\ A,\ A)$, i.e., $(A = \hat{A}A)$.

Definition 4.3

$\forall A \in \Omega^\infty \ \forall v \in D_h$

$A^{(\sim v)} = ((y\sim\text{'}v\text{'})(x;\ y),\ \phi,\ A)$, i.e., $(A^{(\sim v)}A\sim\text{'}v\text{'})$,

where $\sim$ is one of the relational operators $=$, $\neq$, $\geq$, $\leq$, $>$, $<$.

Let O be a special attribute that does not belong to Ω^∞. For the sake of convenience, we add O to Ω_0 by extending Δ to a cartesian product of Δ and $U = (\{0\} \to D_h)$, i.e., $\Delta_{new} = \Delta_{old} \times (\{0\} \to D_h)$. Now we are able to define another important type of adjectives.

Definition 4.4

$\forall A \in \Omega^\infty \quad A^O = ((x=y)(x;\ y),\ A,\ O)$, i.e., $(A^O O = A)$.

We are also allowed to define, for each pair of attributes A and B in Ω^∞, their addition, subtraction, and multiplication as below.

Definition 4.5

$\forall A, \ \forall B \in \Omega^\infty$

$$A+B = (A^O+B^O)O,$$

$$A-B = (A^O-B^O)O,$$

$$A*B = (A^O*B^O)O.$$

Theorem 4.2

$\forall A \in \Omega^\infty \ \forall X \subset \Omega^\infty \quad \underline{r}^\infty((A^O Q)X) = \underline{r}^\infty(AX).$

(proof)
Obvious.

A vocabulary for Δ is a subset V of Σ^* and α, where Σ^* is a set of all the finite strings of an alphabet Σ and α is a function from $(V \cup \Omega^\infty \cup \Theta^\infty)$ to $\Omega^\infty \cup \Theta^\infty$ such that $\forall A \in \Omega^\infty \cup \Theta^\infty\ \alpha(A) = A$. We define, for each subset X of V, a relation over X with respect to $(\Omega, \{\Theta_i\}, \Delta)$ as

$$\langle X \rangle = (X \% \alpha(X))\underline{r}^\infty(\alpha(X)).$$

<u>example 4.1</u>

For an example relation with

$$\Omega_0 = \{\text{driver, licence \#, typist, type speed, salary}\},$$

we are able to define the word "employee" as

$$\alpha(\text{employee}) = \text{driver+typist}.$$

For example, <employee, type speed> can be evaluated as

$$
\begin{aligned}
&\langle \text{employee, type speed} \rangle \\
&= (\text{employee, type speed \% (driver+typist), type speed}) \\
&\qquad \underline{r}^\infty(\text{driver+typist, type speed}) \\
&= (\text{employee, type speed \% (driver+typist), type speed}) \\
&\qquad ((\text{driver+typist}), \text{type speed \% } (\text{driver}^0\text{+typist}^0)0, \\
&\qquad \text{type speed})\underline{r}^\infty((\text{driver}^0\text{+typist}^0)0, \text{type speed}) \\
&= (\text{employee, type speed \% } (\text{driver}^0\text{+typist}^0)0, \text{type speed}) \\
&\qquad (((\text{driver}^0\text{+typist}^0)0, \text{type speed} \\
&\qquad\ \ \% \text{ driver}^0 0, \text{type speed})\underline{r}^\infty(\text{driver}^0 0, \text{type speed}) \\
&\qquad \cup ((\text{driver}^0\text{+typist}^0)0, \text{type speed} \\
&\qquad\ \ \% \text{ typist}^0 0, \text{type speed})\underline{r}^\infty(\text{typist}^0 0, \text{type speed})) \\
&= (\text{employee, type speed \% driver, type speed}) \\
&\qquad \underline{r}^\infty(\text{driver, type speed}) \\
&\ \ \cup(\text{employee, type speed \% typist, type speed}) \\
&\qquad \underline{r}^\infty(\text{typist, type speed}) \\
&= (\text{employee, type speed \% typist, type speed}) \\
&\qquad \underline{r}^\infty(\text{typist, type speed}),
\end{aligned}
$$

while <employee, salary> can be evaluated as

$$(\text{employee, salary \% driver, salary})\underline{r}^{\infty}(\text{driver, salary})$$

$$\cup\,(\text{employee, salary \% typist, salary})\underline{r}^{\infty}(\text{typist, salary}).$$

example 4.2

Let us see another database with

$$\Omega_0 = \{\text{employee, department, salary}\},$$

and define an elementary adjective "rich_man" as

rich_man

= (rich_man(salary) > average(salary; employee/ϕ)).

Then, for example, the query that requests the listing of all the departments that have at least one employee whose salary is more than the average of the company can be simply expressed as

<rich_man(department)>,

while the list of all such employees can be requested by

<rich_man(employee)>.

5. Concluding Remarks

The vocabulary building facility concerned in this paper is a new concept of query semantics. It makes a new approach to the effective enhancement of the database usability. The framework developped in this paper has the following features:

(1) Ad hoc vocabulary building is allowed.

(2) As shown in the examples, even a very complicated request is expressed as a very simple query.

(3) It is not necessary for us to describe a virtual access path or an actual one.

(4) It provides generalized projections and generalized restrictions.

(5) If we consider a vocabulary V as an attribute set, we can construct a new vocabulary V* over V. Since the description of V* includes no elements of Ω^{∞}, the definition of V* is independent from the database.

(6) For each pair of different concepts, the framework provides the means to define their common concept, their difference concept, and their union concept.

If a proper part of our common vocabulary used in daily conversation is adequately built into a vocabulary of the system, then our man-machine communication will become much smoother and more reliable. By having a common vocabulary, a man and a machine can communicate even a very complicated command with only a few words. Our approach will open out new vistas of these possibilities.

Vol. 107: International Colloquium on Formalization of Programming Concepts. Proceedings. Edited by J. Diaz and I. Ramos. VII, 478 pages. 1981.

Vol. 108: Graph Theory and Algorithms. Edited by N. Saito and T. Nishizeki. VI, 216 pages. 1981.

Vol. 109: Digital Image Processing Systems. Edited by L. Bolc and Zenon Kulpa. V, 353 pages. 1981.

Vol. 110: W. Dehning, H. Essig, S. Maass, The Adaptation of Virtual Man-Computer Interfaces to User Requirements in Dialogs. X, 142 pages. 1981.

Vol. 111: CONPAR 81. Edited by W. Händler. XI, 508 pages. 1981.

Vol. 112: CAAP '81. Proceedings. Edited by G. Astesiano and C. Böhm. VI, 364 pages. 1981.

Vol. 113: E.-E. Doberkat, Stochastic Automata: Stability, Nondeterminism, and Prediction. IX, 135 pages. 1981.

Vol. 114: B. Liskov, CLU, Reference Manual. VIII, 190 pages. 1981.

Vol. 115: Automata, Languages and Programming. Edited by S. Even and O. Kariv. VIII, 552 pages. 1981.

Vol. 116: M. A. Casanova, The Concurrency Control Problem for Database Systems. VII, 175 pages. 1981.

Vol. 117: Fundamentals of Computation Theory. Proceedings, 1981. Edited by F. Gécseg. XI, 471 pages. 1981.

Vol. 118: Mathematical Foundations of Computer Science 1981. Proceedings, 1981. Edited by J. Gruska and M. Chytil. XI, 589 pages. 1981.

Vol. 119: G. Hirst, Anaphora in Natural Language Understanding: A Survey. XIII, 128 pages. 1981.

Vol. 120: L. B. Rall, Automatic Differentiation: Techniques and Applications. VIII, 165 pages. 1981.

Vol. 121: Z. Zlatev, J. Wasniewski, and K. Schaumburg, Y12M Solution of Large and Sparse Systems of Linear Algebraic Equations. IX, 128 pages. 1981.

Vol. 122: Algorithms in Modern Mathematics and Computer Science. Proceedings, 1979. Edited by A. P. Ershov and D. E. Knuth. XI, 487 pages. 1981.

Vol. 123: Trends in Information Processing Systems. Proceedings, 1981. Edited by A. J. W. Duijvestijn and P. C. Lockemann. XI, 349 pages. 1981.

Vol. 124: W. Polak, Compiler Specification and Verification. XIII, 269 pages. 1981.

Vol. 125: Logic of Programs. Proceedings, 1979. Edited by E. Engeler. V, 245 pages. 1981.

Vol. 126: Microcomputer System Design. Proceedings, 1981. Edited by M. J. Flynn, N. R. Harris, and D. P. McCarthy. VII, 397 pages. 1982.

Voll. 127: Y.Wallach, Alternating Sequential/Parallel Processing. X, 329 pages. 1982.

Vol. 128: P. Branquart, G. Louis, P. Wodon, An Analytical Description of CHILL, the CCITT High Level Language. VI, 277 pages. 1982.

Vol. 129: B. T. Hailpern, Verifying Concurrent Processes Using Temporal Logic. VIII, 208 pages. 1982.

Vol. 130: R. Goldblatt, Axiomatising the Logic of Computer Programming. XI, 304 pages. 1982.

Vol. 131: Logics of Programs. Proceedings, 1981. Edited by D. Kozen. VI, 429 pages. 1982.

Vol. 132: Data Base Design Techniques I: Requirements and Logical Structures. Proceedings, 1978. Edited by S.B. Yao, S.B. Navathe, J.L. Weldon, and T.L. Kunii. V, 227 pages. 1982.

Vol. 133: Data Base Design Techniques II: Proceedings, 1979. Edited by S.B. Yao and T.L. Kunii. V, 229–399 pages. 1982.

Vol. 134: Program Specification. Proceedings, 1981. Edited by J. Staunstrup. IV, 426 pages. 1982.

Vol. 135: R.L. Constable, S.D. Johnson, and C.D. Eichenlaub, An Introduction to the PL/CV2 Programming Logic. X, 292 pages. 1982.

Vol. 136: Ch. M. Hoffmann, Group-Theoretic Algorithms and Graph Isomorphism. VIII, 311 pages. 1982.

Vol. 137: International Symposium on Programming. Proceedings, 1982. Edited by M. Dezani-Ciancaglini and M. Montanari. VI, 406 pages. 1982.

Vol. 138: 6th Conference on Automated Deduction. Proceedings, 1982. Edited by D.W. Loveland. VII, 389 pages. 1982.

Vol. 139: J. Uhl, S. Drossopoulou, G. Persch, G. Goos, M. Dausmann, G. Winterstein, W. Kirchgässner, An Attribute Grammar for the Semantic Analysis of Ada. IX, 511 pages. 1982.

Vol. 140: Automata, Languages and programming. Edited by M. Nielsen and E.M. Schmidt. VII, 614 pages. 1982.

Vol. 141: U. Kastens, B. Hutt, E. Zimmermann, GAG: A Practical Compiler Generator. IV, 156 pages. 1982.

Vol. 142: Problems and Methodologies in Mathematical Software Production. Proceedings, 1980. Edited by P.C. Messina and A. Murli. VII, 271 pages. 1982.

Vol. 143: Operating Systems Engineering. Proceedings, 1980. Edited by M. Maekawa and L.A. Belady. VII, 465 pages. 1982.

Vol. 144: Computer Algebra. Proceedings, 1982. Edited by J. Calmet. XIV, 301 pages. 1982.

Vol. 145: Theoretical Computer Science. Proceedings, 1983. Edited by A.B. Cremers and H.P. Kriegel. X, 367 pages. 1982.

Vol. 146: Research and Development in Information Retrieval. Proceedings, 1982. Edited by G. Salton and H.-J. Schneider. IX, 311 pages. 1983.

Vol. 147: RIMS Symposia on Software Science and Engineering. Proceedings, 1982. Edited by E. Goto, I. Nakata, K. Furukawa, R. Nakajima, and A. Yonezawa. V. 232 pages. 1983.

This series reports new developments in computer science research and teaching – quickly, informally and at a high level. The type of material considered for publication includes preliminary drafts of original papers and monographs, technical reports of high quality and broad interest, advanced level lectures, reports of meetings, provided they are of exceptional interest and focused on a single topic. The timeliness of a manuscript is more important than its form which may be unfinished or tentative. If possible, a subject index should be included. Publication of Lecture Notes is intended as a service to the international computer science community, in that a commercial publisher, Springer-Verlag, can offer a wide distribution of documents which would otherwise have a restricted readership. Once published and copyrighted, they can be documented in the scientific literature.

Manuscripts

Manuscripts should be no less than 100 and preferably no more than 500 pages in length.
They are reproduced by a photographic process and therefore must be typed with extreme care. Symbols not on the typewriter should be inserted by hand in indelible black ink. Corrections to the typescript should be made by pasting in the new text or painting out errors with white correction fluid. Authors receive 75 free copies and are free to use the material in other publications. The typescript is reduced slightly in size during reproduction; best results will not be obtained unless the text on any one page is kept within the overall limit of 18 x 26.5 cm (7 x 10½ inches). On request, the publisher will supply special paper with the typing area outlined.
Manuscripts should be sent to Prof. G. Goos, Institut für Informatik, Universität Karlsruhe, Zirkel 2, 7500 Karlsruhe/Germany, Prof. J. Hartmanis, Cornell University, Dept. of Computer-Science, Ithaca, NY/USA 14850, or directly to Springer-Verlag Heidelberg.

Springer-Verlag, Heidelberger Platz 3, D-1000 Berlin 33
Springer-Verlag, Tiergartenstraße 17, D-6900 Heidelberg 1
Springer-Verlag, 175 Fifth Avenue, New York, NY 10010/USA

ISBN 3-540-11980-9
ISBN 0-387-11980-9

allowing us to find a solution of G_1, then the same sequence of axioms allows us to find a solution of G_2 as well.

Unfortunately the problem of deciding whether two subgoals are equivalent or not is recursive unsolvable one. However, it does not mean that this notion of equivalency is useless for practical applications. In the method of search reduction described below the difficulties connected with recognition of equivalent subgoals are avoided in the following way:

1) The reducing of unnecessary search is organized as in the method of lemmas (not as in the subgoal extraction method). This simplify the matter considerable because it means that the problem of deciding whether subgoals A_1 and A_2 are equivalent or not arises only when the search for solutions of one of them (say, A_1) is completed. This, in turn, means that the complete deduction search tree for A_1 is finite (otherwise a lemma could not be formed). In this case the problem of equivalency become recursive solvable (although the straightforward deciding algorithm is not suitable for practical applications).

2) A sufficient, but not necessary condition of equivalency is used.

The idea of the method is as follows. In essence we must be able to decide whether a given subgoal A? is an instance of a subgoal equivalent to some known lemma. To do it, after completing the search for solutions of some subgoal G? we will attempt to determine the most general subgoal U_G? equivalent to G? (so G is always an instance of U_G, but U_G, generally speaking, is not an instance of G). Then a lemma consisting of complete set of solutions of U_G? may be formed. The use of such lemmas practically will not differ from the use of standard lemmas. But process of forming lemmas becomes essentially more complicated.

The simplest way to construct U_G is to substitute fresh variables for all subterms from G which have not been examined by the unification algorithm during the search for solutions of G? (really, such subterms can not affect the search process). More thorough analysis shows that even subterms that have been examined by the unification algorithm may be substituted by variables if all these unifications were successful; only subterms which had an influence on the unsuccessful termination of some unification must be kept intact.

Let us describe an algorithm for determining U_G more precisely.

Definition. Let (P, Q) be a pair of non-unifiable atoms. An atom P_1 is said to be a *localization of unification failure* of this pair if P is an instance of P_1 and the pair (P_1, Q) is not unifiable either (of course, P_1 is not determined by these conditions unequivocally).

Algorithms for finding P_1 (note that it is desirable to find the most general atom satisfying these conditions) were studied in connection with so-called "intelligent backtracking" methods. We shall not consider them here (generally speaking, such an algorithm must retain in P_1 only those subterms from P that are "responsible" for unsuccessful termination of the unification and substitute all other subterms by fresh variables).

When a new subgoal G? appears, we define $U_G=f(X_1,\dots X_k)$?, where f is the predicate symbol of G (k-place) and X_1, ..., X_k are variables. During the search for solutions of G? the atom U_G is modified in the following way:

1) If unification of G and P, where P is the head of an axiom, terminated unsuccessfully, then U_G is replaced with unif(U_G, G_1), where G_1 is a localization of the unification failure of the pair (G, P). Note that U_G and G_1 must be unifiable because they have a common instance, namely G.

2) Suppose that $P_1\varphi$? appeared as a result of applying an axiom P_1 & P_2 & ... & $P_n \rightarrow P$ to G?, where φ is the most general substitution unifying G and P; suppose also that the search for solutions of $P_1\varphi$? has been terminated, i.e. $U_{P_1\varphi}$ is definitively determined. Let ψ be the most general substitution such that $P_1\psi$=unif($U_{P_1\varphi}$, P_1), and let Q be the most general atom such that unif(P,Q) is an instance of $P\psi$ ("most general" here means that if Q_1 is another atom with the same property and Q is an instance of Q_1, then Q_1 must be a variant of Q). Then U_G is replaced with unif(U_G, Q).

3) Suppose that an axiom P_1 & P_2 & ... & $P_n \rightarrow P$ was applied to G?, subgoals $P_1\varphi_1$? ,..., $P_k\varphi_k$? (descended from atoms P_1, ..., P_k) appeared, and solutions ψ_1, ..., ψ_k for these subgoals were found (here φ_1 is the most general unifier for P and G, $\varphi_2=\varphi_1\psi_1$, ..., $\varphi_k=\varphi_{k-1}\psi_{k-1}$). Suppose also that the search for solutions of the subgoal $P_{k+1}\varphi_{k+1}$? ($\varphi_{k+1}=\varphi_k\psi_k$) arisen from the atom P_{k+1} was terminated. Then U_G is modified in the same way as it would be modified according to point 2, if the subgoal $P_{k+1}\varphi_{k+1}$? had arisen as a result of applying the axiom $P_{k+1}\tau$ & ... & $P_n\tau \rightarrow P\tau$ to the subgoal G?. Here τ is the most general substitution such that the atoms $P_1\tau$, ..., $P_k\tau$ are instances of $P_1\varphi_2$, ..., $P_k\varphi_{k+1}$ respectively.

Theorem. Let U_G be determined after completing the search for solutions of G? as it was described above. Then

1) G is an instance of U_G.

2) Subgoals G? and U_G? are equivalent.

To form a lemma from the atom U_G we must know a complete set of solutions of U_G?. Generally speaking, it can not be derived from a complete set of solutions of G?. However, it can easily be derived from a complete search tree for G? because complete search trees for equivalent subgoals are isomorphic; we shall not dwell on details here.

It is interesting to compare the proposed method for reducing search with the intelligent backtracking methods (e.g. [2], [10]) which are also intended for reducing unnecessary search during execution of logic programs and use a similar technique (determination of causes of unification failures and so on). The intelligent backtracking methods are based on the following idea. Let a clause $\neg A_1 \vee \neg A_2 \vee \ldots \vee \neg A_n$ be given. Suppose we have found a solution φ of A_1? and so have derived the clause $\neg A_2\varphi \vee \ldots \vee \neg A_n\varphi$. Suppose further that the subgoal $A_2\varphi$? has no solutions. In this situation the Prolog inference engine resumes the search for solutions of A_1?. However, it may appear that even the subgoal A_2? has no solutions. In this case the search for new solutions of A_1? is useless, and the intelligent backtracking methods may try to backtrack to some earlier point (e.g. to the point where a variable binding responsible for an unification failure during the execution of $A_2\varphi$? was produced).

It is easy to see that the proposed method also enables us to attain the similar search reduction. Really, if φ had no influence on the failure in the search for solutions of $A_2\varphi$?, then A_2 is likely to be an instance of $U_{A_2\varphi}$. When it becomes known that $U_{A_2\varphi}$? has no solutions, we can discard all clauses containing an instance of $U_{A_2\varphi}$? (in particular, all clauses containing A_2 or its instance). The use of this "deletion strategy" alone is sufficient to gain no less search reduction than the intelligent backtracking methods can ensure.

In fact, the described method ensures even greater search reduction. First, not only subgoals having no solutions are used for reducing unnecessary search but all other subgoals are used as well. Second, information which is stored in the form of lemmas has a simple logical meaning (complete sets of solutions of several subgoals become known) and can be used throughout the deduction. Information yielded by the intelligent backtracking methods, on the contrary, has local and control nature (in a particular deduction branch backtracking may be performed to a particular point); this information can not be reused in other deduction branches. At last, the proposed method has a wider scope of application because it does not require the process of deduction to be organized in the form of search with backtracking.

However, the overhead associated with this method may be greater than that associated with the intelligent backtracking methods, because we must store all lemmas and determine whether new subgoals are instances of existing lemmas. On the other hand, if one of the methods of the preceding section (e.g. the standard method of lemmas or the subgoal extraction method) is already exploited, then the additional overhead associated with the proposed method is not large.

In conclusion let us notice that a method for reducing search exists which seems to be more interesting than the ones described here, namely, the method which is organized as the subgoal extraction method (not as the method of lemmas) but based on the notion of equivalency described in this section. Such method, preserving completeness of search for solutions, would appear at the same time a powerful means for elimination of loops. More precisely, not only "true loops" but in some cases diverging behaviour of logic programs may also be prevented. Thus, the execution of the example given in the beginning of this section would terminate. In spite of recursive unsolvability of the problem of recognition of equivalent subgoals in case when for none of them the search for solutions is completed, such method may be constructed. The main idea is as follows: we can made assumptions about equivalency of subgoals using some necessary (but not sufficient) condition of equivalency, and reconsider these assumptions every time when two subgoals considered as equivalent proved to be non-equivalent (for more detailes see [9]).

References

1. Clocksin W.F., Mellish C.S. Programming in Prolog. Springer-Verlag, 1981.

2. Cox P.T. Finding backtrack points for intelligent backtracking. In: Implementations of Prolog, 1984, p.216-233.

3. Hirakawa H., Onai R., Furukawa K. Or-parallel optimizing Prolog system: POPS. Lecture Notes in Computer Science, 1986, v.220, p.114-129.

4. Kowalski R. Logic for problem solving. North-Holland, 1979.

5. Loveland D.W. An implementation of the model elimination proof procedure. Journal of the ACM, 1974, v.21, p.124-139.

6. Neiman V.S. Deduction search with single consideration of subgoals. Soviet Math. Dokl., 1986, v.33, p.251-254.

7. Neiman V.S. The system PROVE - implementation of the subgoal extraction method. Preprint Institute of Theoretical Astronomy. Leningrad, 1986 (in Russian).

8. Neiman V.S. Refutation search for Horn sets by subgoal extraction method. Journal of Logic Programming, 1989 (to appear).

9. Neiman V.S. Using partially defined terms in logic inferences. Proc. of "Theory and applications of Artificial Intelligence". Bulgaria, Sozopol, 1988, v.2, p.241-248 (in Russian).

10. Pereira L.M., Porto A. Selective backtracking. In: Logic programming. Academic Press, 1982, p.107-114.

11. Tamaki H., Sato T. OLD resolution with tabulation. Lecture Notes in Computer Science, 1986, v.225, p.84-98.

12. Wolfram D.A., Maher M.J., Lasser J.L. A unified treatment of resolution strategies for logic programs. In: Second Int. Logic Programming Conf., 1984, p.263-276.

CORRECTNESS OF SHORT PROOFS IN THEORY WITH NOTIONS OF FEASIBILITY

V.P. Orevkov
Leningrad Department of Steklov Mathematical
Institute of the Academy of Sciences of the USSR
Fontanka embankment 27, Leningrad 191011, USSR

The present work is a strengthening and generalization of results of Dragalin [1], Gavrilenko [2], Parikh [7] in various directions. We consider the predicate calculus with equality. We get more simple bounds on the complexity of proofs, namely, our bounds depend essentially only on the number of those members in a deduction under consideration which contain the symbol of feasibility, and do not depend on the complexity of formulas in this deduction.

1. Let T be an arbitrary theory, containing Peano's arithmetic PA (in the language with =,<,0,',+ and *). Let θ be some fixed variable free term. T_θ is obtained by adding a new one-place predicate symbol F and the following postulates:

A. Classical or intuitionistic Hilbert-type system for first-order predicate calculus with equality (in the language with F).

B. All non-logical axioms of T (in the language without F).

C. Special axioms:

1) F(0),
2) ∀x(F(x)->F(x')),
3) ∀xy(F(x)->(F(y)->F(x+y))),
4) ∀xy(F(x)->(F(y)->F(x*y))),
5) ∀xy(x<y->(F(y)->F(x))),
6) ~F(θ).

We suppose that our Hilbert-type system contains the rule Gen:

$$\frac{A(x)}{\forall x A(x)},$$

the predicate axioms

∀xA(x)->A(t), A(t)->(Ex)A(x),

∀x(C->A(x))->(C->∀xA(x)), ∀x(A(x)->C)->((Ex)A(x)->C),

where the variable x do not occur free in C, and the axioms for equality

$\forall x(x=x)$, $\forall xy(x=y\to(A(x)\leftrightarrow \underline{A}(y)))$.

Let D be a derivation in T_θ which is written in linear form. Denote by $l_F[D]$ the number of different formulas containing F in D. The function 2^n_i is defined by the recursion equations $2^{n+1}_0=n+1$ and $2^n_{i+1}=2^{(2^n_i)}$.

<u>Theorem 1</u>. Let D be a derivation in T_θ of the formula A, A does not contain F and $n=l_F[D]$. If

$$T \vdash 2^{2n+9}_{3n+4} < \theta$$

then

$$T \vdash A.$$

This theorem is an immediate consequence of the lemma below.

2. Let D be derivation free in the Gentzen-type variant of T_θ. A logical inference L in D is called F-<u>inference</u> if the principal formula of L contains F. A cut in D is called F-<u>cut</u> if the cut formula contains F. Denote by $h_F[D]$ the maximum number of inferences of F in a branch of D (maximum over all branches). D is said to be F-<u>regular</u> if in any branch of D every logical inference above a propositional F-inference is a propositional F-inference.

<u>Lemma 1</u>. Let D be a derivation in T_θ of the formula A, $n=l_F[D]$ and Γ be the list of non-logical and special axioms occurring in D. Then there is a F-cut-free derivation tree D' of the sequent $\Gamma\Rightarrow A$ such that

$$h_F[D'] \leq 2^{2n+9}_{3n+1}$$

<u>Lemma 2</u>. Let D be a F-cut-free derivation tree of the sequent $\Gamma\Rightarrow A$ in the Gentzen-style variants of T_θ, where Γ is list of non-logical and special axioms, and A does not contain F. There is a F-cut-free F-regular derivation tree D' of $\Gamma\Rightarrow A$ such that

$$h_F[D'] \leq 2^{h_F[D]}$$

<u>Lemma 3</u>. Let Σ be a list of atomic formulas, t be a term and Γ be a list of substitution instances of formulas

$(F(x)\to F(x'))$,

$(F(x)\to(F(y)\to F(x+y)))$,

$(F(x)\to(F(y)\to F(x*y)))$,

$(x<y\to(F(y)\to F(x)))$.

If Γ, $F(0)$, $\Sigma\Rightarrow F(t)$ is provable in the predicate calculus with equality and Σ does not contain F then Γ, $\Sigma\Rightarrow t<2^m_2$ is provable in PA, where m is the number of formulas in Γ.

3. Let θ_0 be a term containing 0 and the functional symbols ',+,*. We make the following definition:

$$T_0 \leftrightharpoons PA_{\theta_0}.$$

The <u>rank</u> of the induction axiom

$(A(0)\&\forall x(A(x)\to A(x')))\to A(t)$

is the largest number of occurrences of variable x in the formula A(x). Definitions of the rank of predicates axioms and axioms for equality are analogous. Let D be a derivation in T_0, which is written in linear form. The <u>rank</u> of D is the largest rank of induction axioms, predicates axioms and axioms for equality in D.

Let t be a term. Then h[t] is defined as follows: if t is a variable or a zero-place functional symbol then h[t]=0,

$$h[g(t_1,\dots,t_k)]=1+\max_{1\le i\le k} h[t_i].$$

If A is a formula, let

h[A]=max{h[t]: term t occurs in A}.

Theorem 2. Let D be a derivation in T_0 of a formula A, A does not contain F, r be the rank of D, l be the number of formulas in D and h=h[A]. If

$$PA\vdash 2^m_2<\theta_0,$$

where

$$m=(2r+3)^{5r^2l^2}(h+4)^{2rl}$$

then

$$PA\vdash A$$

In the proof of Theorem 2 we use Theorem 1 and the upper bound on the length of functional periodicity in derivations (cf.[3]).

References

1. Dragalin A.G.: Correctness of inconsistent theories with notions of feasibility. Lect. Notes Comp. Sci. (1985), v.108, 58-79
2. Gavrilenko Yu.V. (Гавриленко Ю.В.): Монотонные теории достижимых чисел. Докл. АН СССР, 1(1984), т.276, 18-22
3. Makanin G.S.: (Маканин Г.С.) Проблема разрешимости уравнений в свободной полугруппе. Матем. сб., 2(1977), т.103, 147-236
4. Orevkov V.P.: (Оревков В.П.) Верхние оценки удлинения выводов при устранении сечений. Зап. научн. семин. ЛОМИ, (1984), т.137, 87-98
5. Orevkov V.P.: (Оревков В.П.) Верхние оценки удлинения выводов при устранении сечений в исчислении предикатов с равенством. Всесоюзн. конференция по прикл. логике. Тезисы докладов. Новосибирск, (1985), 167-168
6. Orevkov V.P.: (Оревков В.П.) Оценки сложности вывода в теории достижимых чисел. УШ Всесоюзн. конференция по математической логике. Тезисы докладов. Москва, (1986), с.146
7. Parikh R.: Existence and feasibility in arithmetic. J. Symbolic Logic, 3(1971), v.36, 494-508

A Formulation of the Simple Theory of Types (for Isabelle)

Lawrence C. Paulson
Computer Laboratory, University of Cambridge
Pembroke Street, Cambridge CB2 3QG, England

Abstract

Simple type theory is formulated for use with the generic theorem prover Isabelle. This requires explicit type inference rules. There are function, product, and subset types, which may be empty. Descriptions (the η-operator) introduce the Axiom of Choice. Higher-order logic is obtained through reflection between formulae and terms of type *bool*. Recursive types and functions can be formally constructed.

Isabelle proof procedures are described. The logic appears suitable for general mathematics as well as computational problems.

1 Introduction

Isabelle is a theorem prover for various logics, including several first-order logics, Martin-Löf's Type Theory, and Zermelo-Fraenkel set theory [28,29]. Each new logic is formalized within Isabelle's meta-logic. New types and constants express the syntax of the logic, while new axioms express its inference rules.

The present formulation of simple type theory (also called higher-order logic) may interest logicians. It also illustrates Isabelle applied to an area where hard choices must be made. The simple theory of types is *too* simple; it can be enriched in numerous ways, as hinted by Church [5]. The traditional formalization, with implicit type constraints, goes beyond Isabelle's view of syntax. Isabelle supports ML-style type inference with unification; the formulation offers a limited degree of polymorphism.

There are several reasons for implementing higher-order logic in Isabelle.

Gordon and others have used higher-order logic, with great success, for hardware verification [6,14]. They have developed a theorem prover called HOL, based on LCF. Another implementation of higher-order logic is TPS [2]. How well does Isabelle perform against specialized systems like these?

Zermelo-Fraenkel set theory [36] is intended as a foundation of mathematics, but is inconvenient for formal proof — even set theorists use intuition and diagrams. Yet set theory is the basis of the Z specification language [35]. Philippe Noël has conducted extensive set theory proofs in Isabelle [25]. Type theory is also intended as a foundation for mathematics, and seems nearly as powerful. How does it compare with set theory as a practical formal language?

Isabelle's meta-logic is a fragment of Church's version of type theory [28]. It is natural to ask whether the meta-logic can be formalized in itself. Actually, the object version of higher-order logic has to be much larger than the meta-logic because it is intended for expressing all kinds of mathematics. The meta-logic only has to express other formal systems.

The Isabelle implementation is definitely not intended for teaching Church's notation. Here TPS is the champion, with a special character set for Church's subscripted Greek letters. However, most modern authors write $x : \alpha$ and $\alpha \rightarrow \beta$ rather than x_α and $\beta\alpha$. Church's axiom system is now antiquated, largely dating back to *Principia Mathematica*. There are improved formulations but most use the Hilbert style. Natural deduction is far superior for automated proof.

Sections 2 and 3 of the paper introduce Isabelle and the theory of types. Sections 4 to 10 discusses some issues in the Isabelle formulation, while Section 11 presents the formulation itself. The remaining sections describe proof procedures and offer some conclusions. The computer file containing the rules is included as an appendix.

2 Overview of Isabelle

Isabelle represents its object-logics within a fragment of intuitionistic higher-order logic including implication, universal quantifiers, and equality [28]. The implication $\phi \Longrightarrow \psi$ means 'ϕ implies ψ', and expresses logical entailment. The quantification $\bigwedge x \,.\, \phi$ means 'ϕ is true for all x', where x has a fixed type, and expresses generality in object-rules and axiom schemes. The equality $a \equiv b$ means 'a equals b', and expresses definitions.

The meta-logic includes the typed λ-calculus, which is convenient for formalizing the syntax of object-logics, particularly variable binding. Provisos of quantifier rules (of the sort 'x not free in the assumptions') are enforced by meta-level quantification.

Like in LCF [30], backwards proofs are developed using tactics and tacticals, which are implemented using Standard ML. But an inference rule in LCF is a function from the premises to the conclusion, while in Isabelle it is an axiom in the meta-logic stating that the premises imply the conclusion.

Since Isabelle axioms are essentially Horn clauses, the proof techniques draw ideas from PROLOG. Huet's higher-order unification procedure [18] takes account of α, β, and η-conversions during unification. Higher-order unification can return multiple or infinitely many results. While the general problem is undecidable, the procedure works well in Isabelle.

3 A brief history of type theory

Bertrand Russell invented the theory of types to resolve the paradoxes in the foundations of mathematics.

One pillar of type theory is that functions differ from individuals. There are also functions of functions, etc., giving a hierarchy of types. To start, there is a type of individuals (Church's ι) and a type of propositions (Church's o). If α and β are types then so is $\alpha \rightarrow \beta$, the type of functions from α to β. A statement is meaningless unless it obeys the type constraints: a function of type $\alpha \rightarrow \beta$ can only be applied an object of type α. The logical constants are functions over the type of propositions. In $\forall x$ and $\exists x$ the variable x must range over some type.

Russell's other pillar was the *vicious circle principle*, which concerned statements like 'all propositions are either true or false' [37, page 37]. If this were itself a proposition then it would refer to itself, a possibly dangerous circularity. The vicious circle principle forbade such 'impredicative propositions' through a system of orders. A statement about all propositions of order n was itself a proposition of order $n+1$. Whitehead and Russell showed how this ramified type theory resolved the paradoxes, but it was too weak to justify classical mathematics. They were forced to assume the Axiom of Reducibility, squashing the orders down to one.

Simple type theory remains when the vicious circle principle is abandoned. Although sets must be introduced in a strict hierarchy, propositions need not be. The idea of orders appears today in the *universes* of Martin-Löf's Type Theory [22], where propositions are represented by types. The terminology persists: simple type theory is called *higher-order logic* because it permits unrestricted quantification over propositions of all orders.

The main achievement of Church [5] is a precise formulation of the syntax. (Gödel calls the vague syntax in *Principia* 'a considerable step backwards as compared with Frege' [13, page 448].) Church formalizes syntax, including quantifiers, in the typed λ-calculus. His technique is now standard in generic theorem proving.

See Hatcher [16] and Gödel [13] for further discussion of the history of these type theories, and Andrews [1] for the formal development.

4 Fundamental issues in type theory

The following sections discuss basic issues in type theory: subtypes, description operators, empty types, polymorphism, and higher-order reasoning. Here, we consider the semantics and syntax.

We begin with basic notation and conventions for syntactic variables:

- *types* are Greek letters α, β, γ, ...
- *bound variables* are x, y, z
- *terms* are a, b, c, d, ...; the ordered pair of a and b is $\langle a, b\rangle$; the relation $a : \alpha$ means 'a has type α'
- *formulae* are P, Q, R, ...; the true formula is $\top$; the false formula is $\bot$

4.1 Types as sets

Type theory is intended as the foundation of mathematics, but it has a simple interpretation in set theory. Types denote sets, abstractions denote set-theoretic functions, and the typing relation denotes set membership. The present formulation retains this semantics. There is no clear alternative: most mathematical reasoning involves sets.

Many theorem provers perform type checking with parsing, but Isabelle cannot do this for its object-logics. Type errors are detected much later: during proofs. Proofs must include explicit type checking using type inference rules, like in Martin-Löf's Type Theory. Type symbols appear as extra arguments to constants.

Ill-typed terms can be written. Their value can vary among models because it is not determined by the axioms. There is no special 'undefined' value. Similarly, an ill-typed formula has some truth value. If this seems unsatisfactory, observe that a traditional theory of Peano arithmetic specifies no value for division by zero, yet the term $a/0$ denotes some number in each model, for each a.

Of course, there are alternative semantics. Fourman and Scott [11,33] can reason about whether a/b exists, but their *existence predicate* involves some complexity. Their logic has a topos semantics, which is a categorical generalization of set theory. Martin-Löf's Type Theory [22] has a constructive, operational semantics. An ambitious type theory can even be based on classical sets: Borzyszkowski et al. formalize general products, some domain theory, and types of types [10]. If the present logic seems pedestrian compared with these, remember that it claims to express the Simple Theory of Types.

4.2 Variable binding and substitution

Isabelle has a typed λ-calculus at the meta-level to deal with operators, variable binding, and substitution uniformly.[1] All forms of variable binding — abstraction, quantifiers, descriptions — are expressed through meta-level abstraction. Compound expressions like $\mathrm{fst}(a)$ and $P \,\&\, Q$ are expressed through meta-level application. Further details are discussed elsewhere [28]. It may be simpler to regard all compound forms, variable-binding or not, as primitive. When a term is written as $b(x)$, this can be regarded as setting off the occurrences of x, so that $b(a)$ nearby indicates substitution. In fact, $b(x)$ is a meta-level application, and substitution takes place by meta-level β-reduction.

Church represents syntax in the object-level typed λ-calculus. His formulation defines λ-abstraction and application, then uses these to express quantifiers and descriptions.

Meta-abstraction works better than object-abstraction in Isabelle. It is also more modular. With meta-abstraction always available, different fragments of the logic can be understood independently of its internal notion of function.

A meta-level function may not correspond to any object-level function. For example, the pairing function is defined for all values of all types. It is defined over the whole object-level domain. It cannot be an object-level function, but could be represented by a family of object-level functions of various types. This must be borne in mind when comparing the present formulation with Church's.

Notation for object-level functions. Abstraction is written $\lambda x : \alpha \,.\, b(x)$. Application is written with the explicit 'apply' operator (‘), as in f ‘ a. The apply operator is not used in general discussions of simple type theory.

5 Subtypes

A subtype is a collection of the elements of a type that share some common property. Typically, a subtype defines an abstract type from a type of representations. The abstract type contains just the elements that represent abstract objects. Any predicate $P(x)$ over a type α defines a subtype $\{x : \alpha \,.\, P(x)\}$. Subtypes usually make type checking undecidable, for checking whether a belongs to $\{x : \alpha \,.\, P(x)\}$ requires proving $P(a)$.

Consider defining the sum type $\alpha + \beta$. The left injection of a can be represented by the pair of abstractions

$$\langle(\lambda x : \alpha \,.\, a = x), (\lambda y : \beta \,.\, \bot)\rangle$$

[1] This works like Martin-Löf's system of arities [26].

while the right injection of b can be represented by

$$\langle(\lambda x : \alpha . \bot), (\lambda y : \beta . b = y)\rangle$$

If formulae are terms of type *bool*, both injections have type $(\alpha \to bool) \times (\beta \to bool)$. The subtype containing just the injections is the sum type $\alpha + \beta$.

Because a subtype may depend upon bound variables, we must consider introducing dependent types: general products and sums. These cause no semantic difficulties, but seem unnecessary (see also Dana Scott [33]). The term

$$\lambda z : \alpha . \lambda y : \{x : \alpha . R(z, x)\} . y$$

has no legal type. Its type could be $\prod_{z:\alpha} . \{x : \alpha . R(z, x)\} \to \alpha$ if we added dependent types. However

$$\lambda z : \alpha . \exists y : \{x : \alpha . R(z, x)\} . P(y)$$

has type $\alpha \to bool$ because the body of the abstraction is a formula.

Traditionally a term has a unique type, but each element of a subtype also belongs to its parent type. Gödel [13, page 466] describes uniqueness of types as a suspect principle, noting that it precludes reasoning about types. Such reasoning is a necessity for Isabelle. Allowing types to overlap causes no difficulty in the set-theoretic semantics.

Gordon's HOL has a different treatment of subtypes (see Melham [23]). To keep uniqueness of types and decidable type checking, conversion functions distinguish elements of a subtype from elements of the parent type. Determining that the conversion functions are applied correctly still requires theorem proving. Subtypes are defined by top-level commands, so there is no question of dependent types. Subtypes must be non-empty.

6 Descriptions

Descriptions, present in type theory from the beginning, name an object by a defining property. The unique description $\iota x : \alpha . P(x)$ means 'the x satisfying $P(x)$'. For example, $\sqrt{a}$ is $\iota x : nat . x^2 = a$. The inference rule verifies that there exists a unique value that satisfies $P(x)$:

$$\frac{\exists x : \alpha . P(x) \;\&\; (\forall y : \alpha . P(y) \supset y = x)}{P(\iota x : \alpha . P(x))}$$

Descriptions can also embody the Axiom of Choice. Hilbert's ϵ-operator, written $\epsilon x : \alpha . P(x)$, means 'some x satisfying $P(x)$'. The rule drops the requirement of uniqueness.

$$\frac{\exists x : \alpha . P(x)}{P(\epsilon x : \alpha . P(x))}$$

Replacing the premise $\exists x : \alpha \,.\, P(x)$ by the two premises $a : \alpha$ and $P(a)$ would impose the stronger requirement (especially in classical logic) of exhibiting the term a.

Gordon's HOL uses Hilbert's ϵ-operator, while Church [5] formalizes both forms of description. See Leisenring [21] for a full discussion of Hilbert's ϵ-operator.

6.1 Descriptions in *Principia Mathematica*

Whitehead and Russell (in Chapter III of *Principia*) argue that descriptions are meaningless by themselves. They give translations to eliminate descriptions from statements. In their view *The author of Waverley was a poet* means Waverley was written by some poet, not that Sir Walter Scott was a poet. This is because if *The author of Waverley* denotes Sir Walter Scott, then *Scott is the author of Waverley* means the same as *Scott is Scott*, which cannot be intended. Girard calls this the question of *sense vs denotation* [12]. The question applies broadly, not just to descriptions: does $2 + 2 = 4$ mean simply $4 = 4$?

If there is no object meeting the description, its meaning is problematical. To Whitehead and Russell, a statement like *The present King of France is bald* is simply false, while the meaning of *The present King of France is not bald* depends upon the scope of the *not*.

The modern view is that a description denotes some object satisfying the given property, if there is one. Otherwise it is undefined — however we understand this.

6.2 The Axiom of Choice and classical logic

In higher-order logic, the Axiom of Choice implies the excluded middle. The argument, due to Diaconescu, is sketched by D. Scott [33]. Let *two* be the type whose values are 0 and 1. To derive $P \vee \neg P$ for some formula P, define type *set* as the following subtype of $two \rightarrow bool$:

$$set \equiv \{q : two \rightarrow bool \,.\, \exists x : two \,.\, q(x) \,\&\, ((\forall x : two \,.\, q(x)) \leftrightarrow P)\}$$

Note that q_0 and q_1 belong to type *set*, where

$$q_0 \equiv \lambda x : two \,.\, (x = 0) \vee P$$
$$q_1 \equiv \lambda x : two \,.\, (x = 1) \vee P$$

Informally, q_0 and q_1 correspond to the sets $\{0, 1?\}$ and $\{1, 0?\}$, where 0? and 1? are included just if P holds. Each element of *set* contains some $x : two$. By the Axiom of Choice (Hilbert's ϵ-operator) there is a corresponding function $f : set \rightarrow two$:

$$f \equiv \lambda q : set \,.\, \epsilon x : two \,.\, q(x)$$

Whether $f(q_0) = f(q_1)$ holds or not is decidable, for it is an equality between natural numbers. And this equality decides P:

- If $f(q_0) = f(q_1)$ then P. By the rule for descriptions, both $q_0(f(q_0))$ and $q_1(f(q_1))$ hold, namely $(f(q_0) = 0) \vee P$ and $(f(q_1) = 1) \vee P$. There are four subcases, of which three imply P and one implies $0 = 1$.
- If $f(q_0) \neq f(q_1)$ then $\neg P$. Assuming that P holds implies $q_0 = \lambda x : two\,.\,\top = q_1$, and so $f(q_0) = f(q_1)$, contradiction.

7 Empty types

Traditional formulations of higher-order logic require that all types are non-empty. This hardly matters for Church, who has only two basic types (propositions and individuals). With many basic types, the requirement becomes unnatural. If the subtype $\{x : \alpha.P(x)\}$ depends on free variables, it could sometimes be empty. Empty types require careful formulation of quantifiers and descriptions.

Isabelle's meta-logic uses implicit type checking and does not admit empty meta-types. Empty types are not needed at the meta-level.

7.1 Church's formulation of quantifiers

Church postulates a supply of variables $x_\alpha, y_\alpha, z_\alpha, \ldots$ for each type α. Typical quantifier rules are (with the usual variable restrictions)

$$\frac{P(x_\alpha)}{\forall x_\alpha\,.\,P(x_\alpha)}\ \forall\text{-intr} \qquad\qquad \frac{\forall x_\alpha\,.\,P(x_\alpha)}{P(a_\alpha)}\ \forall\text{-elim}$$

$$\frac{P(a_\alpha)}{\exists x_\alpha\,.\,P(x_\alpha)}\ \exists\text{-intr} \qquad\qquad \frac{\exists x_\alpha\,.\,P(x_\alpha) \qquad \begin{matrix}[P(x_\alpha)]\\ Q\end{matrix}}{Q}\ \exists\text{-elim}$$

In the $\forall$-intr rule x_α is a variable of type α. In the $\forall$-elim rule, a_α is any term of type α, even a variable. This rule and $\exists$-intr are unsound if α is empty, with many false consequences:

$$\neg(\forall x_\alpha\,.\,\bot) \qquad \exists x_\alpha\,.\,\top \qquad (\forall x_\alpha\,.\,P(x_\alpha)) \supset (\exists x_\alpha\,.\,P(x_\alpha))$$

These look like trivial theorems of first-order logic, but a first-order domain may not be empty.

7.2 Quantifier rules admitting empty types

Explicit type checking admits empty types — almost by accident. The quantifier rules are

$$\frac{\begin{array}{c}[x:\alpha]\\ P(x)\end{array}}{\forall x:\alpha\,.\,P(x)}\ \forall\text{-intr} \qquad\qquad \frac{\forall x:\alpha\,.\,P(x) \qquad a:\alpha}{P(a)}\ \forall\text{-elim}$$

$$\frac{P(a) \qquad a:\alpha}{\exists x:\alpha\,.\,P(x)}\ \exists\text{-intr} \qquad\qquad \frac{\exists x:\alpha\,.\,P(x) \qquad \begin{array}{c}[x:\alpha,\ P(x)]\\ Q\end{array}}{Q}\ \exists\text{-elim}$$

There is no supply of typed variables; instead, $\forall$-intr and $\exists$-elim discharge the assumption $x:\alpha$. The rules $\forall$-elim and $\exists$-intr demand a proof of $a:\alpha$; if a is some variable y, the proof will depend on the assumption $y:\alpha$. If there is no closed term of type α then $\exists x:\alpha\,.\,\top$ has no proof, but $\forall y:\alpha\,.\,\exists x:\alpha\,.\,\top$ does:

$$\frac{\dfrac{\top \qquad [y:\alpha]}{\exists x:\alpha\,.\,\top}}{\forall y:\alpha\,.\,\exists x:\alpha\,.\,\top}$$

7.3 Descriptions and empty types

Because it is always defined, Hilbert's ϵ-operator gives every type α the element $\epsilon x:\alpha.\top$. Another form of description, the η-operator, permits empty types. If there is no $x:\alpha$ satisfying $P(x)$ then $\eta x:\alpha\,.\,P(x)$ is *undefined* — its value and type are unspecified. This typing rule for descriptions makes type checking undecidable.

Hilbert's ϵ-operator can express the quantifiers: for example, $\exists x:\alpha\,.\,P(x)$ as $P(\epsilon x:\alpha\,.\,P(x))$. This does not work with the η-operator, for if $\exists x:\alpha\,.\,P(x)$ is false then $P(\eta x:\alpha\,.\,P(x))$ is meaningless.

The ϵ-operator can be defined through the η-operator if α is non-empty, because $Q\supset\exists x:\alpha\,.\,P(x)$ implies $\exists x:\alpha\,.\,Q\supset P(x)$ under classical logic.[2] Putting $\exists y:\alpha\,.\,P(y)$ for Q, the body of the η can always be satisfied:

$$\epsilon x:\alpha\,.\,P(x) \quad\equiv\quad \eta x:\alpha\,.\,(\exists y:\alpha\,.\,P(y))\supset P(x)$$

7.4 Alternative formulations

The above quantifier and description rules are adopted for the present formulation of simple type theory. Here are two other ways — both based on topos theory — of admitting empty types.

[2]Classical logic obtains by Diaconescu's argument, for η implies the Axiom of Choice.

- Fourman [11] and Dana Scott [33] formalize the notion of existence: a term can have a valid type and yet be undefined. A type is empty if it has no defined elements. The $\forall$-elimination rule can only be applied to a defined term. The description $\iota x : \alpha . P(x)$ exists only if $P(x)$ is satisfied by a unique value, but always has type α.

- Lambek and P. J. Scott [20, page 130] present quantifier rules that maintain a list of the typed variables on which the conclusion depends.

8 Polymorphism

The 'typical ambiguity' in *Principia* is a form of polymorphism where type symbols in expressions are simply not shown. Type checking in ML is a formal version of the same thing: types are inferred but not shown. Isabelle does not (at present) allow any hiding of syntax. This calls for another kind of polymorphism, where certain constants have no type symbols at all. Let us consider how to minimize type symbols in elements of function, product, and sum types. To do this safely, we must remember the semantics.

8.1 Functions

The notation for abstraction could have type symbols for the function's domain and range:

$$\lambda_{\alpha,\beta} x . b(x) \quad : \quad \alpha \to \beta$$

where $b(x) : \beta$ for $x : \alpha$.

The type symbol α is essential. Functions like $\lambda x . 0$ and $\lambda x . x$ cannot be interpreted as sets without specifying the set of values x may take. In domain theory a function can be defined over the universal domain. But we are in set theory, where an operation 'on everything' is not a function. An element of $\alpha \to \beta$ may (hereditarily) contain all the elements of α and β, so these must be sets.

The type symbol β is superfluous, however, by set theory's Axiom of Replacement. If α denotes a set then $\{b(x) \mid x : \alpha\}$ is also a set. Deleting β improves the notation:

$$\lambda x : \alpha . b(x) \quad : \quad \alpha \to \beta$$

Martin-Löf's Type Theory has polymorphic functions like $\lambda x . x$, but its semantics is operational: its functions are algorithms, not graphs. Nor can we omit the type symbols as a syntactic convenience, hoping they could be replaced in principle. Anne Salvesen [32] presents proofs in Martin-Löf's Type Theory that fail when type symbols are added. Her arguments are general, and should apply here as well.

Note that only the 'contravariant' aspect of functions — the domain of application — requires a type label. Product and sum types do not need any type labels.

8.2 Products

For products, a pairing constructor with type symbols is

$$\text{Pair}_{\alpha,\beta}(a)(b) : \alpha \times \beta \qquad (a : \alpha, b : \beta)$$

This could abbreviate the abstraction

$$\lambda x : \alpha \,.\, \lambda y : \beta \,.\, (x = a) \,\&\, (y = b) \quad : \quad \alpha \to \beta \to \mathit{bool}$$

which contains the type symbols α and β. But pairs need not carry type symbols. In set theory, pairs formed by the operation $\langle a, b\rangle = \{\{a\}, \{a, b\}\}$ do not depend on what sets contain a and b. The Isabelle formulation includes a polymorphic pairing operator $\langle a, b\rangle$.

Many other authors, for various reasons, take pairing as a primitive of the typed λ-calculus [12,20,33].

8.3 Sums

The disjoint sum has left and right injections:

$$\text{Inl}_{\alpha,\beta}(a) : \alpha + \beta \qquad (a : \alpha)$$

$$\text{Inr}_{\alpha,\beta}(b) : \alpha + \beta \qquad (b : \beta)$$

The type symbols are again unnecessary. In set theory, the injections $\text{Inl}(a) = \langle\{a\}, \emptyset\rangle$ and $\text{Inr}(b) = \langle\emptyset, \{b\}\rangle$ depend only on the values of a and b, not on the sets that contain them.

In the current version of the logic, a monomorphic (type labelled) disjoint sum is derived as shown in Section 5. Taking polymorphic injections as primitive seems to be needless extra complexity, for it does not greatly improve the notation.

8.4 Comparison with other type systems

The Edinburgh Logical Framework, a type theory for representing formal systems, can express the implicit type checking of Church's higher-order logic [15]. In this case all constants are fully decorated with type symbols, and terms contain much redundant type information.

Gordon's HOL system, like LCF, uses polymorphic type checking. Its type variables, written *, **, etc., are syntactic variables ranging over types. The identity combinator I might have the polymorphic type

```
I : * -> *
```

Because a constant's type is part of its name, the HOL constant `I` stands for a family of constants I_α, satisfying the schematic typing

$$I_\alpha : \alpha \rightarrow \alpha$$

For example, `I(I)` abbreviates $I_{(\alpha\rightarrow\alpha)\rightarrow(\alpha\rightarrow\alpha)}(I_{\alpha\rightarrow\alpha})$. This is not self-application: it involves two different instances of I.

Under the set-theoretic semantics, an identity function on all types could not exist. Coquand [8] has shown that polymorphic higher-order logic is inconsistent. A constant like $I_\alpha : \alpha \rightarrow \alpha$ is here called *monomorphic* (following Salvesen [32]) because of its type label.

9 Higher-order reasoning

Quantification over propositions is what makes simple type theory 'higher-order'. First-order logic allows quantification over individuals; second-order logic allows quantification over properties of individuals; third-order logic allows quantification over properties of properties of individuals; and so forth. Higher-order (or ω-order) logic allows all these quantifications. A formula is simply a term of type *bool*.

The logic programming language λProlog, though based on higher-order logic, forbids quantification over *bool* [24]. So it is really first-order logic extended with typed λ-expressions. The meta-logic of Isabelle avoids quantification over *bool* to simplify the theory, but this restriction is not enforced.

Quantification over propositions permits many different formulations of higher-order logic. Absurdity ($\bot$) is definable as $\forall p : \mathit{bool} \,.\, p$, the proposition that implies all propositions. Conjunction and disjunction are definable by

$$\begin{aligned} P \,\&\, Q &\equiv \forall r : \mathit{bool} \,.\, (P \supset (Q \supset r)) \supset r \\ P \vee Q &\equiv \forall r : \mathit{bool} \,.\, (P \supset r) \supset ((Q \supset r) \supset r) \end{aligned}$$

Andrews [1] presents a formulation based on equality. For example, the universal quantifier is defined in terms of truth ($\top$) as

$$\forall x : \alpha \,.\, P(x) \;\equiv\; (\lambda x : \alpha \,.\, P(x)) = (\lambda x : \alpha \,.\, \top)$$

Representing formulae by terms of type *bool* is inconvenient in Isabelle. Explicit type inference (to ensure that all theorems have type *bool*) encumbers proofs. Type checking can be minimized by formulating each rule such that the conclusion is well-typed provided its premises are. Then type checking only takes place when assumptions are discharged. In a trial implementation, even this much checking was inefficient.

Now formulae are a separate syntactic class. The present formulation defines first-order logic. It then adds *reflection* — isomorphisms between formulae and terms of type *bool* — to obtain higher-order logic.

- $term(P)$ is a term (of type *bool*) if P is a formula
- $form(b)$ is a formula if b is a term

A predicate (or *class*) on α is just a function of type $\alpha \rightarrow bool$. Class formation is λ-abstraction over a formula, while the membership predicate is function application. Let us introduce some class-theoretic notation:

$$\{\!|x : \alpha \,.\, P(x)|\!\} \equiv \lambda x : \alpha \,.\, term(P(x))$$
$$a \in S \equiv form(S \text{ ` } a)$$

Class theory is the main vehicle for mathematical reasoning in *Principia.*

The higher-order logic of Fourman and D. Scott [11,33] also distinguishes between terms and formulae. The primitive types are products and powersets; functions are represented by their graphs, like in set theory. Reflection functions can be defined through class abstraction and the membership predicate: abstraction creates a class (a term) from a formula, while membership in a class is a formula.

10 Recursive data types

Recursive types, like the natural numbers, lists, and trees, are an active research area. The wellordering types of Martin-Löf's Type Theory are general transfinite trees [22]. The Nuprl system, although largely based on Martin-Löf, uses positive recursive type definitions [7]. Boyer and Moore's 'shell principle' introduces recursive structures [4]. LCF can define recursive types using domain theory [30]. Recursive types can also be constructed in simple type theory.

The natural numbers can be constructed in various ways, assuming an Axiom of Infinity. In *Principia*, the number 2 is the class of all pairs of some type α. In Church, 2 is $\lambda f : \alpha \rightarrow \alpha \,.\, \lambda x : \alpha \,.\, f(fx)$. Both definitions are cumbersome and entail different types of natural numbers for each type α. The Isabelle formulation postulates a type *nat* of natural numbers satisfying induction and primitive recursion.

Melham [23] describes one treatment of recursive types, defining lists in terms of natural numbers, and trees in terms of lists. He has implemented this in Gordon's HOL system, which uses Church's logic.

The Isabelle formulation uses a different treatment inspired by Huet [19]. For a given set of constructors it involves two steps:

1. Find a type rich enough to represent all possible constructions.

2. Restrict to the subtype inductively generated by the constructors.

Let us define $list(\alpha)$, the type of lists over α. The representing type is $nat \times \alpha \to bool$, where the list $[x_0, x_1, \ldots, x_n]$ is represented by the class of pairs

$$\{\!|\langle 0, x_0\rangle, \langle 1, x_1\rangle, \ldots, \langle n, x_n\rangle|\!\}$$

The constructors are nil_α and $\mathrm{cons}_\alpha(a, l)$. (They must be labelled with type α because classes have type labels.) The empty list is represented by the empty class. To put an element in front of a list, $\mathrm{cons}_\alpha(a, l)$ increments m in all pairs $\langle m, x\rangle$ in l, then adds the pair $\langle 0, a\rangle$.

$$\begin{aligned}
\mathrm{nil}_\alpha &\equiv \{\!|u : nat \times \alpha \,.\, \bot|\!\} \\
\mathrm{cons}_\alpha(a, l) &\equiv \{\!|u : nat \times \alpha \,.\, u = \langle 0, a\rangle \\
&\qquad \vee (\exists m : nat \,.\, \exists x : \alpha \,.\, \langle m, x\rangle \in l \;\&\; u = \langle \mathrm{Succ}(m), x\rangle)|\!\}
\end{aligned}$$

The representing type includes many non-lists. Tarski's theorem (see Huet [19]), which asserts that every monotone function over a complete lattice has a least fixed point, can be used to define the subtype of lists. The monotone function takes a class F of lists and returns the class of all lists obtained by a further application of the constructors:

$$\begin{aligned}
&\{\!|l : nat \times \alpha \to bool \,.\, l = \mathrm{nil}_\alpha \\
&\qquad \vee (\exists x : \alpha \,.\, \exists l' : nat \times \alpha \to bool \,.\, l' \in F \;\&\; l = \mathrm{cons}_\alpha(x, l'))|\!\}
\end{aligned}$$

Trees with labelled edges can also be represented by classes of pairs. For lists, the number in each pair gives the position of an element. The position of an element in a tree can be given by a list of edges. If the trees have countable branching, the representing type could be $list(nat) \times \alpha$. Trees are sometimes represented like this in set theory.

Tarski's theorem also handles recursively defined classes. For example, the reflexive/transitive closure of the relation R is inductively generated from the identity relation by composition with R. This too is the least fixed point of a monotone function.

11 The formulation of simple type theory

Because of type inference there are two forms of judgement: 'formula P is true', written simply P, and 'term a has type α', written $a : \alpha$. Type assertions cannot be combined by logical connectives — which would not be in the spirit of type theory — because $a : \alpha$ is not a formula.

Appendix A is the Isabelle rule file, including a few uninteresting rules omitted below.

11.1 Equality

The formula $a =_\alpha b$ means that a and b are equal and have type α. There is no deep reason for having a typed equality relation. A proof that a equals b must involve showing that a and b have some type α, and this type information could be useful later.[3]

The reflexivity, symmetry, and substitution rules are

$$\frac{a : \alpha}{a =_\alpha a} \qquad \frac{a =_\alpha b}{b =_\alpha a} \qquad \frac{a =_\alpha b \qquad P(b)}{P(a)}$$

The type information can be extracted. If $a = b$ then both terms have the relevant type.

$$\frac{a =_\alpha b}{a : \alpha} \qquad \frac{a =_\alpha b}{b : \alpha}$$

11.2 Types

Functions

These rules for abstraction and application are typical of type inference systems: see Chapter 15 of Hindley and Seldin [17]. Applications are written with an explicit operator: $f \,`\, a$.

$$\frac{\begin{matrix}[x : \alpha] \\ b(x) : \beta\end{matrix}}{(\lambda x : \alpha \,.\, b(x)) : \alpha \to \beta} \qquad \frac{f : \alpha \to \beta \qquad a : \alpha}{f \,`\, a : \beta}$$

We have β and η-conversion:

$$\frac{a : \alpha \qquad \begin{matrix}[x : \alpha] \\ b(x) : \beta\end{matrix}}{(\lambda x : \alpha \,.\, b(x)) \,`\, a =_\beta b(a)} \qquad \frac{f : \alpha \to \beta}{\lambda x : \alpha \,.\, f \,`\, x =_{\alpha \to \beta} f}$$

[3]In Martin-Löf's Type Theory, equality can only be understood with respect to some type, so the relation is typed for semantic reasons.

In η-conversion, variable x may not be free in f. All rules that discharge the assumption $x : \alpha$ are subject to the proviso that x is not free in the conclusion or other assumptions. This will be taken for granted below.

Finally, there is a rule for the construction of equal abstractions. It does not follow from the substitution rule above because x is bound in the conclusion.

$$\frac{\begin{array}{c}[x : \alpha] \\ b(x) =_\beta c(x)\end{array}}{(\lambda x : \alpha \,.\, b(x)) =_{\alpha \to \beta} (\lambda x : \alpha \,.\, c(x))}$$

Products

The pair of a and b is written $\langle a, b \rangle$; the projections are fst and snd. These constants contain no type symbols.

Type assignment rules for pairing and the projections are

$$\frac{a : \alpha \qquad b : \beta}{\langle a, b \rangle : \alpha \times \beta} \qquad \frac{p : \alpha \times \beta}{\mathrm{fst}(p) : \alpha} \qquad \frac{p : \alpha \times \beta}{\mathrm{snd}(p) : \beta}$$

Conversion (equality) rules for pairing and the projections are

$$\frac{a : \alpha \qquad b : \beta}{\mathrm{fst}(\langle a, b \rangle) =_\alpha a} \qquad \frac{a : \alpha \qquad b : \beta}{\mathrm{snd}(\langle a, b \rangle) =_\beta b}$$

The elimination rule for products resembles a rule of Martin-Löf's Type Theory:

$$\frac{p : \alpha \times \beta \qquad \begin{array}{c}[x : \alpha,\ y : \beta] \\ Q(\langle x, y \rangle)\end{array}}{Q(p)}$$

It implies $\langle \mathrm{fst}(p), \mathrm{snd}(p) \rangle =_{\alpha \times \beta} p$ for $p : \alpha \times \beta$.

Subtypes

The type checking of subtypes involves the truth of $P(a)$ and is therefore undecidable.

$$\frac{a : \alpha \qquad P(a)}{a : \{x : \alpha \,.\, P(x)\}}$$

The elimination rules say that if $a : \{x : \alpha \,.\, P(x)\}$ then $a : \alpha$ and $P(a)$.

$$\frac{a : \{x : \alpha \,.\, P(x)\}}{a : \alpha} \qquad \frac{a : \{x : \alpha \,.\, P(x)\}}{P(a)}$$

Natural numbers

The type of natural numbers is called *nat*. The Axiom of Infinity is expressed in the most convenient form: through the existence of functions defined by primitive recursion.

The typing rules for 0 and successor are

$$0 : nat \qquad \frac{a : nat}{\mathrm{Succ}(a) : nat}$$

The typing rule for rec is

$$\frac{a : nat \qquad b : \beta \qquad \begin{array}{c}[x : nat,\ y : \beta]\\ c(x,y) : \beta\end{array}}{\mathrm{rec}(a, b, xy\ .\ c(x,y)) : \beta}$$

The operator $\mathrm{rec}(a, b, xy\ .\ c(x,y))$, where x and y are bound in $c(x,y)$, expresses primitive recursion. In the meta-level typed λ-calculus c is a function, so $\mathrm{rec}(a, b, xy\ .\ c(x,y))$ will henceforth be abbreviated to $\mathrm{rec}(a, b, c)$. The conversion rules for rec are

$$\frac{b : \beta \qquad \begin{array}{c}[x : nat,\ y : \beta]\\ c(x,y) : \beta\end{array}}{\mathrm{rec}(0, b, c) =_\beta b} \qquad \frac{a : nat \qquad b : \beta \qquad \begin{array}{c}[x : nat,\ y : \beta]\\ c(x,y) : \beta\end{array}}{\mathrm{rec}(\mathrm{Succ}(a), b, c) =_\beta c(a, \mathrm{rec}(a, b, c))}$$

Because rec binds variables, it requires its own substitution rule:

$$\frac{a =_{nat} d \qquad b =_\beta e \qquad \begin{array}{c}[x : nat,\ y : \beta]\\ c(x,y) =_\beta f(x,y)\end{array}}{\mathrm{rec}(a, b, c) =_\beta \mathrm{rec}(d, e, f)}$$

The mathematical induction rule is

$$\frac{a : nat \qquad Q(0) \qquad \begin{array}{c}[x : nat,\ Q(x)]\\ Q(\mathrm{Succ}(x))\end{array}}{Q(a)}$$

11.3 Logic

Implication and universal quantification are taken as primitive; the other logical constants are defined through them.

The rules for implication are

$$\frac{\begin{array}{c}[P]\\ Q\end{array}}{P \supset Q} \qquad \frac{P \supset Q \qquad P}{Q}$$

The rules for universal quantification are

$$\frac{\begin{array}{c}[x : \alpha] \\ P(x)\end{array}}{\forall x : \alpha \, . \, P(x)} \qquad \frac{\forall x : \alpha \, . \, P(x) \qquad a : \alpha}{P(a)}$$

The following rule gives classical logic (which follows anyway from the Axiom of Choice).

$$\frac{\begin{array}{c}[\neg P] \\ P\end{array}}{P}$$

Reflection

The operator term(P) maps a formula to a term of type *bool*, while form(a) maps such a term to a formula. Since there is no way to decide whether a formula is true or false, term(P) is non-constructive. The truth value of form(a) is specified only where a has type *bool*.

The typing rule says that term(P) has type *bool*, even if P is ill-typed!

$$\text{term}(P) : bool$$

Isomorphism rules state that term and form preserve truth:

$$\frac{a : bool}{\text{term}(\text{form}(a)) =_{bool} a} \qquad \frac{P}{\text{form}(\text{term}(P))} \qquad \frac{\text{form}(\text{term}(P))}{P}$$

Although *form*(a) is syntactically a formula for all terms a, it preserves truth only if a has type *bool*.

Also, term and form preserve equivalence:

$$\frac{\begin{array}{cc}[P] & [Q] \\ Q & P\end{array}}{\text{term}(P) =_{bool} \text{term}(Q)}$$

The analogous property for *form* — that $a =_{bool} b$ and form(b) imply form(a) — follows by substitution.

Definitions of other connectives

These definitions of other connectives yield their usual properties. The terms False and True have type *bool*; the absurdity formula ($\bot$) is form(False).

$$\begin{aligned}
\text{False} &\equiv \text{term}(\forall p : bool\,.\,\text{form}(p)) \\
\text{True} &\equiv \text{term}(\forall p : bool\,.\,\text{form}(p) \supset \text{form}(p)) \\
P \,\&\, Q &\equiv \forall r : bool\,.\,(P \supset Q \supset \text{form}(r)) \supset \text{form}(r) \\
P \vee Q &\equiv \forall r : bool\,.\,(P \supset \text{form}(r)) \supset (Q \supset \text{form}(r)) \supset \text{form}(r) \\
\exists x : \alpha\,.\,P(x) &\equiv \forall r : bool\,.\,(\forall x : \alpha\,.\,P(x) \supset \text{form}(r)) \supset \text{form}(r) \\
\neg P &\equiv (P \supset \text{form}(\text{False})) \\
P \leftrightarrow Q &\equiv (P \supset Q) \,\&\, (Q \supset P)
\end{aligned}$$

Descriptions

The η-operator is adopted, which assumes an Axiom of Choice and is only defined if some suitable object exists.

$$\frac{\exists x : \alpha\,.\,P(x)}{(\eta x : \alpha\,.\,P(x)) : \alpha} \qquad \frac{\exists x : \alpha\,.\,P(x)}{P(\eta x : \alpha\,.\,P(x))}$$

Two descriptions are equal if they are defined and the formulae are equivalent. The second premise ensures that the description is defined.

$$\frac{\begin{array}{c}[x : \alpha] \\ P(x) \leftrightarrow Q(x)\end{array} \qquad \exists x : \alpha\,.\,P(x)}{(\eta x : \alpha\,.\,P(x)) =_\alpha (\eta x : \alpha\,.\,Q(x))}$$

11.4 Definitions of types

These include the empty type *void*, the singleton type *unit*, and the union type $\alpha + \beta$.

$$void \equiv \{p : bool\,.\,\text{form}(\text{False})\} \qquad unit \equiv \{p : bool\,.\,p =_{bool} \text{True}\}$$

The sum type consists of all left injections and right injections.

$$\begin{aligned}
\alpha + \beta \equiv \{w : (\alpha \to bool) \times (\beta \to bool).\quad &(\exists x : \alpha\,.\,w = \text{Inl}(\alpha, \beta, x)) \\
\vee\quad &(\exists y : \beta\,.\,w = \text{Inr}(\alpha, \beta, y))\}
\end{aligned}$$

Injections are defined in a standard way as pairs of classes [33].[4]

$$\begin{aligned}
\text{Inl}(\alpha, \beta, a) &\equiv \langle \lambda x : \alpha\,.\,\text{term}(a =_\alpha x), \lambda y : \beta\,.\,\text{False}\rangle \\
\text{Inr}(\alpha, \beta, b) &\equiv \langle \lambda x : \alpha\,.\,\text{False}, \lambda y : \beta\,.\,\text{term}(b =_\beta y)\rangle
\end{aligned}$$

[4] The definition of $\alpha + \beta$ used by Melham [23] does not work if either type is empty.

The operator $\text{when}(\alpha, \beta, \gamma, p, c, d)$ performs case analysis on a sum type, where c and d are meta-level functions.[5]

$$\begin{aligned}\text{when}(\alpha,\beta,\gamma,p,c,d) \equiv \eta z:\gamma.\quad & (\forall x:\alpha\,.\,p =_{\alpha+\beta} \text{Inl}(\alpha,\beta,x) \supset z =_\gamma c(x))\\ \& \quad & (\forall y:\beta\,.\,p =_{\alpha+\beta} \text{Inr}(\alpha,\beta,y) \supset z =_\gamma d(y))\end{aligned}$$

These operators have type labels because they are defined by terms containing type symbols. All variables on the right side in a definition must be present on the left.

Basic laws like $\text{when}_{\alpha,\beta,\gamma}(\text{Inl}_{\alpha,\beta}(a), c, d) = c(a)$ are proved in the Isabelle theory. The operator is computable despite being defined by description.

11.5 Class Theory

Class theory includes the relations *membership* and *subclass* and the operations *union*, *intersection*, and *powerset*. Union and intersection are also defined for a class of classes. The class abstraction $\{\!|x:\alpha\,.\,P(x)|\!\}$ abbreviates $\lambda x:\alpha\,.\,term(P(x))$. Operators defined by class abstraction have the type label α as an extra argument, so none of these are infix operators in Isabelle's concrete syntax.

$$\begin{aligned}
a \in S &\equiv form(S\,`\,a)\\
S \subseteq_\alpha T &\equiv \forall z:\alpha\,.\,z \in S \supset z \in T\\
S \cup_\alpha T &\equiv \{\!|z:\alpha\,.\,z \in S \lor z \in T|\!\}\\
S \cap_\alpha T &\equiv \{\!|z:\alpha\,.\,z \in S \,\&\, z \in T|\!\}\\
\bigcup_\alpha F &\equiv \{\!|z:\alpha\,.\,\exists S:\alpha \to bool\,.\,S \in F \,\&\, z \in S)|\!\}\\
\bigcap_\alpha F &\equiv \{\!|z:\alpha\,.\,\forall S:\alpha \to bool\,.\,S \in F \supset z \in S)|\!\}\\
\mathcal{P}_\alpha(S) &\equiv \{\!|T:\alpha\,.\,T \subseteq_\alpha S|\!\}
\end{aligned}$$

12 Sample proofs in Isabelle

A logic is traditionally illustrated by sample proofs. Theorems proved using Isabelle include basic facts, lemmas used in proof procedures, Tarski's Theorem, and well-founded recursion. Proof procedures exist for first-order logic, rewriting, and class theory.

12.1 Simple proof procedures

The rewriting package is based on the one for Martin-Löf's Type Theory, as are the sample proofs in elementary number theory. Using rewriting and induction, arithmetic is developed up to the theorem $a \bmod b + (a/b) \times b = a$.

[5] Conventional notation is $\text{when}_{\alpha,\beta,\gamma}(p, x\,.\,c(x), y\,.\,d(y))$, where x and y are bound variables.

Reflection works well in higher-order reasoning. Natural deduction rules for the logical constants are easily derived from their higher-order definitions. A standard example of higher-order logic is Cantor's Theorem that every set has more subsets than elements, which can be expressed as follows:

$$\neg\Big(\exists g : \alpha \to (\alpha \to bool) \,.\, \forall f : \alpha \to bool \,.\, \exists j : \alpha \,.\, f = g \text{ ` } j\Big)$$

(There is no onto function from α to $\alpha \to bool$.) While TPS [2] can prove Cantor's Theorem automatically, Isabelle must be guided towards the proof.

The proof procedures for first-order logic work directly with the natural deduction rules, as sketched in Chapter 2 of my book [30]. Although none of the procedures is complete or fast, they can prove many examples automatically:

$$\Big(\exists y : \alpha \,.\, \forall x : \alpha \,.\, J(y,x) \leftrightarrow \neg J(x,x)\Big) \supset \neg\Big(\forall x : \alpha \,.\, \exists y : \alpha \,.\, \forall z : \alpha \,.\, J(z,y) \leftrightarrow \neg J(z,x)\Big)$$

Similar proof procedures for class theory reason about unions, intersections, subsets, etc. This example is proved automatically:

$$\frac{F : (\alpha \to bool) \to bool \qquad G : (\alpha \to bool) \to bool}{\bigcap_\alpha (F \cup G) = (\bigcap_\alpha F) \cap (\bigcap_\alpha G)}$$

Classes are also used to construct a type of lists and derive structural induction.

12.2 Well-founded induction and recursion

Well-founded recursion is a general method of defining total recursive functions, while well-founded induction reasons about functions so defined. These principles, which hold for every well-founded relation, play a central role in the Boyer/Moore logic [4]. They have been derived using Isabelle.

Given a relation $R : \alpha \times \alpha \to bool$, let us write $y \, R \, x$ instead of $\langle y, x\rangle \in R$ and abbreviate 'R is well-founded' as $\mathrm{wf}_\alpha(R)$. Classically, R is well-founded if there are no infinite descending chains $\cdots x_3 \, R \, x_2 \, R \, x_1 \, R \, x_0$. The following definition is more convenient:

$$\begin{aligned} \mathrm{wf}_\alpha(R) \;\equiv\; & \Big(\forall x : \alpha \,.\, (\forall y : \alpha \,.\, y \, R \, x \supset y \in S) \supset x \in S\Big) \\ & \supset (\forall x : \alpha \,.\, x \in S) \end{aligned}$$

This easily yields well-founded induction:

$$\frac{\mathrm{wf}_\alpha(R) \qquad a : \alpha \qquad \begin{matrix} [x : \alpha] \\ \forall y : \alpha \,.\, y \, R \, x \supset y \in P(x) \end{matrix}}{P(a)}$$

A recursive function f is well-founded along R if $f(x)$ depends only on $f(y)$ such that $y\,R\,x$. This condition can be stated using subtypes. Type α_{Rx} is the restriction of α to predecessors of x under R.

$$\alpha_{Rx} \equiv \{y : \alpha \,.\, y\,R\,x\}$$

The body of the recursive function has the form $H(x, f)$ where $x : \alpha$ is the argument and $f : \alpha_{Rx} \to \beta$ handles recursive calls. Type checking ensures that f is only called below x. The resulting recursive function is applied to argument a by $\mathrm{wfrec}_{\alpha,\beta}(R, H, a)$.

$$\frac{\mathrm{wf}_\alpha(R) \qquad R : \alpha \times \alpha \to \mathit{bool} \qquad a : \alpha \qquad \begin{array}{c}[x : \alpha,\ f : \alpha_{Rx} \to \beta] \\ H(x, f) : \beta\end{array}}{\mathrm{wfrec}_{\alpha,\beta}(R, H, a) : \beta}$$

Under the same premises, wfrec satisfies the recursion equation

$$\mathrm{wfrec}_{\alpha,\beta}(R, H, a) =_\beta H(a, \lambda x : \alpha_{Ra} \,.\, \mathrm{wfrec}_{\alpha,\beta}(R, H, x))$$

Some observations: The variables x and f are subject to the usual 'not free in' conditions. The abstraction $\lambda x : \alpha_{Ra} \,.\, \mathrm{wfrec}_{\alpha,\beta}(R, H, x)$ restricts the function to arguments below a. Here H is a meta-level function (wfrec binds variables); if we had dependent types, H could be an object-function of type $\prod_{x:\alpha}(\alpha_{Rx} \to \beta) \to \beta$.

Defined by a description, wfrec takes the union of all graphs of functions that satisfy the recursion equation below some $x : \alpha$. Its typing rule holds because this union forms the graph of a function on α. Observe how type checking can involve substantial proof. With the help of a few extra lemmas, the equality rule is then proved.

This work follows Suppes's treatment of transfinite recursion in set theory [36]. Operator wfrec is defined once and for all, and its properties proved, for all well-founded relations in the logic. It is far stronger than my work in Martin-Löf's Type Theory [27], which considers certain ways of constructing well-founded relations and their corresponding recursion operators.

13 Conclusions

Programs are typically verified within a special logic of computation. Although several such logics have been successful, they sometimes restrict abstract mathematical reasoning — needed even for computational proofs.

- The *Logic for Computable Functions* (LCF) embeds a typed λ-calculus, where types denote domains, into first-order logic [30]. LCF is good for reasoning about nonterminating processes, but termination proofs can become a chore (in my opinion [31]). The restriction to domains and continuous functions has serious consequences [34].

- Martin-Löf's Type Theory is based on computation [22,26]. By the interpretation of propositions-as-types, a type can express a complete program specification. Developments and applications are proceeding rapidly [3]. However, the theory does not admit classical set-theoretic arguments. Unwanted proof objects in types cause complications [32].
- Boyer and Moore use quantifier-free first-order logic with well-founded induction and recursion [4]. Although this combination gives unique simplicity and power, it is hard to do without quantifiers.

Simple type theory may be suitable for reasoning about computation. It offers a rich collection of computable functions, including general recursive and higher-order functions, but it is not restricted to computable functions. Subtypes and classes can express program specifications. The main question is how to recognize when a function is computable.

Some people will wonder whether classical logic is appropriate. Why not use intuitionistic higher-order logic instead? Simply remove the double-negation law and the Axiom of Choice (replacing the η-operator by ι). Although I have an interest in constructive logic, this suggestion requires a stronger argument. Intuitionism is a deep and evolving subject. There is little agreement about whether intuitionistic higher-order logic, with its impredicative quantification, is constructive.

The Calculus of Constructions, by Coquand and Huet [9], is also intended for reasoning about programs. In use it is very like simple type theory: the Isabelle proof of Tarski's theorem follows Huet's [19]. However, it interprets propositions-as-types and has a clear notion of computation. Experiments with the Calculus and the Isabelle formulation of type theory will make an interesting comparison.

Acknowledgement. Isabelle was developed under grant GR/E 0355.7 from the Science and Engineering Research Council. Mike Fourman made many valuable remarks, especially about descriptions. Thanks also to Thomas Melham, Dale Miller, Philippe Noël, and Jan M. Smith for advice.

References

[1] Peter B. Andrews. *An Introduction to Mathematical Logic and Type Theory: To Truth Through Proof.* Academic Press, 1986.

[2] Peter B. Andrews, Dale A. Miller, Eve L. Cohen, and Frank Pfenning. Automating higher-order logic. In W. W. Bledsoe and D. W. Loveland, editors, *Automated Theorem Proving: After 25 Years*, pages 169–192, American Mathematical Society, 1984.

[3] Roland Backhouse, Paul Chisholm, Grant Malcolm, and Erik Saaman. Do-it-yourself type theory. *Formal Aspects of Computing*, 1:19–84, 1989.

[4] Robert S. Boyer and J Strother Moore. *A Computational Logic.* Academic Press, 1987.

[5] Alonzo Church. A formulation of the simple theory of types. *Journal of Symbolic Logic*, 5:56–68, 1940.

[6] Avra J. Cohn. A proof of correctness of the VIPER microprocessor: the first level. In Graham Birtwistle and P. A. Subrahmanyam, editors, *VLSI Specification, Verification and Synthesis*, pages 27–71, Kluwer Academic Publishers, 1988.

[7] R. L. Constable, S. F. Allen, H. M. Bromley, W. R. Cleaveland, J. F. Cremer, R. W. Harper, D. J. Howe, T. B. Knoblock, N. P. Mendler, P. Panagaden, J. T. Sasaki, and S. F. Smith. *Implementing Mathematics with the Nuprl Proof Development System.* Prentice-Hall International, 1986.

[8] Thierry Coquand. An analysis of Girard's paradox. In *Symposium on Logic in Computer Science*, pages 227–236, IEEE Computer Society Press, 1986.

[9] Thierry Coquand and Gèrard Huet. The calculus of constructions. *Information and Computation*, 76:95–120, 1988.

[10] Andrzej Borzyszkowski et al. *Towards a Set-Theoretic Type Theory.* Technical Report, Polish Academy of Sciences, Institute of Computer Science, Gdańsk, 1988.

[11] Michael P. Fourman. The logic of topoi. In J. Barwise, editor, *Handbook of Mathematical Logic*, pages 1053–1090, North-Holland, 1977.

[12] Jean-Yves Girard. *Proofs and Types.* Cambridge University Press, 1989. Translated by Yves LaFont and Paul Taylor.

[13] Kurt Gödel. Russell's mathematical logic. In Paul Benacerraf and Hilary Putnam, editors, *Philosophy of Mathematics: Selected Readings*, Cambridge University Press, 1983. Essay first published in 1944.

[14] Michael J. C. Gordon. HOL: a proof generating system for higher-order logic. In Graham Birtwistle and P. A. Subrahmanyam, editors, *VLSI Specification, Verification and Synthesis*, pages 73–128, Kluwer Academic Publishers, 1988.

[16] William S. Hatcher. *The Logical Foundations of Mathematics.* Pergammon Press, 1982.

[17] J. Roger Hindley and Jonathon P. Seldin. *Introduction to Combinators and λ-Calculus.* Cambridge University Press, 1986.

[18] G. P. Huet. A unification algorithm for typed λ-calculus. *Theoretical Computer Science*, 1:27–57, 1975.

[19] Gérard Huet. Induction principles formalized in the Calculus of Constructions. In *Programming of Future Generation Computers*, pages 205–216, Elsevier, 1988.

[20] J. Lambek and P. J. Scott. *Introduction to Higher Order Categorical Logic.* Cambridge University Press, 1986.

[21] A. C. Leisenring. *Mathematical Logic and Hilbert's ϵ-Symbol.* MacDonald, 1969.

[22] Per Martin-Löf. Constructive mathematics and computer programming. In C. A. R. Hoare and J. C. Shepherdson, editors, *Mathematical Logic and Programming Languages*, pages 167–184, Prentice-Hall International, 1985.

[23] Thomas F. Melham. Automating recursive type definitions in higher order logic. In Graham Birtwistle and P. A. Subrahmanyam, editors, *Current Trends in Hardware Verification and Automated Theorem Proving*, pages 341–386, Springer-Verlag, 1989.

[24] G. Nadathur. *A Higher-Order Logic as the Basis for Logic Programming.* PhD thesis, University of Pennsylvania, 1987.

[25] Philippe Noël. *Experimenting with Isabelle in ZF Set Theory.* Technical Report, University of Cambridge Computer Laboratory, 1989. Preprint.

[26] Bengt Nordström, Kent Petersson, and Jan Smith. *Programming in Martin-Löf's type theory. An introduction.* Oxford University Press, 1989. In press.

[27] Lawrence C. Paulson. Constructing recursion operators in intuitionistic type theory. *Journal of Symbolic Computation*, 2:325–355, 1986.

[28] Lawrence C. Paulson. The foundation of a generic theorem prover. *Journal of Automated Reasoning*, 5:363–397, 1989.

[29] Lawrence C. Paulson. Isabelle: the next 700 theorem provers. In P. Odifreddi, editor, *Logic and Computer Science*, Academic Press, 1989? In press.

[30] Lawrence C. Paulson. *Logic and Computation: Interactive proof with Cambridge LCF.* Cambridge University Press, 1987.

[31] Lawrence C. Paulson. Verifying the unification algorithm in LCF. *Science of Computer Programming*, 5:143–170, 1985.

[32] Anne Salvesen. *On Information Discharging and Retrieval in Martin-Löf's Type Theory.* PhD thesis, University of Oslo, 1989. Report 803, Norwegian Computing Center, Oslo.

[33] Dana Scott. Identity and existence in intuitionistic logic. In M. P. Fourman, editor, *Applications of Sheaves*, pages 660–696, Springer-Verlag, 1979. Lecture Notes in Mathematics 753.

[34] Stefan Sokołowski. Soundness of Hoare's logic: an automatic proof using LCF. *ACM Transactions on Programming Languages and Systems*, 9:100–120, 1987.

[35] J. M. Spivey. *Understanding Z: A Specification Language and its Formal Semantics.* Cambridge University Press, 1988.

[36] Patrick Suppes. *Axiomatic Set Theory.* Dover, 1972.

[37] A. N. Whitehead and B. Russell. *Principia Mathematica.* Cambridge University Press, 1962. Paperback edition to *56, abridged from the 2nd edition (1927).

A Appendix: The Isabelle Rule File

```
(*  Title: HOL/ruleshell
    Author: Lawrence C Paulson, Cambridge University Computer Laboratory
    Copyright   1989  University of Cambridge

Rules of Higher-order Logic (Type Theory)

!!!After updating, rebuild  ".rules.ML"  by calling make-rulenames!!!
*)

signature HOL_RULE =
  sig
  structure Thm : THM
  val sign: Thm.Sign.sg
  val thy: Thm.theory
(*INSERT-RULESIG -- file produced by make-rulenames*)
  end;


functor HOL_RuleFun (structure HOL_Syntax: HOL_SYNTAX and Thm: THM
sharing HOL_Syntax.Syntax = Thm.Sign.Syntax) : HOL_RULE =
struct
structure Thm = Thm;

val thy = Thm.enrich_theory Thm.pure_thy "HOL"
    (["term","form","type"], HOL_Syntax.const_decs, HOL_Syntax.syn)
[
  (*** Equality ***)

  ("refl",  "[| a: A |] ==> [| [ a = a : A ]  |]"),

  ("sym",  "[| [ a = b : A ]  |] ==> [| [ b = a : A ]  |]"),

  (*Equal terms are well typed -- all rules must enforce this! *)
  ("eq_type1",  "[| [ a = b : A ] |] ==> [| a: A |]"  ),

  ("eq_type2",  "[| [ a = b : A ] |] ==> [| b: A |]"  ),

  ("subst",
    "[| [ a = c : A ]  |]  ==>  [| P(c) |]  ==>  [| P(a) |]"),
```

```
(*** TYPES ***)

(** Functions **)

("Lambda_type",
  "(!(x)[| x: A |] ==> [| b(x) : B |])  ==>    \
\    [| lam x:A. b(x) : A->B |]"  ),

("Lambda_congr",
  "(!(x)[| x: A |] ==> [| [ b(x) = c(x) : B ] |])  ==>    \
\    [| [ lam x:A. b(x) = lam x:A. c(x) : A->B ] |]"  ),

("apply_type",
  "[| f: A->B |] ==> [| a: A |]  ==> [| f`a : B |]"  ),

("beta_conv",
  "[| a : A |] ==> (!(x)[| x: A |] ==> [| b(x) : B |])  ==> \
\    [| [ (lam x:A.b(x)) ` a = b(a) : B ] |]"  ),

("eta_conv", "[| f: A->B |] ==> [| [ lam x:A. f`x = f : A->B ] |]"  ),


(** Products **)

("pair_type", "[| a: A |] ==>  [| b: B |]  ==>  [| <a,b> : A*B |]"  ),

("prod_elim",
  "[| p : A*B |]  ==> \
\    (!(x,y)[| x: A |] ==> [| y: B |] ==> [| Q(<x,y>) |])  ==> \
\    [| Q(p) |]"  ),

("pair_inject",
  "[| [ <a,b> = <c,d> : A*B ] |] ==>  \
\    ([| [ a = c : A ] |] ==> [| [ b = d : B ] |] ==> [| R |])  ==>  \
\    [| R |]"  ),

(*fst and snd could be defined using descriptions...they are not to avoid
  excessive type labels -- which is the point of defining products here. *)

("fst_type",  "[| p: A*B |] ==> [| fst(p) : A |]"  ),
("snd_type",  "[| p: A*B |] ==> [| snd(p) : B |]"  ),

("fst_conv",  "[| a: A |] ==>  [| b: B |] ==> [| [ fst(<a,b>) = a: A] |]"  ),
("snd_conv",  "[| a: A |] ==>  [| b: B |] ==> [| [ snd(<a,b>) = b: B] |]"  ),

("split_def", "split(p,f) == f(fst(p), snd(p))"  ),


(** Subtypes **)

("subtype_intr",  "[| a: A |] ==> [| P(a) |] ==> [| a : {x:A.P(x)} |]"  ),

("subtype_elim1",   "[| a: {x:A.P(x)} |] ==> [| a:A |]"),
("subtype_elim2",   "[| a: {x:A.P(x)} |] ==> [| P(a) |]"),


(** Natural numbers **)

("Zero_type",  "[| 0: nat |]"  ),
("Succ_type",  "[| a: nat |] ==> [| Succ(a) : nat |]"  ),

("rec_type",
  "[| a : nat |] ==> \
\    [| b : C |]  ==> \
\    (!(x,y)[| x: nat |] ==> [| y: C |] ==> [| c(x,y): C |])  ==>  \
\    [| rec(a,b,c) : C |]"  ),
```

```
  ("rec_congr",
    "[| [ a = a' : nat ] |] ==> \
\    [| [ b = b' : C ] |]  ==> \
\    (!(x,y)[| x: nat |] ==> [| y: C |] ==> \
\            [| [ c(x,y) = c'(x,y): C ] |]) ==>  \
\    [| [ rec(a,b,c) = rec(a',b',c') : C ] |]"  ),

  ("rec_conv0",
    "[| b: C |] ==> \
\    (!(x,y)[| x: nat |] ==> [| y: C |] ==> [| c(x,y): C |])  ==>  \
\    [| [ rec(0,b,c) = b : C ] |]"  ),

  ("rec_conv1",
    "[| a : nat |] ==> \
\    [| b : C |]  ==> \
\    (!(x,y)[| x: nat |] ==> [| y: C |] ==> [| c(x,y): C |]) ==>  \
\    [| [ rec(Succ(a),b,c) = c(a, rec(a,b,c)) : C ] |]"  ),

  ("nat_induct",
    "[| a: nat |] ==> [| Q(0) |] ==>    \
\    (!(x)[| x: nat |] ==> [| Q(x) |] ==> [| Q(Succ(x)) |]) ==>    \
\    [| Q(a) |]"  ),


  (*** Logic ***)

  (** Implication and quantification *)

  ("classical",  "([| ~P |] ==> [| P |])  ==> [| P |]"),

  ("imp_intr",
    "([| P |] ==> [| Q |])  ==>  [| P-->Q |]"),

  ("mp",
    "[| P-->Q |] ==> [| P |]  ==> [| Q |]"),

  ("all_intr",
    "(!(x)[| x: A |] ==> [| P(x) |])  ==>  [| ALL x:A.P(x) |]"),

  ("spec",
    "[| ALL x:A.P(x) |] ==> [| a : A |]  ==> [| P(a) |]"),



  (** Reflection *)

  ("term_type", "[| term(P) : bool |]"  ),

  ("term_conv", "[| p: bool |] ==> [| [ term(form(p)) = p : bool ] |]"  ),

  ("form_intr", "[| P |] ==> [| form(term(P)) |]"),

  ("form_elim", "[| form(term(P)) |] ==> [| P |]"),

  ("term_congr",
    "([| P |] ==> [| Q |]) ==> ([| Q |] ==> [| P |]) ==>   \
\    [| [ term(P) = term(Q) : bool ] |]"),


  (** Reduction predicate for simplification. *)

  (*does not verify a:A!  Sound because only trans_red uses a Reduce premise*)
  ("refl_red", "Reduce(a,a)"  ),

  ("red_if_equal", "[| [ a = b : A ] |] ==> Reduce(a,b)"),

  ("trans_red", "[| [ a = b : A ] |] ==> Reduce(b,c) ==> [| [ a = c : A ] |]"),
```

```
(** Definitions of other connectives*)

("False_def", "False == term(ALL p:bool.form(p))"),
("True_def",  "True == term(ALL p:bool.form(p)-->form(p))"),
("conj_def",  "P&Q == ALL r:bool. (P-->Q-->form(r)) --> form(r)"),

("disj_def",
 "P|Q == ALL r:bool. (P-->form(r)) --> (Q-->form(r)) --> form(r)"),

("exists_def",
 "(EXISTS x:A. P(x)) ==  ALL r:bool. (ALL x:A. P(x)-->form(r)) --> form(r)"),

("not_def",  "~P == (P-->form(False))"),
("iff_def",  "P<->Q == (P-->Q) & (Q-->P)"),


(** Conditionals *)

("cond_def", "cond(A,p,a,b) == PICK x:A.(form(p)  & [x=a:A]) | \
\                                         (~form(p) & [x=b:A])" ),

(** Descriptions *)

("Pick_type", "[| EXISTS x:A.P(x) |] ==> [| (PICK x:A.P(x)) : A |]"),

("Pick_congr",
  "(!(x)[| x: A |] ==> [| P(x) <-> Q(x) |]) ==>    \
\   [| EXISTS x:A.P(x) |]  ==>  [| [ PICK x:A.P(x) = PICK x:A.Q(x) : A ] |]"),

("Pick_intr", "[| EXISTS x:A.P(x) |] ==> [| P(PICK x:A.P(x)) |]"),


(** Definitions of Classes*)
("member_def", "a<:S == form(S`a)"),
("subset_def",  "subset(A,S,T) == ALL z:A. z<:S --> z<:T"),
("un_def",   "un(A,S,T) == lam z:A. term(z<:S | z<:T)"),

("int_def",   "int(A,S,T) == lam z:A. term(z<:S & z<:T)"),
("union_def",
    "union(A,F) == lam z:A. term(EXISTS S:A->bool. S<:F & z<:S)"),
("inter_def",
    "inter(A,F) == lam z:A. term(ALL S:A->bool. S<:F --> z<:S)"),
("pow_def",
    "pow(A,S) == lam T:A. term(subset(A,T,S))"),


(** Definitions of types*)

(*the types "void" and "unit"*)
("void_def", "void == {p: bool. form(False)}"),
("unit_def", "unit == {p: bool. [p=True:bool]}"),

(*unions: the type A+B *)
("plus_def",
  "A+B == {w: (A->bool) * (B->bool). \
\ (EXISTS x:A. [w = Inl(A,B,x) : (A->bool) * (B->bool)]) | \
\ (EXISTS y:B. [w = Inr(A,B,y) : (A->bool) * (B->bool)]) }"),

("Inl_def", "Inl(A,B,a) == <lam x:A.term([ a = x : A ]), lam y:B.False>"),
("Inr_def", "Inr(A,B,b) == <lam x:A.False, lam y:B.term([ b = y : B ])>"),
("when_def",
  "when(A,B,C,p,c,d) == PICK z:C.  \
\ (ALL x:A. [ p = Inl(A,B,x) : A+B ] --> [ z = c(x) : C ]) & \
\ (ALL y:B. [ p = Inr(A,B,y) : A+B ] --> [ z = d(y) : C ])"  )];
end;
```

ON CONNECTIONS BETWEEN CLASSICAL AND CONSTRUCTIVE SEMANTICS

Starchenko S.S. Voronkov A.A.
Institute of Mathematics
630090 Novosibirsk-90
USSR

Earliest investigations of constructive semantics of the first order logic were carried out by Kolmogorov [1] for propositional language and Kleene [2] for arithmetical language. Later some similar semantics have been developed for interpretation of intuitionistic predicate calculus, intuitionistic arithmetic and variants of constructive analysis and set theory [3-8]. Further developments are connected with realizability-like semantics based on the theory of constructive models [9]. These are n-realizability [10,11], constructive truth in the sense of [12] and constructive truth in the sense of [13]. There are many reasons to investigate such semantics. At the first place they shed some light on the foundations of logic. At the second place many interesting theorems on these semantics were proved that connect together theory of constructive models, model theory and constructive logic [12,13]. At the third place the semantics have applications in the theory of program synthesis since they reflect constructive understanding of data types in programming language.

It is well known that data in programs can have various structures (data types) and program execution can be considered as processing of the data. Thus to extract programs from proofs at least one thing is needed: to build calculi describing properties of data types. One of possible approaches is to consider data type as a model and to base appropriate calculus on the theory of this model. For example if the data type in hand is natural numbers then for an appropriate calculus can be taken the intuitionistic (Heyting) arithmetic.

But when more sophisticated objects than natural numbers are considered, it is not a very simple matter to construct suitable calculus. Some attempts to give general definitions were undertaken in [10-15]. All of these definitions are based on semantics very close to Kleene's realizability [2]. Satisfability of a formula in such semantics can be considered as a constructive analogue of a truth of the formula. From the program synthesis' viewpoint constructively true formulae are those that can be taken as axioms of corresponding constructive calculus.

The investigations of constructive truth revealed many interesting mathematical problems [12.13]. In this paper we shall investigate connections between constructive and classical truth.

The main questions for us are the following:

1) When we can use in constructive proofs classically true formulae with no constructive foundation of them? More exactly. it means finding the classes of formulae (or classes of models) for which constructive truth coincides with classical one.

2) To carry out how much can diverse the classical theory of model from its constructive theory. In other words do there exist models, the theories of which are "good" from the classical viewpoint but "bad" constructively. Such models are not very suitable for applications in program synthesis since very little part of their properties can be used in corresponding constructive calculus.

According to the two kinds of problems above we divided the article into two sections (2 and 3). Section 1 contains basic definitions. New results are given with more or less complete proofs. Those published in [13] are given without proofs.

This paper was written when authors were in Karaganda State University. We are grateful to all employees of the higher algebra department (and especially Vladimir D. Ten) who made our work very fruitful.

1. Basic definitions

Let $\underline{M} = \langle M, \sigma \rangle$ be a model of signature $\sigma = \langle P_0, P_1, \ldots, f_0, f_1, \ldots \rangle$ and ν be an enumeration of $\underline{M}$. i.e. ν maps the set of natural numbers on $\lambda \circ M$. The pair $\langle \underline{M}, \nu \rangle$ is called an <u>enumerated model</u> iff the set

$\{\langle i,j \rangle : j$ is a Gödel number of $\langle x_0, \ldots x_n \rangle$ with $f_i(\nu x_1, \ldots, \nu x_n) = \nu x_0\}$ is decidable.

Let Σ be an expansion of σ by the constants $\nu_0, \nu_1, \ldots$. An enumerated model $\langle \underline{M}, \nu \rangle$ is called <u>constructive</u> iff the set

$$\{\langle i, j_1, \ldots, j_k \rangle : \underline{M} \models P_i(\nu j_1, \ldots, \nu j_k)\}$$

is decidable.

An enumerated model $<\underline{M},\nu>$ is called strongly constructive iff the set of all sentences of Σ true in $\underline{M}$ is decidable.

Now we are going to define constructive semantics of formulae. For the sake of uniformity we change their original names. Defining below n-realizability corresponds to n-realizability from [11], p-realizability corresponds to constructive truth in generalized models [12], v-realizability and c-realizability are constructive truth and classical realizability from [13].

We begin with the definition of n-realizability. For every sentence A of signature Σ we construct an enumerated set $\underline{S}_A=<S_A,\nu_A>$ and the relation $s\ \underline{n}\ A$, where $s\in S_A$ according to the following laws:

1) For atomic A let $\underline{S}_A \equiv \underline{1}$ (one-element enumerated set) and $s\ \underline{n}\ A$ iff A is true.

2) $\underline{S}_{A\&B} \equiv \underline{S}_A \underline{x} \underline{S}_B$ (the direct product of two enumerated sets $\underline{S}_A$ and $\underline{S}_B$ [16]) and $<s_1,s_2>\underline{n}\ A\&B$ iff $s_1\underline{n}\ A$ and $s_2\underline{n}\ B$.

3) $\underline{S}_{AvB}\equiv \underline{S}_A+\underline{S}_B$ (the direct sum of enumerated sets $\underline{S}_A$ and $\underline{S}_B$ [16]).

Let inl and inr are natural mappings of $\underline{S}_A$ and $\underline{S}_B$ respectively in $\underline{S}_A+\underline{S}_B$. Then $s\ \underline{n}\ AvB$ iff s is of the form $inl(s_1)$ such that $s_1\underline{n}\ A$ or s is of the form $inr(s_2)$ with $s_2\ \underline{n}\ B$.

4) $\underline{S}_{A->B}\equiv Mor_p(\underline{S}_A,\underline{S}_B)$ (the enumerated set of all partial morphisms from $\underline{S}_A$ to $\underline{S}_B$ [16]). $s\ \underline{n}\ A->B$ iff for every a such that $a\ \underline{n}\ A$, $s(a)\ \underline{n}\ B$ holds.

5) $\underline{S}_{\sim A}\equiv \underline{1}$ and $s\ \underline{n}\ \sim\underline{A}$ iff for every $s_1\in\underline{S}$ $s_1\ \underline{n}\ A$ does not hold.

6) $\underline{S}_{(Ex)A(x)} \equiv <\underline{M},\nu> \underline{x}\ \underline{S}_{A(\nu 0)}$ and $<\nu m,s>\ \underline{n}\ (Ex)A(x)$ iff $s\ \underline{n}\ A(\nu m)$.

7) $\underline{S}_{\forall xA(x)}\equiv Mor_p(<\underline{M},\nu>,\underline{S}_{A(\nu 0)})$ and $s\ \underline{n}\ \forall xA(x)$ iff for every $m\in n$ $s(\nu m)\ \underline{n}\ A(\nu m)$ holds.

The sentence A is called n-realizable in an enumerated model $<\underline{M},\nu>$ iff there exists s such that $s\ \underline{n}\ A$. We omit $<\underline{M},\nu>$ when its identity is clear.

The definition of p-realizability differs from n-realizability in the following:

1) Everywhere "enumerated set" should be replaced by the set N of natural numbers, the pair $<n_1, n_2>$ by its Gödel number, instead of inl(n) and inr(n) the Gödel numbers <0,n> and <1,n> are taken and "morphism" should be replaced by a "Kleene number of the partial recursive function".

2) For atomic formulae A the relation n $\underline{p}$ A is supposed to be given and to satisfy the following properties:

a) A is true iff there exists an n with n $\underline{p}$ A;

b) if P is a predicate symbol and the values of terms $t_1, \ldots, t_n$ are equal to the values of terms $t_1', \ldots, t_n'$ then n $\underline{p}$ $P(t_1, \ldots, t_n)$ implies n $\underline{p}$ $P(t_1', \ldots, t_n')$.

The definition of v-realizability is similar to the definition of n-realizability, but it needs an auxillary definition of c-realizability. The latter differs from n-realizability in the following:

Everywhere "enumerated set" should be replaced by Scott's information system [17], an enumerated set $\underline{1}$ should be replaced by one-element information system with the same designation and instead of enumerated set $<\underline{M}, \nu>$ an information system corresponding to $<\underline{M}, \nu>$ should be taken (for the complete definition see [13]).

An element a v-realizes formula A iff a $\underline{c}$ A and a is a computable element [13,15,18].

For the formula A, enumerated model $<\underline{M}, \nu>$ and $r \in \{n,p,v\}$ we write $<\underline{M}, \nu> \models_r A$ iff A is r-realizable in $<\underline{M}, \nu>$.

At the conclusion of this section we sketch how realizability-like semantics can be applied to program synthesis. Let Π be a constructive proof of the closed formula (Ex)A(x). Using Π we find an element a which realizes this formula. By the definition of, say, v-realizability it has the form <n,b> where b $\underline{v}$ $A(\nu n)$. Thus the element νn "meets specification A(x)".

Reasoning by analogy, if a realizes $\forall \bar{x}(Ey)A(\bar{x},y)$ then we can effectively find an algorithm which for any tuple $\bar{x}$ of elements from the model $\underline{M}$ gives an y such that $A(\bar{x},y)$ is realizable. Thus this algorithm can be considered as a program meeting specification $A(\bar{x},y)$ where $\bar{x}$ are input variables, y is output variable.

2 On coincidence of the constructive truth with the classical one

In this section we study the following question: under what conditions constructive truth coincides with the classical truth.

In [13] we pointed out that all strong constructive models possess this property and in general inverse is not true. Here we give a characterization of such models not appealing to the definition of realizability. Let us formulate some auxiliary statements first.

Let L be well known (originating with Kolmogorov) operator on the set of formulae defined as follows:

1) $L(A)={\sim}{\sim}A$ for atomic A.

2) $L(A\&B)=L(A)\&L(B)$

3) $L(AvB)=L(A)vL(B)$

4) $L(A\text{->}B)={\sim}(L(A)\&{\sim}L(B))$

5) $L({\sim}A)={\sim}L(A)$

6) $L((Ex)A)={\sim}\forall x{\sim}L(A)$

7) $L(\forall xA)=\forall xL(A)$

Lemma 1. Let $\langle \underline{M}, \nu\rangle$ be an enumerated model of signature σ and A be a sentence of this signature. Then $\underline{M}\models A$ iff $\langle \underline{M}, \nu\rangle\models_r L(A)$ where $r\in\{n,p,v\}$.

Let μ be a Gödel numbering of the set of all sentences of σ.

Corollary. Under conditions of Lemma 1

$$\mu^{-1}(Th(\underline{M}))\leq_1 \mu^{-1}(Th_r(\underline{M},\nu)).$$

Thus the theory of model cannot be more complicated than its constructive theory.

Let us fix the enumerated model $\langle \underline{M}, \nu\rangle$ of signature σ. We say that the formula A of signature σ is decidable iff the set $\{\langle m_1,\dots,m_n\rangle : \underline{M}\models A(\nu m_1,\dots,\nu m_n)\}$ is decidable.

Lemma 2. The following are equivalent.

1)The formula A is decidable.

2)$\langle \underline{M}, \nu\rangle\models_r L(A)v{\sim}L(A)$.

Theorem 3. For $r\in\{n,v\}$ the following are equivalent:

1)Any formula A of signature σ is decidable.

2)$Th(\underline{M})$ coincides with $Th_r(\underline{M},\nu)$.

Proof. (2=>1) Assume that there exists an undecidable formula A. By Lemma 2 the formula $L(A)$ v ${\sim}L(A)$ is not constructively true. But obviously it is classically true.

(1=>2) Let any formula A be decidable. By induction on the length of A we prove the following statement: Let all free variables of A belong to $x_1,\dots,x_n$. Then there exists an f such that if

$A(\vartheta m_1,\ldots,\vartheta m_n)$ is true then $f(\vartheta m_1,\ldots,\vartheta m_n)\,\underline{v}\,A(\vartheta m_1,\ldots,\vartheta m_n)$. To prove it we simply define such f:

1) For atomic A f_A is the constant function having on any element the unique element of $\underline{1}$ as the value.

2) $f_{B\&C}(\bar{x})=\langle f_B(\bar{x}),f_C(\bar{x})\rangle$.

3)
$$f_{BvC}(\bar{x}) = \begin{cases} inl(f_B(\bar{x})) & \text{if } M\models B(\bar{x}) \\ inr(f_C(\bar{x})) & \text{if } M\models C(\bar{x})\,\&\,{\sim}B(\bar{x}) \\ \text{Undefined otherwise} \end{cases}$$

4) $f_{B\to C}(\bar{x})$ is the functional $\lambda y[f_C(\bar{x})]$.

5) $f_{\sim B}$ is as for atomic formulae.

6) $f_{(Ey)B(y)}=\langle\vartheta(\mu yB(\bar{x},y)),f_B(\bar{x},\mu yB(\bar{x},y))\rangle$, where $\mu yB(\bar{x},y)$ means "the y with the least n-number such that $B(\bar{x},y)$ holds".

7) $f_{\forall yB(y)}(\bar{x})$ is the functional $\lambda y[f_{B(y)}(\bar{x})]$.

Computability of f_A is trivial. The property $f_A(m)\,\underline{v}\,A(m)$ follows from the definition of realizability.

<u>Corollary.</u> If $\langle\underline{M},\vartheta\rangle$ is a strongly constructive model then $Th(\underline{M})$ coincides with $Th_r(\underline{M},\vartheta)$ for $r\in\{n,v\}$.

<u>Theorem 3'.</u> Let (M,n) be an enumerated model of signature σ. Then $Th(\underline{M})$ coincides with $Th_p(\underline{M},\vartheta)$ iff the following two conditions are satisfied:

1) Every formula of signature σ is decidable.

2) For any predicate symbol P of σ there exists a partial recursive function f such that for any natural numbers $m_1,\ldots,m_n$ $\underline{M}\models P(\vartheta m_1,\ldots,\vartheta m_n)$ implies $f(m_1,\ldots,m_n)\ \underline{p}\ P(\vartheta m_1,\ldots,\vartheta m_n)$.

Proof trivially follows from the proof of Theorem 3 and the fact that the latter condition is equivalent to p-realizability of the formula

$$\forall x_1\ldots\forall x_n({\sim}{\sim}P(x_1,\ldots,x_n)\to P(x_1,\ldots,x_n)).$$

<u>Corollary.</u> Let $\langle\underline{M},\vartheta\rangle$ be a strong constructive model. Then the following are equivalent:

1) $Th(\underline{M})$ coincides with $Th_p(\underline{M},\vartheta)$.

2) See the condition 2) from theorem 3'.

Above we have characterized models for which two notions of truth are equivalent. Now we study the question for what formulae are these two notions equivalent independently of enumerated model.

For n-realizability there is a very simple family of such formulae:

Theorem 4. Let $\underline{M}$ be a model and formula A have no occurences of disjunction and existential quantifier. Then for any enumeration ν of $\underline{M}$ we have $\underline{M} \models A$ iff $\langle \underline{M}, \nu \rangle \models_n A$.

In the case of v-realizability we come to the well known class of formulae: Harrop formulae. This class of formulae was discovered in [19] in connection with investigations of disjunctive property and E-property of intuitionistic predicate calculus with auxiliary axioms. Harrop had shown that if we add to intuitionistic predicate calculus any set of Harrop formulae as axioms then the resulting calculus satisfies these properties. Later Goad used this feature of Harrop formulae in the system of program synthesis [20].

Theorem 4'. Let $\langle \underline{M}, \nu \rangle$ be an enumerated model and a be a Harrop formula. Then $\underline{M} \models A$ iff $\langle \underline{M}, \nu \rangle \models_v A$.

For p-realizability we do not know any analogue of theorems 4 and 4'. At the conclusion of section we note that theorems here show that the definitions of realizability in which realizations of atomic formulae do not depend on formula are more fruitful.

3 Examples of model with decidable theory and undecidable constructive theory

In corollary to Lemma 1 we show that the theory of a model is always simpler than its constructive theory. In this section we will show that there exist models with very simple (decidable) theory and constructive theory as complicated as you need. Thus the using of constructive semantics makes the first order language more expressive.

Let $R \subsetneq N$ be an infinite set of natural numbers which contains 0 and has infinite complement. On the sum $M=N^2$ we define function f(x) as follows:

$$f(\langle m,n \rangle) = \begin{cases} \langle m+1,0 \rangle & \text{if } n=0 \\ \langle 0,0 \rangle & \text{otherwise} \end{cases}$$

We also define on M the relation E(x,y) as follows: for $a_1, a_2 \in M$, $a_1 = \langle m_1, n_1 \rangle$, $a_2 = \langle m_2, n_2 \rangle$ relation $E(a_1, a_2)$ holds iff one of the following three conditions are satisfied:

(i) $m_1 = m_2$, $m_1 \in R$

(ii) $m_1 = m_2$, $n_1 \in R$, $n_2 \in R$.

(iii) $m_1 \notin R$, $m_2 \notin R$, $n_1 \notin R$, $n_2 \notin R$.

Let $B_0(x)$ and $B_1(x)$ be the following formulae in the language $\langle =, f, E \rangle$:

$$B_0(x) \equiv (Ey_1 y_2)(\sim y_1 = y_2 \& x = f(y_1) \& x = f(y_2));$$

$$B_1(x) \equiv \sim B_0(f(x))$$

If $C(x)$ is a formula and $n \in N$ then $(E!x)C(x)$ and $(E>n\ x)C(x)$ are as usual shorthands for

$$(Ex)C(x) \& \forall x_1 x_2 (C(x_1) \ \& \ C(x_2) \to x_1 = x_2) \text{ and}$$

$$(Ex)_0 \ldots x_n \left(\underset{0 \le i \le n}{\&} C(x_i) \& \underset{i \ne j}{\&} \sim x_i = x_j \right)$$

It is easy to prove the following lemma:

Lemma 5. In the model $\underline{M} = \langle M, f, E, = \rangle$ the following propositions are true:

$$C_0 \equiv \forall x E(x,x) \& \forall xy (E(x,y) \to E(y,x) \&$$

$$\forall xyz (E(x,y) \& E(y,z) \to E(x,z));$$

$$C_1 \equiv (E!x) B_0(x)$$

$$C_2 \equiv \forall x (C_0(x) \to C_1(x))$$

$$C_{3,n} \equiv (E>n\ z) \forall x (E(x,z) \to \sim B_1(x)), \ n \in N;$$

$$C_4 \equiv \forall z (B_1(z) \to B_1(f(z))) \ \& \ \forall z (B_1(z) \& \sim B_0(z) \to (E!y)(f(y) = z));$$

$$C_{5,n} \equiv (Ey_0 \ldots y_n n (\underset{i \ne j}{\&} \sim E(y_i, y_j)), \ n \in N;$$

$$C_{6,n} \equiv \forall y (E>n\ x) E(y,x), \ n \in N;$$

$$C_7 \equiv \forall z ((Ex)(E(x,z) \& B_1(x)) \to (E!x)(E(x,z) \& B_1(x)));$$

$$D_{n,m} \equiv \forall z\,\tilde{}\,(fn(z)=fm(z)),\ n<m\in N.$$

Let S be the following set of sentences:

$\{C_0, C_1, C_2, C_4, C_7\} \cup \{C_{3,n}, C_{5,n}, C_{6,n} : n\in N\} \cup \{D_{n,m} : n<m\in N\}$ and let $T=Th(\underline{M})$.

From Lemma 6 it follows that $T|-S$. The inverse is also true:

Lemma 6. $S|-T$.

Proof. Using Taimanov-Ehrenfeuht method one can show that if $N|=S$ then $N|=T$. Thus $S|=T$.

It is easy to see that the set S is decidable in any effective enumeration of the set of all formulae. Thus from Lemma 6 it follows

Corollary. T is decidable.

Let n be any effective enumeration of N (e.g. Cantor's). Let us consider the enumeration model $<\underline{M},\nu>$.

Lemma 7. Let D be the set of Gödel numbers of decidable formulae of $<\underline{M},\nu>$. Then $R\leq_1 D$.

Proof. We can assume that R is not decidable. Consider the sequence of formulae

$F_n(x) \equiv (Ey)(B_o(y)\&E(fn(y),x))$.

Let

$X_n \equiv \{\nu^{-1}(a):\underline{M}|=F_n(a)\}$.

By definition of E and F we have

$$X_n = \begin{cases} \nu^{-1}(\{<n,m>:m\in R\}) \text{ if } n\in R \\ \nu^{-1}(\{<n,m>:m\in N\}) \text{ otherwise} \end{cases}$$

So X_n is decidable iff $n\notin R$.

From Lemmas 1 and 7 it follows

Theorem 8. Let $r\in\{n,p,v\}$. Then for every set R of natural numbers there exists an enumerated model $<\underline{M},\nu>$ such that $Th(\underline{M})$ is decidable and $R\leq_1 \mu^{-1}(Th_r(\underline{M},\nu))$.

The following definition of constructive truth of formula in a model can be taken: A is constructively true in M iff $<\underline{M},\nu>:=_r A$ holds for any enumeration of $\underline{M}$ (see [12]). It is interesting to generalize our results for the constructive truth in a model. But our construction is not suitable for this.

References

1. Kolmogorov, A.N.: Zur Deutung der intuitionistische Logic. Math. Z. (1932), v.35
2. Kleene S.C.: On the interpretation of intuitionistic number theory. J Symb. Logic, (1945), v.10, 109-124
3. Kreisel G.: Interpretation of analysis by means of constructive functionals of finite types. Constructivity of Mathematics, North Holland, (1959), 101-128
4. Gödel K.: Uber eine noch nicht benutzte Erweiterung des finiten Standpunktes. Dialectica, 3/4, (1958), v.12, 280-287
5. Kleene S.C.: Vesley R.Y. The foundations of intuitionistic mathematics. North Holland, (1956)
6. Howard W.A.: The formulae-as-types notion of construction. Festschrist on the Occasion of H.B.Curry's 80th Birthday, Academic Press, (1980), 479-490
7. Diller J.: Modified realization and the formulae-as-types notion, Ibid, 491-502
8. McCarthy Ch.: Realizability and recursive set theory. Ann. Pure and Appl. Log., 2(1986), 32, 153-184
9. Ershov Yu.L.: Decidability problems and constructive models. Moscow, Nauka, (1980) (In Russian)
10. Nepeivoda N.N., Sviridenko D.I.: On the theory of program synthesis. Trudy Math. Inst., 2(1982), 159-175
11. Voronkov A.A.: Automatic program synthesis and realizability. Matematicheskie Osnovy Raspoznavaniya Obrasov, Erevan, Acad. Sci. of Armenian SSR, (1985), 38-40 (In Russian)
12. Plisko V.E.: On language with constructive logical connectives. Soviet Math. Doklady, 1(1987), v.296
13. Voronkov A.A: Model theory based on constructive notion of truth. Trudy Math. Inst., Novosibirsk, (1988), v.8, 25-42 (In Russian)
14. Voronkov A.A.: Logic programs and their synthesis. Preprint of Math. Inst., Novosibirsk, 23(1986), p.32 (In Russian)
15. Voronkov A.A.: Synthesis of logic programs. Ibid, 24(1986), p.42.
16. Ershov Yu.L.: Theory of enumerations. Moscow, Nauka, 1977 (In Russian)
17. Scott D.S.: Domains for denotational semantics Lect. Notes in Comp. Sci. 1982, v.140, 577-612
18. Kamimura T., Tang J.: Effectively given spaces. Theoretical Comp. Sci. (1984)

19. Harrop R.: Concerning formulas of the types A->BvC, A->(Ex)B(x) in intuitionistic formal system. J. Symb. Logic, 1(1962), v.27, 27-32

20. Goad C.A.: Computational uses of the manipulation of formal proofs. Stanford Univ. Tech. Report, CS-80-819, (1980), p.122

FLOW-DIAGRAMS WITH SETS

Alexey P. Stolboushkin

Program Systems Institute of the USSR Academy of Sciences, P.O. box 11, Pereslavl-Zalessky 152140 Soviet Union

ABSTRACT.

We investigate flow-chart programs equipped with sets to keep intermediate results of computations. It is shown that such programs form the weakly semi-universal class, i.e. are able to distinguish finite interpretations from infinite. It is not known whether this class is semi-universal, but if we in addition equip programs with level 2 sets, the resulting class is proved to be semi-universal, i.e. every generalized computable function in every infinite interpretation can be computed by some programs of the class. Some other properties of these two classes are proved.

0. INTRODUCTION.

In the comparative schematology as well as in the theory of Dynamic Logics the flow-charts are investigated which are additionally equipped with stacks, arrays etc. (see [T], [H], [KSU]). We investigate here the flow-chart schemes with (finite) sets in the same sense, i.e. as additional program memory. Of course these flow-chart schemes allow instructions of some new type to process the sets. We will call them **set instructions**. We consider also finite sets of finite sets (level 2 sets) as well as level 2 set instructions.

The investigation of flow-chart schemes with sets is interesting to theoretical point of view. Intuitively, sets are closely related to so called boolean arrays (see [U]), namely, n-dimensional boolean arrays in some sense correspond to level n sets. But it is not known whether flow-charts with one-dimentional boolean arrays is weakly semi-universal (in contrast with the theorem 1 below).

In Section 2 we show (Theorem 1) that the class of flow-chart schemes with level 1 sets is weakly semi-universal. The question about its semi-universality remains open. I conjecture it is not semi-universal. If so, we have a wonderful example of a programming formalism which is weakly semi-universal but not semi-universal.

1. Preliminaries

I give here necessary definitions only. The reader is referred to [KSU] for the full list of definitions as well as for the full bibliography.

By a **signature** we mean a sequence Ω of function, relation and constant symbols, each nonconstant symbol with a nonnegative arity. If not stated otherwise we assume that the signatures under consideration are always finite, and contain the symbol = of equality. We say that a signature is unary iff all its function symbols are unary. We assume that Ω is always nontrivial, i.e. it contains at least one function symbol, and that the equality symbol ''='' does always occur in Ω.

The definitions of first-order language over the signature Ω is well-known. In particular, we use standard definitions of expressions and formulae over the signature Ω.

A Ω-**structure** is an arbitrary set A together with an appropriate interpretation of symbols of Ω. A Ω-**interpretation** is a pair (A,a) where A is a Ω-structure and a is a sequence of elements in A which generate A. This definition is motivated by the observation that all values which occur in a computation of a program must belong to the structure generated by the input values.

Let now for every natural n we fix the set of variables of level n. First-order language may handle only level 0 variables.

Definition 1. First-order expressions will be called level 0 expressions, while quantifier-free first-order formulae will be called level 0 formulae. All the variables in level 0 expressions are free.

Let the definitions of level i formulae and level i expressions be given. Let t be a level i expression. Then

1. Every level $i+1$ variable is a level $i+1$ expression.

2. $\emptyset^{i+1}$ is a level $i+1$ expression (denoting the empty set of the level $i+1$; we will occasionally omit $i+1$).

3. $\{t\}$ is a level $i+1$ expression (denoting the level $i+1$ singleton containing only the element denoted by t).

4. Let $u_1,\ldots,u_n$ be some free variables of t, of levels $0\le j_1,j_2,\ldots,j_n\le i$ resp., and let $T_1,T_2,\ldots,T_n$ be expressions of levels $j_1+1,j_2+1,\ldots,j_n+1$. Let φ be a level $i+1$ formula. Then $\{t\,|\,u_1\in T_1,u_2\in T_2,\ldots,u_n\in T_n,\varphi\}$ is a level $i+1$ expression. It denotes the level $i+1$ set formed by the following way. We consider such values of $u_1,\ldots u_n$ that both the tuple of these values belongs to $Val(T_1)\times\ldots\times Val(T_n)$ and φ holds on these values of $u_1,\ldots u_n$. Then we form the level $i+1$ set of the values of t under all such tuples of values of $u_1,\ldots u_n$. φ and t may contain other variables being *parameters*. All occurences of $u_1,\ldots u_n$ in that expression, except of possible occurences in $T_1,T_2,\ldots,T_n$, will be called closed occurences.

5. If t_1,t_2 are level $i+1$ expressions then $(t_1\cup t_2)$, $(t_1\cap t_2)$, and $(t_1\setminus t_2)$ are too (theirs correspond to standard set-theoretic operations).

6. If φ is a level i formula, it is a level $i+1$ formula too.

7. Let u be a level i variable, t_1 be a level $i+1$ expression, and let φ be a level i formula. Then $(\exists u\in t_1)\varphi$ and $(\forall u\in t_1)\varphi$ are level $i+1$ formulae. Their semantics are clear. Such an occurence of the variable u will be called a **closed** occurence.

8. If φ, ψ are level $i+1$ formulae, then $(\varphi\&\psi)$, $(\varphi\vee\psi)$, $(\neg\varphi)$ are too.

9. There are neither other level $i+1$ formulae nor other level $i+1$ expressions.

10. Every occurence of every variable in level $i+1$ expression will be called free iff it is not closed (see (4), (7) above). Let t_1 be a level $i+1$ expression. A variable u will be called **free in** t_1 iff there is a free occurence of u in t_1.

We will use some common abbreviations, such as parenthesis omitting, the signs $\subset$, $\le$, $\notin$ etc.

DEFINITION 2. An expression t will be called a **pure** level i expression iff t is a level i expression and t contains either no subformulae, or level $i-1$ subformulae only.

A **(deterministic) flow-diagram** (or "flow-chart scheme", or "flow-chart program") over Ω, is a finite directed graph of nodes labelled by instructions. The **instructions** may take one of the following forms:

assignments $x:=c$, or $x:=y$, or $x:=f(x_1,\dots,x_n)$, or

tests $r(x_1,\dots,x_n)?$, or

start, or *stop*.

Each node labelled by an assignment has one outcoming edge. There are two edges outcoming from each test node - one for "yes" and one for "no", while *stop* has no outcoming edges at all. The number of edges incoming to nodes is arbitrary, except of course of the *start* node which has no incoming edges. (We assume that there is only one such node). In the above, $x, y, x_1, \dots, x_n$ are (simple) variables, and f, r, c are respectively function, relation and constant symbols in Ω.

Some of the variables in a program may be considered to be input or output variables. If (A,a) is a Ω-interpretation such that the length of a is equal to the number of input variables in a program S, then we may define in an obvious way the **computation** of S in (A,a) and the output of this computation (if it exists). Thus, for any Ω-structure A, S will define a partial function S_A from A^k into A^m (where k,m represent the numbers of input and output variables), given by assigning to any a in A^k the output of the computation of S in the interpretation generated in A by a. The class of all deterministic flow-diagrams over Ω is denoted by DFD_Ω, or DFD if Ω is fixed.

A flow-chart is called to be **loop-free** iff it contains no cycles perhaps except of those of lenght 1 (such a cycle corresponds to a "diverge" instruction).

Now we are going to extend this formalism by adding various features:

(A) Nondeterminism

Nondeterministic flow-diagrams are allowed to contain a choice instruction "*or*" (two outcoming edges). The semantics is defined as in the deterministic case, but S_A is now a relation $S_A \subseteq A^k \times A^m$ rather than a partial function, since there are possible many computations on the same input. The class of nondeterministic flow-diagrams is denoted by NFD_Ω or NFD.

(B) Sets

(Deterministic) flow-chart schemes with (level 1) sets (denoted by $FDSet_1$) are equipped with level 1 sets and have additional instructions of the form:

(assignments) $u:=t$, where u is a variable of the level 1, and t is a level 1 expression.

(low level) $x:=t$, where x is a level 0 variable and t is a level 1 expression. This instruction has two outcoming arrows, and the 2nd arrow is always come to the same node (the cycle of the length 1). If the value of t is a singleton, i.e. contains one element only, then this sole element will be assign to x and the scheme come through the 1st arrow. Otherwise x won't be changed and the scheme come through the 2nd arrow (diverges).

Additionally flow-chart schemes with level 1 sets may use arbitrary level 1 formulae as tests.

The definition of **flow-chart schemes with level 2 sets** ($FDSet_2$) is of the same sort as with level 1 sets, and additionally introduces the assignments of the level 2 and low-level from 2 to 1. Arbitrary level 2 formulae are admitted as tests in $FDSet_2$ schemes.

The definitions of **pure** $FDSet_1$ **and pure** $FDSet_2$ require that only pure level 1 (level 2) expressions may be used in schemes, but in pure $FDSet_2$ arbitrary level 1 expressions are admitted (this is natural for the level 2).

We say that programs S and S' are equivalent iff, for every A, it holds that $S_A=S'_A$. A class of programs K is reducible to another class K' (in symbols: $K\le K'$) iff, for any program in K, there exists an equivalent one in K'. (Results of the form $K\le K'$ may depend on a signature. We will put "$K_\Omega\le K'_\Omega$" to state that the inequality holds for a given Ω. If not set otherwise, the notation "$K\le K'$" will mean "for every Ω, $K_\Omega\le K'_\Omega$"). We write $K\equiv K'$ to denote that $K\le K'$ and $K'\le K$, and $K<K'$ to denote that $K\le K'$ but not $K'\le K$. (The notation $K<K'$ without any reference to a signature will mean "for all Ω, $K_\Omega\le K'_\Omega$ and for some Ω, $K_\Omega<K'_\Omega$").

Let S be a program over a signature Ω and let A be a Ω-structure. We say that an **element** b **occurs in a computation of** S **in** (A,a) iff b occurs in a or it is assigned to a (simple) variable in some step of the computation. The **program** S **can search every finite** Ω**-interpretation** (A,a) (a is assumed to be of a length equal to the number of input variables in S) iff the following conditions hold:

- for each finite (A,a) and each finite (converging) computation of S in (A,a), all elements of A occur in the computation.

- for each finite (A,a) there is at least one finite computation of S in (A,a).

If a class K of programs contains for all k a program S_k that stops on every finite Ω-interpretation (A,a) with a in A^k, and does not stops in every infinite, then we say that K **is weakly**

semi-universal. K is said to be **semi-universal** iff for each k, there is S_k in K, such that S_k defines a unique enumeration of elements in every finite interpretation (A,a) with a in A^k, i.e., S can search any interpretation, and in addition the order of elements of A given by their first occurrences in all computations of S in (A,a) is the same.

If a class is semi-universal then it must contain a program S being able to enumerate in a unique order every (not necessarily finite) interpretation, by defining a successor function.

Let K be a class of programs over Ω. The **logic of programs in** K, denoted $L(K)$, is defined as classical first-order logic over Ω, with use of the additional construct $\langle\ \rangle$, of the following meaning: for a program S and a formula α, $\langle S\rangle\alpha$ is true in (A,a) iff there is a finite computation of S on a such that the output satisfies α.

It is an easy exercise to prove that for any formula α of $L(K)$ there is an equivalent formula β, such that all the occurrences of $\langle\ \rangle$ in β are in the context $\langle S\rangle true$.

We say that a logic L_1 **is reducible to another logic** L_2 (in symbols: $L_1 \le L_2$) iff for any formula α in L_1 there is an equivalent formula β in L_2.

2. Results

An obvious result is:

Theorem 1. *The class $FDSet_1$ is weakly semi-universal.*

Sketch of proof. Usind just one set (of the level 1), we can easily define the following "pumping" process. At first time, we define our set as the set of all generators of the given interpretation. At any following time, we define new value of our set as the union of its old value and all "atomic" expressions over elements of the same old value. If, at some step, our set doesn't increase, then we stops in the step. We need in one additional set to check whether our set increases.

■

Usually the result about weakly semi-universality of a deterministic formalism automatically gives us the result about its semi-universality. This is not the case of flow-diagrams with sets, because the set-instructions are not of sequential type and so do not define any ordering of generated values of the interpretation.

More precisely, the above process defines an equivalency relation on elements of the interpretation as follows.

Let $(A,a_1,\dots,a_n)$ be an interpretation. We define L_n, $n\geq 0$, by induction as follows:

1. $L_0=\{a_1,\dots,a_n\}$.
2. $L_{n+1} = \bigcup_{t\in T} \{t(e_1,\dots,e_{m_t}) \mid e_1,\dots,e_{m_t} \in \bigcup_{k=0}^{n} L_k\} \setminus \bigcup_{k=0}^{n} L_k$,

where T is the set of all atomic expressions of the signature.

So each L_n is just a class of this equivalency, and $a\equiv b$ iff $a,b\in L_n$ for some n.

Let, for some fixed interpretation (A,a) we have got a flow-chart sub-scheme (i.e. marked sub-graph) CH with one incoming and one outcoming arrows, which sub-scheme defines a "choice", in the following sense. Let $L\subseteq L_n$ for some n. Then, having that L as the value of certain level 1 variable in the incoming arrow, sub-scheme CH gives some element of L as the value of certain (simple) variable in the outcoming arrow); and CH always finishes for any L as above.

It is easy to prove

LEMMA 2. *$FDSet_1$ ($FDSet_2$) is semi-universal on the class I of interpretations iff for any $n>0$ there exists flow-chart sub-scheme CH_n in the class $FDSet_1$ ($FDSet_2$) such that CH_n defines the above discussed "choice" on any $(A,a)\in I$, $a\in A^n$.*

■

I don't know whether $FDSet_1$ is a semi-universal class or not. I conjecture it isn't. To my mind this is an interesting open question. If my conjecture is true, $FDSet_1$ is an example of the programming formalism which is weakly semi-universal but not semi-universal.

Never the less, $FDSet_1$ is semi-universal for a large class of interpretations. The following theorem is proved by P.Urzyczyn.

THEOREM 3. *$FDSet_1$ is semi-universal on unoids.*

■

REMARK. *Low-level assignments are not necessary in this theorem.*

The well-known definitions of pebble games and of the pebble complexity are given, for example, in [K]. Let us consider only binary function symbols and the pebble game with potentially infinite number of pebbles. Every expression is presented as a directed graph. The *height* of a node is the length of the longest way to the node from the leafs (this height corresponds to the nesting level of the sub-expression labelling this node). Then

LEMMA 4. *The pebble game on arbitrary expression can be always organized by such a way that, in an arbitrary step, almost all active pebbles have the different heights, except of perhaps four pebbles.*

We leave the proof to the reader.

■

THEOREM 5. *$FDSet_1$ is universal on infinite structures.*

Sketch of proof. With no loss of generality we may consider only one-generated interpretations of the sole binary signature function symbol. The signature also contains a *constant* a whose value is the generator. Such interpretations are called *groupoids*.

Since these groupoids are infinite, that "pumping" process gives us the possibility to use counters. Thus $FDSet_1$ has on considered groupoids the universal control power (see [KSU] for details).

The only thing we need in to finish the proof is such a sub-scheme E that, for some fixed numeration $t_1, t_2, \ldots$ of closed expressions, E, given an arbitrary natural number n and an arbitrary one-generated groupoid G, either gives the value of t_n in G, if t_n is unable to be abbreviated in G, or gives the reply "t_n can be abbreviated in G", otherwise. So, the required E has got to have one incoming, and two outcoming arrows.

A (level 0) expression without variables is called closed. We say, that a given closed expression t can be abbreviated in a given G, iff either t or one of its subexpressions is equivalent in G to some another one from one of preceeding equivalence class (see the definition of equivalence classes L_m above).

The difficulty is, of course, the pebble complexity of elements, in general unlimited (see [K] for details). We may use a level 1 variable to keep intermediate elements, but for getting an element from the variable we have to identify it exactly.

We could easily get the element from the value of the level 1 variable if this element belongs to L_n for some n, and no another element of L_n belongs to this value. Such strict way of identification admits us to construct a singleton containing only the element we need in, and then we may apply the level-decreasing assignment instruction.

Thus, using the Lemma 4, we can compute the value of the expression by the above way, if only this expression is unable to be abbreviated in the considered groupoid. This actually proves the theorem. Technical details are left to the reader.

■

Remark. *We cannot eliminate level-decreasing assignments in the theorem 5, because of pebble complexity arguments.*

So, it is not only clear, if $FDSET_1$ is semi-universal to the class of finite interpretations of a polyadic signature.

We know (Theorem 3), that for monadic signatures $FDSet_1$ is semi-universal. Thus, it is easy to see that, in the sense of [TU], [HP], and [ST]):

Theorem 6. *$FDSet_1$ and $L(FDSet_1)$ correspond to the complexity class LINEARSPACE.*

■

If we introduce the nondeterministic analog of $FDSet_1$ - $NFDSet_1$, we obtain a description of *NLINEARSPACE*.

Let us now consider the class $FDSet_2$. It turns out

Theorem 7. *$FDSet_2$ is semi-universal.*

Sketch of proof. Once again consider only the case of one-generated groupoids. Every element of a groupoid (perhaps except of the generator) can be obtained, in general by different ways, as a multiplication of two other elements. A natural terminology here is "father" and "mother". Hence an element can have a lot of fathers and (much more surprisingly!) a lot of mothers, and thereto an element can be both its father and its mother in different marriages (multiplications) as well as in the same.

Let us inductively define a linear order in every one-generated groupoid as follows:

1. If $a \in L_n$, $b \in L_m$ and $n < m$, then $a < b$.
2. Let $a, b \in L_n$ for some n, i.e. a and b be equivalent.

Basis. $n=0$. The generator is the sole element of L_0.

Induction step. Let us consider the set $Fathers_{n-1}(a)$ and $Mothers_{n-1}(a)$, where $Fathers_{n-1}(a)$ is defined as intersection of $\bigcup_{m=0}^{n-1} L_m$ with all fathers of a, $Mothers_{n-1}(a)$ - as the intersection of $\bigcup_{m=0}^{n-1} L_m$ with all mothers of a.

Let us also consider the sets $Fathers_{n-1}(b)$ and $Mothers_{n-1}(b)$. It is easy to see that these four sets are nonempty.

Due to the induction hypothesis, we can find the least element of $Fathers_{n-1}(a)$, say $Father_{n-1}(a)$, and the least element of $Fathers_{n-1}(b)$, say $Father_{n-1}(b)$.

If $Father_{n-1}(a) < Father_{n-1}(b)$, then we let $a<b$.

If $Father_{n-1}(b) < Father_{n-1}(a)$, then we let $b<a$.

Let us consider the case when $Father_{n-1}(a)=Father_{n-1}(b)=c$. Then we define two new sets:

$$Mothers'_{n-1}(a)=\{d\in Mothers_{n-1}(a) \mid c\cdot d=a\},$$

$$Mothers'_{n-1}(b)=\{d\in Mothers_{n-1}(b) \mid c\cdot d=b\}.$$

Once again we can find the least elements of $Mothers'_{n-1}(a)$, say $Mother_{n-1}(a)$, and the least element of $Mothers'_{n-1}(b)$, say $Mother_{n-1}(b)$.

If $Mother_{n-1}(a) < Mother_{n-1}(b)$, then we let $a<b$.

If $Mother_{n-1}(b) < Mother_{n-1}(a)$, then we let $b<a$.

If $Mother_{n-1}(b) = Mother_{n-1}(a) = d$, then $a=b=c\cdot d$.

PROPOSITION 8. *The above definition correctly defines a linear order in every one-generated groupoid.*

■

Let us return to the proof of Theorem 7. Due to the Lemma 2, it is enough to consider the case when we have got some $L\subseteq L_n$ for some n, and to choose an element of L in the sense of the above defined linear order.

The proof of Theorem 7 obviously follows from Lemma 9 below, because we may first find the least father and then, if necessary, the least mother too.

■

LEMMA 9. *Let G be an one-generated groupoid with the generator g, and let $L \subseteq L_n$ for some n. If the set L_{n+1} is nonempty, then there exists an $FDSet_2$ subscheme computing the least element of L.*

Proof. So let us consider $L \subseteq L_n$, $n>0$, $L \neq \emptyset$ (if $n=0$, L contains only the groupoid's generator). Using a level 1 variable, we obviously can simulate the counters from 0 up to $n+1$. We organize these counters as follows:

0 corresponds to $\mathbf{0} = \bigcup_{m=0}^{n+1} L_m$,

$i+1$ corresponds to $\mathbf{i+1} = \mathbf{i} \setminus L_i = \bigcup_{m=i+1}^{n+1} L_m$.

These values can be easily computed by $FDSet_1$ subscheme using g and L.

To compute the least element of L, we first compute its fathers, grand-fathers, etc.

Namely, let us consider the sequence

$L = Fathers_n(L)$, $Fathers_{n-1}(L)$, $Fathers_{n-2}(L)$, ..., $Fathers_0(L)$,

where

$$Fathers_{n-i-1}(L) = \{a \in \bigcup_{m=0}^{n-i-1} L_m \mid (\exists b \in \bigcup_{m=0}^{n-i-1} L_m)(a \cdot b \in Fathers_{n-i}(L))\}.$$

It is easy to see that in some step i, $0 \leq i \leq n$, the set $Fathers_{n-i}(L)$ contains the generator g. If L itself contains g, then we found the minimal element of L. Else we denote that g as $Father_{n-i}(L)$ and form the level 2 set

$$Fathers = \{\{Father_{n-i}(L)\} \cup \mathbf{n-i+1}\}.$$

We also form the level 2 set

$$Goals = \{L \cup \mathbf{n+1}\}.$$

We remind the reader that $\mathbf{i} = \bigcup_{k=i}^{n+1} L_k$. Thus, neither $Father_{n-i}(L)$ and $\mathbf{n-i+1}$, nor L and $\mathbf{n+1}$ has a nonempty intersection.

Let $L' = \{a \in \bigcup_{k=0}^{n-i-l} L_k \mid Father_{n-i}(L) \cdot a \in Fathers_{n-i+1}(L)\}$, where l is the least natural number such that L' is nonempty. Now we substitute the set L' instead of L and once again repeat the above procedure, including the building of the sequence of fathers of levels, already, $n-i, n-i-1, \ldots, 0$.

Again, in some step j, $Fathers_{n-i-j}(L')$ contains g and we fix that number $n-i-j$. If L' itself contains g, then we found the minimal element of L'. Else we add the set $\{Father_{n-i-j}(L')\} \cup$ **n-i-j+1** to the level 2 set *Fathers*, and add the set $L' \cup$ **n-i+1** to the level 2 set *Goals*. Then we once again repeat the procedure with set $L'' = \{a \in \bigcup_{k=0}^{n-i-j-l} L_k \mid Father_{n-i-j}(L') \cdot a \in Fathers_{n-i-j+1}(L')\}$, where l is the least natural number such that L'' is nonempty.

And so on. After a number of repetitions, we have that g belongs to L'''''. Thus we found the least element of that L'''''. Now we do the back step.

Namely, let for simplicity L'' contain g. Then we compute the element $d = Father_{n-i-j}(L') \cdot g$. Next, we extract the previous L' from the level 2 set *Goals*. To do that, we use its *indice* (i.e., the set **n-i+1**). This is the least indice among the element of *Goals*. We leave details to the reader. If now our d belongs to the extracted goal (in this case, to L'), then we found the least element of L' (d), and we do the following back step, extracting the left compositor for d from the level 2 set *Fathers*, and computing d'. Otherwise, i.e. if d does not belong to the last extracted goal, we return ourselves to the "forward process", adding the set $\{d\} \cup$ **n-i-j+2** to the *Fathers*, while $L' \cup$ **n-i+2** to *Goals*, and focusing at the set $L_1'' = \{a \in \bigcup_{k=0}^{n-i-j-l+1} L_k \mid d \cdot a \in Fathers_{n-i-j+2}(L')\}$, where l is the least natural number such that L_1'' is nonempty.

The above process will be finished after some number of steps and will give us the least element of L.

Reader can easily observe that neither level 2 set *Fathers* nor level 2 set *Goals* contains two element with coinciding indices in some step of that process.

■

So $FDSet_2$ is of semi-universal power.

Let us now consider the dynamic logic $L(FDSet_2)$ of that class. Every formula φ of $L(FDSet_2)$ for every n defines the class $\mathfrak{F}_n(\varphi)$ of finite interpretations such that $(A,a)\in\mathfrak{F}_n(\varphi)$ iff φ is true in (A,a), $a\in A^n$. We will call $\mathfrak{F}_n(\varphi)$ a **spectrum** of φ, if n is the number of free variables of φ.

The following theorem can be proved for arbitrary signature.

THEOREM 10. *For any natural k there is a natural numeration of finite interpretations (A,a), $a\in A^k$, such that, for any recursive $M\subseteq\{0,1,\ldots\}$, $M\in\bigcup_c DSPACE(c\cdot 2^n)$ iff M is the index set of the spectrum $\mathfrak{F}_k(\varphi)$ of some formula φ of $L(FDSet_2)$ with k free variables.*

■

3. CONCLUDING REMARKS

All the results formulated in Section 2 remain true for **pure** $FDSet_1$ and **pure** $FDSet_2$ (see corresponding proofs). So, because of semi-universality of $FDSet_2$, we have:

THEOREM 11. *$FDSet_2$ $(L(FDSet_2))$ is reducible into pure $FDSet_2$ (pure $L(FDSet_2)$).*

■

From the same arguments

THEOREM 12. *$FDSet_1$ $(L(FDSet_1))$ is reducible into pure $FDSet_1$ (pure $L(FDSet_1)$) on unoids.*

■

THEOREM 13. *$FDSet_1$ $(L(FDSet_1))$ is reducible into pure $FDSet_1$ (pure $L(FDSet_1)$) on infinite interpretations.*

■

From the theorems 6 and 9, we have:

THEOREM 14. *$FDSet_1 < FDSet_2$ and $L(FDSet_1) < L(FDSet_2)$.*

■

We remind the reader that the question whether $FDSet_1$ is semi-universal or not, remains open.

Acknowledgements

A part of this work was done during author's visit in Warsaw University during June, 1987. I thank Warsaw University for pleasant possibility to work, and Jurek Tiuryn and Paweł Urzyczyn for fruitful discussions. I'm especially grateful to Paweł Urzyczyn, in discussions with him certain ideas of the paper were born. I thank M.A.Taitslin for helpful comments.

References

[T] **Tiuryn,J.**, *An introduction to first-order dynamic logic*. Research report, Washington State University, Pullman, 1984.

[H] **Harel,D.**, *First-order dynamic logic*, LNCS **68**, Springer-Verlag, Berlin, 1979.

[U] **Urzyczyn,P.**, *Logics of programs with boolean memory*, Fundamenta Informaticae, **11**, № 1 (1988).

[U1] **Urzyczyn,P.**, Private communications, 1987.

[HP] **Harel,D.**, and **Peleg,D.**, *On static logic, dynamic logic and complexity classes*, Inform. Control, **60**, 1984.

[K] **Kfoury, A.J.** *Pebble game and logics of programs*. In "Harvey Friedman's Research on the Foundation of Mathematics", (L.A.Harrington et al., Eds.), Elsevier Science Publishers B.V. (North-Holland), 1985.

[KSU] **Kfoury,A.J.**, **Stolboushkin,A.P.**, and **Urzyczyn,P.** *Some open questions in the theory of program schemes and their logics*, Uspekhi Matematicheskikh Nauk, **44**, 1(265), 1989 (Russian).

[TU] **Tiuryn,J.**, and **Urzyczyn,P.** *Some relationships between logic of programs and complexity theory*. In "IEEE Annu. Symp. Found. Comp. Sci.", 1983.

[ST] **Stolboushkin,A.P.**, and **Taitslin,M.A.** *Dynamic logics*. In "Cybernetics and computer machinery" (V.A.Melnikov, Ed.), Nauka Science Publ., Moscow, 1986 (Russian).

THE RESOLUTION PROGRAM, ABLE TO DECIDE SOME SOLVABLE CLASSES

Tanel Tammet
Institute of Cybernetics
Estonian Academy of Sciences
Akadeemia tee 21, 200108, Tallinn, Estonia

Abstract

In the first part of the paper we describe our theorem-prover implementing a certain modification of the resolution strategy of N.Zamov. Zamov's strategy decides a number of solvable classes, including Maslov's Class K (this class contains most well-known decidable classes like Gödel's Class, Skolem's Class, Monadic Class). We describe several experiments performed with the theorem-prover.

The second part of a paper consists of a proof that a presented modoification of Zamov's strategy decides a wide class of formulas with functional symbols, called E+.

The work described here has been guided by G.Mints. We would also like to thank N.Zamov for helpful discussions.

Introduction

Since the first-order predicate calculus is undecidable, the complete proof methods (for example, the resolution method and Gentzen's method) are guaranteed to terminate only in case the answer is positive, e.g. the formula is refutable (in case of the resolution method). If it is not, the algorithm will in general loop, since (except the positive result) there can't be any effective criteria for stopping the search.

Algorithms which are guaranteed to stop and provide correctly not only the positive but also the negative answers (decision algorithms), exist for the solvable subclasses of first order predicate calculus only. Well-known solvable classes are, for example, Gödel's Class, Skolem's Class, Maslov's Class (see [2]). Familiar decision algorithms for these sub-classes are specialized and in general based on testing a formula on a model of (super)exponential size.

So it is quite desirable to modify a general method like resolution in such a way that it became, besides being a complete proof procedure, also a decision procedure for some solvable classes. N.Zamov [11], [12] has shown that using a special ordering of literals, it is possible to guarantee termination of the resolution method for all formulas (including invalid ones) of a wide decidable class, the so-called Maslov's Class K [7] (see also [12]). This class contains most well-known decidable classes, like Gödel's Class (prefix $(Ex_1)(Ex_2)(\forall y_1)\ldots(\forall y_m)$), Skolem's Class (prefix $(Ex_1)\ldots(Ex_k)(\forall y_1)\ldots(\forall y_m)$ and every literal contains either one variable y_i or all variables $x_1,\ldots,x_k$), Monadic Class.

The strategy proposed by Zamov is complete in the general case, requires a small modification of the resolution method and proves in experiments to be much more efficient than building the countermodel of a formula. Since it reduces the search space to the finite set of disjuncts, in addition of being able to prove the nonvalidity of formulas, it normally proves the validity of the formulas in solvable classes much faster than the familiar modifications of the resolution method.

Experiments with our theorem-prover

We have written a resolution-based theorem prover which implements Zamov's resolution strategy, a variant of Maslov's inverse method [6] (see also [12]) and a lot of well-known resolution strategies. The program has a menu-driven user interface and is designed for an interactive use. It is written in LISP and runs on any IBM-PC or compatible. The program has decided all the first-order formulas from Church [1] (some of them are valid, some are not), the longest runtimes for these problems were less than a minute on the IBM-PC-AT. The efficiency of using a decision strategy is noted by a fact that without such a strategy neither our prover nor version 0.9 of W.McCune's prover OTTER (which doesn't incorporate Zamov's strategy) couldn't refute the negation of most complicated valid formula in [1] (both provers exhausted all available memory).

Our prover has solved most first-order predicate calculus problems (Schubert's steamroller, for example, was solved in a minute) found in automated deduction literature. Other test examples include predicate logic translations of formulas in modal and intuitionistic propositional logic. For example, the predicate logic translation (via Kripke semantics) of the modal formula $\Box\Diamond(\Box\Diamond p \,\&\, \Box\Diamond q) \to (\Box\Diamond p \,\&\, \Box\Diamond q)$

presented by M.Fitting as a problem too complicated for some (natural deduction style) theorem provers built specially for modal logic, was proved in about 3 minutes.

One of the applications (Zamov's strategy was of no use for this application) was a computer-aided proof of a general theorem in the theory of relevant implication -> : usual axioms $R_{\rightarrow}$ for -> remain complete for -> if the standard deduction rules modus ponens (A, A->B / B) and substitution (B / Bs for any substitution s) are replaced by the condensed detachment rule D (A, A'->B / Bs), where s is a most general unifier of A, A' (see [3], [10]). This theorem was posed as an open question in [3]. The axioms $R_{\rightarrow}$ are following:
(B) (a->b)->((c->a)->(c->b)), (C) (a->(b->c))->(b->(a->c)), (I) a->a, (W) (a->(a->b))->(a->b).

In [3] it was also left open whether ((a->a)->a)->a is deducible from the axioms $R_{\rightarrow}$ by the rule D, thus G. Mints proposed to investigate the problem. We modelled it using our theorem-prover in the following well-known manner. Implicational formulae A are represented by terms t_A. If x is a propositional variable, then t_x=x (treated now as an individual variable). Formula A->B is represented by a term $t_{A\rightarrow B}=i(t_A, t_B)$. We use single monadic predicate P to represent derivability. Axioms A are represented by literals $P(t_A)$, e.g. P(i(a,a)) represents axiom (I): a->a. Rule D is represented as a clause MP={ -P(i(x,y)),-P(x), P(y)}. New formulae are derived using resolution rule without factoring, since factoring is a limited version of substitution rule. At the first attempt we used encoding

$$P(B),\ P(C),\ P(I),\ P(W),\ MP,\ -P(t_{((a\rightarrow a)\rightarrow a)\rightarrow a})$$

and ran our program in hope of obtaining the derivation of ((a->a)->a)->a from $R_{\rightarrow}$. This did not come true: the hyperresolution derivation of contradiction was obtained very quickly but corresponded to a propositional derivation using substitution rule in essential way. Next we ran the program to obtain all consequences of MP, P(B), P(C), P(I), P(W) by hyperresolution. In fact it simply generated all D-consequences of our axioms. After some efforts the desired formula was derived. The main idea of the proof found became the main idea of G.Mints's hand-made proof of the condensed-detachment completeness of $R_{\rightarrow}$ (for details see [9]).

Computer programs with the capacity to decide some decidable classes have been written in seventies by A.Sochilina [15] and W.Joyner [4]. Sochilina's program didn't give the user an opportunity to choose between different proof strategies and the program is not supported any more. Joyner's program used a close, but a weaker method than Zamov's strategy.

First formulation of Zamov's strategy

At first we bring the presentation of the strategy as given by N.Zamov in [12]. We will call it Z1-strategy.

Definition. The term t covers the term s if at least one of the following conditions is satisfied:

1) t=s
2) $t=f(t_1,\ldots,t_n)$, s is a variable and $s=t_i$ for some i, $i=1,\ldots,n$
3) $t=f(t_1,\ldots,t_n)$, $s=g(t_1,\ldots,t_m)$, $n\geq m\geq 0$.

Definition. The relation $>_1$ between literals is defined as follows:

$A >_1 B$ if for every argument t_2 of the literal B there exists an argument t_1 of the literal A such that t_1 covers t_2 or t_2 is a subterm of the term t_1.

Notice that the relation $>_1$ between literals is not a strict ordering relation, since it leaves some pairs of literals uncomparable and for other pairs A, B it can be the case that $A>_1 B$ and $B>_1 A$. Nevertheless the $>_1$ relation is transitive and is also preserved under substitution: if $A>_1 B$, then $As >_1 Bs$ for any substitution s.

Z1-strategy is defined by a restriction on a resolution rule (we use a formulation of resolution method where resolution rule and factorization rule are separated. See [5] for differences between definitions of resolution method). We assume that the resolution method we are using incorporates the strategy of eliminating subsumed and tautologous disjuncts.

Restriction on resolution rule: only the maximal (in the sense of ordering $>_1$) literals in the disjunct are allowed to be resolved upon.

Example. In the disjunct {A(f(x,g(x)),y), B(g(x),g(x)), C(x,0), D(y,x), E(g(x),z)} only the literals A(f(x,g(x)),y), E(g(x),z) and C(x,0) are allowed to be resolved upon.

N.Zamov has shown (see [12], [12]) that Z1-strategy is complete (since the $>_1$ ordering is a kind of pi-orderings introduced by J.Maslov, see [8]) and for the formulas of Maslov's Class K, only terms of the certain kind will be generated. These terms cannot contain proper subterms with the same or bigger arity, therefore the set of generated terms is finite. As the length of generated disjuncts is shown to be bounded by a certain function from the size of generated terms, the search space is finite and the Z1-strategy solves Maslov's Class K. We give the definiton of Class K in one of the following sections.

A modification of Zamov's strategy

In the following we are going to use a certain modification of Zamov's strategy, the completeness of which is also proven by the fact that it is a pi-strategy (see [8]). We will call the modification Z2-strategy.

Definition. Let A and B be literals. By maxdepth(x,L) we denote the maximal depth of occurrence of the variable x in literal L. Let $\bar{x}$ be a set of variables in A, $\bar{y}$ be a set of variables in B. We say that A $>_2$ B iff at least one of the following is true:

1) $\bar{y} \subseteq \bar{x}$ and $\forall y_i \in \bar{y}(\text{maxdepth}(y_i,B) \leq \text{maxdepth}(y_i,A))$ and $Ey_j \in \bar{y}(\text{maxdepth}(y_j,B) < \text{maxdepth}(y_j,A))$.

2) $\bar{y}$ is a proper subset of $\bar{x}$ and neither A nor B contain function symbols with arity > 0.

3) $\bar{y}$ is empty, $\bar{x}$ is not empty.

The new ordering $>_2$ is transitive and is preserved under substitution (in the weak sense, though: if A $>_2$ B then it is guaranteed that Bs $>_2$ As cannot be the case, but it is not guaranteed that As $>_2$ Bs).

Z2-strategy is defined by a restriction on a resolution rule analogous to the restriction used for Z1-strategy:

Restriction on the resolution rule: only the maximal (in the sense of new ordering $>_2$) literals in a disjunct are allowed to be resolved upon.

Example. In the disjunct {A(f(f(x,x),y),1), B(2,g(y,x)), C(y,z), D(1,f(2,1))} only the literals A(f(f(x,x),y),1) and C(y,z) are allowed to be resolved upon.

It is possible to show that unlike Z1-strategy, Z2-strategy solves the Initially-extended Class K (this class contains, for example, the Initially-extended Gödel's Class with prefix $(Ex_1)\ldots(Ex_n)(\forall y_1)(\forall y_2)(Ez_1)\ldots(Ez_m)$).

Maslov's Class K

The prefix of the literal L in a prenex predicate formula F is obtained from the prefix of F by discarding quantifiers for those variables, which do not occur in L.

The definition we are going to present is given in terms of refutability, not provability as in [7]:

F is said to belong to the Class K if it does not contain function symbols and there are variables $x_1,\ldots,x_n$ in a set of variables of F such that every x_i $(1 \le i \le n)$ is situated in the prefix of F to the left of all existential quantifiers and for every prefix P of a literal in F at least one of the following is true:

1) P consists of the single general quantor ∀.
2) P ends with the existential quantor E.
3) P is of the form $(\forall x_1)(\forall x_2)\ldots(\forall x_n)$.

Maslov's Class K is introduced and shown to be solvable in [7]. See [11], [12] for an overview of Class K and the proof of its decidability by Z1-strategy.

Class E+

We say that the first-order predicate calculus formula F in the prenex form belongs to the Class E+ if the prefix of F consists only of the general quantors (we present Class E+ in terms of refutability) and the matrix of F has the following properties:

A) if a term t in some literal L contains variables, then it contains all the variables present in L.

B) if two literals L and L' in some disjunct D share a variable, all the variables in L and L' are shared.

The class just presented contains the class of formulas with function symbols where every atom contains only one variable; this class has been called N_1 and E (if given in terms of derivability, formulas in E can generally be assumed to be of the form (Ex)M). The solvability of E has been shown in various ways, see [13] and [14]. E contains, for example, the skolemized version of Initially-extended Ackermann Class (the class of formulas without function symbols and with the prefix $(Ex_1)\ldots(Ex_n)(\forall y)(Ez_1)\ldots(Ez_m)$).

Since we consider the class we'll investigate here to be a "natural" superclass of E, we have named it E+.

Example: The following formula belongs to class E+:

∀xyzuv{ [P(f(f(1)),g(3,g(y,x))) V P(u,f(u))] &
[-P(f(l(x,y,z)),z) V G(2) V G(v) V -G(u)] &
[E(g(l(z,x,y),f(2)))] &
[R(g(v,g(v,u)),g(u,v)) V R(f(2),g(f(f(1)),g(u,v)))]}.

In the following we will show that Z2-strategy decides Class E+', which is the same as Class E+, except for the property B, which in E+' is replaced by a property B':

B') if a literal L in some disjunct D contains variables, then it contains all variables present in D.

If Z2-strategy decides E+', it is clear that Z2-strategy with splitting decides the class E+, since every formula in E+ can be split into a finite set of formulas in E+'.

Splitting is a following well-known rule: if a disjunct D in a set of disjuncts S can be split into two parts A and B such that A and B do not share variables, then S is contradictory iff both (S-D)+A and (S-D)+B are.

Notice that instead of using splitting, the decision algorithm for E+ could be obtained from Z2-strategy by restricting factorization in the following way: if literals A and B in a disjunct D do not share variables, then it is not allowed to generate a factorization (Ds)-(Bs) (where s=MGU(A,B) is the most general unifier of A and B and for our case here s must be obtained by actually unifying A with B) in case (Ds)-(Bs) does not subsume D.

In the following f[l] (where f is some letter and l denotes either some term or a set of terms) will denote a term containing either l or all elements of l as proper subterms. We will use a letter s for substitutions and write ts for a result of an application of s to an expression t. Generally s is a set $\{t_1/x_1,\dots,t_n/x_n\}$ where x_i are variables, t_i are terms. MGU(g,t) denotes the most general unifier of expressions g,t (g and t may be terms, atoms or literals). We'll say that a substitution s binds two terms a and b iff as=bs.

In order to simplify the proof we assume that if $MGU(g,t)=\{t_1/x_1,\dots,t_n/x_n\}$ exists, it is always such that each t_i $(1\le i\le n)$ has an occurrence in g or t. Thus the application of substitution is generally an iterative process.

<u>Lemma 1</u> If literals L and L' both have a property A (in the sense of the definition of E+), then MGU(L,L') does not contain a subset $\{t[\bar{y}]/x,\ g[\bar{x}]/y\}$, where $\bar{x}$ is a set of variables in L, $\bar{y}$ is a set of variables in L', y is an element of $\bar{y}$, x is an element of $\bar{x}$.

<u>Proof:</u> Obvious, since $\{t[\bar{y}]/x,\ g[\bar{x}]/y\}$ is a cyclical substitution, thus it cannot be a value of MGU(L,L').
END OF PROOF.

<u>Lemma 2</u> If disjuncts D and D' both have a property B' (in the sense of the definition of Class E+') and every literal in D and D' has a property A, then all disjuncts inferred from D and D' using Z2-strategy, also have the property B' and every literal in them has a property A.

Proof: 1) Property B': If B'(D), then B'(Ds), where D is any disjunct and s is any substitution. If the disjunct D is obtained from disjuncts D_1 and D_2 with property B' by Z2-strategy, resolving upon literal L_1 in D_1 and literal R_1 in D_2, then L_1 contains all variables in D_1 and R_1 contains all variables in D_2.

2) Property A: Obviously factorization step preserves A (the value of MGU, used for factorizing D, cannot have elements of the kind t/x, where t contains variables, since then by properties A and B' t must also contain x). Next we will consider a resolution step.

In order that property A were not preserved, such a substitution s must be applied to some literal L in D (let $\bar{x}=\{x_1,\dots,x_k\}$ be a set of variables in L) that s contains t/x_i $(1\le i\le k)$, ts contains variables and L contains some variable $x_j \ne x_i$ such that Ls also contains x_j and ts does not contain x_j.

$s=MGU(L_1, L_2)$, where by property B' and Z2-strategy L_1 contains variables $x_1,\dots,x_k$. Let $\bar{y}=\{y_1,\dots,y_r\}$ be a set of variables in L_2. Suppose s contains $t[\bar{y}]/x_i$ $(1\le i\le k)$ and L contains a variable $x_j \ne x_i$. We'll examine whether Ls can contain x_j. Consider following cases:

1) x_j occurs in some term $g[\bar{x}]$ in L and s contains an element $g[\bar{x}]/y_n$ for some variable y_n in L_2. Due to lemma 1 this is impossible.

2) s contains x_j/y_n for some y_n in $\bar{y}$. Then either ts contains x_j or, if s binds x_j to some ground term, Ls doesn't contain x_j.

3) s contains y_n/x_j for some y_n in $\bar{y}$. Then Ls doesn't contain x_j.

4) s contains l/x_j, where l is a ground term. Then Ls does not contain x_j.

5) s contains $g[\bar{y}]/x_j$. Then Ls does not contain x_j.

END OF PROOF.

Lemma 3 If disjuncts D_1 and D_2 both have a property B' and every literal in them has a property A, then no disjunct inferred from D_1 and D_2 with a factorization or resolution step of Z2-strategy can contain variables deeper than the depth of the deepest variable in D_1 and D_2.

Proof: Let D be a disjunct inferred from D_1 and D_2 using Z2-strategy. D can contain some variable deeper than any variable in D_1 and D_2 only in case a substitution used for inferring D contains such an element $t[\bar{y}]/x_i$ (where $\bar{y}$ is some set of variables) that some literal Ls in D is such that x_i occurs in L (L is the original of Ls in D_1 or D_2) in some term g and s does not bind all variables in $\bar{y}$ to ground terms. We'll show that this is impossible.

1. Inference by factorization. Since $B'(D_1)$ and A holds for every literal in D_1, factorization cannot give such a substitution s that contained $t[\bar{y}]/x_i$ (then $\bar{y}$ should contain x_i, which is impossible).

2. Inference by resolution. Let L, L_1 D_1, L_2 D_2. D_1 contains variables $x_1, \ldots, x_k$, D_2 contains variables $y_1, \ldots, y_r$. Let $s = MGU(L_1, L_2)$. Since we use Z2-strategy, variable x_i cannot occur deeper in L than in L_1. Let $f[x_i]$ be the term in L_1 where x_i has its deepest occurence in L_1. We can assume that D_2 does not contain variables deeper than D_1. Due to Z2-strategy, $f[x_i]$ has the deepest occurrences of variables in D_1, D_2.

Suppose s contains the needed element $t[\bar{y}]/x_i$. Then during the computation of s the term $f[x_i](t[\bar{y}]/x_i) = f[t[\bar{y}]]$ must have been unified with some term l in L_2. Consider the following cases.

1) l is a variable. Then l cannot be unified with $f[t[\bar{y}]]$, since y contains l.

2) l is a ground term. Then every variable in y is bound to some ground term, thus s cannot contain such an element $t[\bar{y}]/x_i$ that variables in $\bar{y}$ were not bound to some ground term.

3) $l = f[g]$, g is either a variable or a ground term and g occurs in the term $f[g]$ on the same place as $t[\bar{y}]$ first occurs in $f[t[\bar{y}]]$. If g is a variable, then l cannot be unified with $f[t[\bar{y}]]$ since $\bar{y}$ contains g. If g is a ground term, then every variable in $\bar{y}$ is bound to some ground term, thus s cannot contain such an element $t[\bar{y}]/x_i$ that variables in $\bar{y}$ were not bound to some ground term.

4) $l = f[t'[\bar{y}]]$, where $t'[\bar{y}]$ occurs in $f[t'[\bar{y}]]$ on the same place as $t[\bar{y}]$ first occurs in $f[t[\bar{y}]]$. Suppose that $t[\bar{y}]$ and $t'[\bar{y}]$ are unifiable. In case $t[\bar{y}]$ contains non-ground functional terms as proper subterms, then iterate our analysis (notice that if A(L) for some literal L, then A(t) for any subterm of L, in case we treat t as a literal) until such a term $t^*[\bar{y}]$ is reached which does not contain non-ground functional terms as proper subterms.

Therefore, in order that s existed and Ls contained some variables in y, L_2 must contain such a term $f[t'[\bar{y}]]$ that the maximal depth of variables in $f[t'[\bar{y}]]$ were not smaller than the maximal depth of variables in $f[t[\bar{y}]]$. Since $f[x_i]$ is assumed to contain the deepest occurrence of variables in D_1 and D_2, $f[t[\bar{y}]]$ contains variables deeper than the maximal depth of variables in D_1 and D_2, which is impossible. Thus no variable in Ds may occur deeper than the depth of the deepest variable in D_1, D_2.

END OF PROOF.

<u>Corollary 1</u> Resolution method with Z2-strategy cannot infer such a disjunct D from the formula F in class E+' that any variable occured deeper in D than the maximal depth of variables in F.

<u>Proof:</u> immediate from lemmas 2 and 3.
END OF PROOF.

Let F be some formula in Class E+'. In the following we will denote the depth of the deepest occurence of a variable in F by VDmax. Let T be a set of all non-ground functional terms in F (if T is empty, the proof of the forthcoming theorem 1 is trivial). Let L be some literal in F. Since T is a finite set, there must be a finite set L* of such literals which can be obtained from L by applying substitutions of the kind $\{el(T)/x_1,\ldots,el(T)/x_k\}$ (where $x=\{x_1,\ldots,x_k\}$ is some set of variables, el(T) denotes an arbitrary element of T) to L so that no variable occurs deeper in L* than VDmax. (T and L* are defined up to renaming the variables).

<u>Example:</u> F=∀uxy{[A(f(x)) V L(x,f(1))] & [-P(f(g(u,y)),u)]}.
Then VDmax=2, T={f(x), f(g(u,y)), g(u,y)}.
If L=A(f(x)), then L*={A(f(x)), A(f(g(x,y))), A(f(f(x)))}.
If L=-P(f(g(u,y)),u), then L*={-P(f(g(u,y)),u)}.

<u>Definition.</u> Let L be some literal such that we can compute L*. VNmax(L) denotes the maximal number of different variables in literals in L*.

<u>Example:</u> Let F be the same as in the previous example.
VNmax(A(f(x))=2.
VNmax(A(f(f(x)))=1.
VNmax(A(f(1)))=0.
VNmax(-P(f(g(u,y)),u)=2.
VNmax(A(f(f(1)))=0.
VNmax(A(x))=4. {Notice A(g(g(u",y"),g(u',y'))))}.

<u>Lemma 4</u> Let F be a formula in E+'. The resolution method with Z2-strategy cannot infer such a disjunct $(L_1' \vee \ldots \vee L_n')$ from F that for some L_i' $(1 \le i \le n)$ $VNmax(L_i') > VNmax(L_i)$, where L_i is the original of L_i'.

<u>Proof:</u> immediate from the corollary 1 and the definition of VNmax.
END OF PROOF.

Lemma 5 Let disjuncts D_1 and D_2 have the property B' and each literal in D_1, D_2 have the property A. Let the resolution method with Z2-strategy infer such a new disjunct $D=(L_1' \vee \ldots \vee L_n')$ from D_1 and D_2 that D contains a term deeper than the deepest term in D_1, D_2. Then:

$$\forall i (1 \le i \le n) \begin{cases} \text{VNmax}(L_i')=0 & \text{if VNmax}(L_i)=0 \\ \text{VNmax}(L_i')<\text{VNmax}(L_i) & \text{if VNmax}(L_i)\neq 0 \end{cases}$$

where L_i is the original of L_i' in D_1 or D_2, VNmax is computed in respect to the original formula in E+' from which D_1, D_2 were inferred.

Proof: Due to lemma 3 the only possibility for some L_i' to contain a term deeper than the deepest term in D_1 and D_2 is substituting some ground term t for such a variable x_i in L_i that x_i occurs in some functional term $f[x_i]$ in L_i. Then x_i will occur in no term in D, thus the VNmax of all literals in D_1 which contain variables (VNmax≠0) is bigger than VNmax of the corresponding literals in D.

Next we'll consider the disjunct D_2, which did not contain x_i. Let x_i have a deepest occurence in L_1 in a term $g[x_i]$. Notice that after applying a substitution $\{t/x_i\}$ to the literal L_1 in D_1 (assume L_1 is resolved upon in D_1) the term $g[x_i]$ will turn into $g[t]$, which by assumption is a deeper term than the deepest one in D_1, D_2 (since we use Z2-strategy, x_i has a deepest occurrence in L_1) . Let l be the term which occurs in R_1 (assume R_1 is resolved upon in D_2) on the place of the first occurence of $g[x_i]$ in L_1. Consider the following cases:

1) $l=g[t]$. This is impossible, since $g[t]$ is deeper than the deepest term in D_1, D_2.

2) $l=g[y_j]$. Then literals in D which correspond to literals in D_2 will contain a ground term on place of a variable y_j, thus VNmax of these literals in D will be less than VNmax of their originals in D_2.

3) l is a variable. Then literals in D which correspond to literals in D_2 will contain a ground term on place of a variable l thus VNmax of these literals in D will be less than VNmax of their originals in D_2.

END OF PROOF.

Theorem 1 The resolutiom method with Z2-strategy is a decision algorithm for Class E+'.

Proof: Let F be a formula in Class E+'. F has a certain VDmax. Due to corollary 1 no disjunct inferred from F using resolution method with Z2-strategy can contain variables deeper than VDmax. All literals in F have a certain VNmax. Lemmas 4 and 5 demonstrate that if a depth of a literal becomes bigger than a certain given bound for F, then VNmax of a literal will be correspondingly smaller. By lemma 4 VNmax of a literal cannot grow. Thus such a bound on the depth of terms must exist that all literals of this depth have a VNmax equal to 0, thus they do not contain variables and their depth cannot grow.

Therefore resolution with Z2-strategy cannot infer literals deeper than the beforementioned depth bound. Thus the number of inferrable literals (up to renaming variables) is also bounded. Then, due to lemma 3, length of the inferrable disjuncts is bounded, thus the number of inferrable disjuncts is bounded.
END OF PROOF.

References

1. Church, A. Introduction to mathematical logic I. (Princeton University Press, New Jersey, 1956).
2. Dreben, B., Goldfarb, W.D. The decision problem: solvable classes of quantificational formulas. (Addison-Wesley, Reading, 1979).
3. Hindley, R., Meredith, D. Principal type-schemes and condensed detachment. Preprint, October 1987.
4. Joyner, W.H. Resolution strategies as decision procedures. J. ACM 23 (3)(1976), 396-417.
5. Leitsch, A. On different concepts of resolution. Zeitschr. f. math. Logik und Grundlagen d. Math. 35 (1989) 71-77.
6. Maslov, S.Ju. An inverse method of establishing deducibility in the classical predicate calculus. Dokl. Akad. Nauk. SSSR 159 (1964) 17-20=Soviet Math. Dokl. 5 (1964) 1420, MR 30 #3005.
7. Maslov, S.Ju. The inverse method for establishing deducibility for logical calculi. Trudy Mat. Inst. Steklov 98 (1968) 26-87= Proc. Steklov. Inst. Math. 98 (1968) 25-96, MR 40 #5416; 43 #4620.
8. Maslov, S.Ju. Proof-search strategies for methods of the resolution type. Machine Intelligence 6 (American Elsevier, 1971) 77-90.
9. Mints, G.E, Tammet, T. Experiments in proving formulas of non-classical logics with a resolution theorem-prover. To appear.
10. Wos, L., Overbeek, R., Lusk, E. Boyle, J. Automated reasoning: introduction and applications. (Prentice-Hall, New Jersey, 1984).
11. Zamov, N.K., On a bound for the complexity of terms in the resolution method. Trudy Mat. Inst. Steklov 128 (1972), 5-13.
12. Zamov, N.K., Maslov's inverse method and decidable classes. Annals of Pure and Applied Logic 42 (1989), 165-194.
13. Маслов С.Ю, Минц Г.Е. Теория поиска вывода и обратный метод. Доп. к русскому переводу: Чень,Ч., Ли, Р. Математическая логика и автоматическое доказательство теорем. (Наука, М., 1983) 291-314.
14. Оревков В.П. Один разрешимый класс формул классического исчисления предикатов с функционалными знаками. Сб: II симпозиум по кибернетике (тезисы), Тбилиси 1965, 176.
15. Сочилина А. В. О программе, реализующей алгоритм установления выводимости формул классического исчисления предикатов. Семиотика и информатика 12 (1979).

A STRUCTURAL COMPLETENESS THEOREM FOR A CLASS OF CONDITIONAL REWRITE RULE SYSTEMS

Sergey G. Vorobyov
Program Systems Institute of the
USSR Academy of Sciences
152140 Pereslavl-Zalessky
USSR

ABSTRACT. *For a class of quantifier-free logical theories, axiomatized by conditinal equivalences, we prove a completeness result of the form: if a theory T from the class generates the uniquely terminating conditional rewrite rule system, and a partition* $T_1 \cup T_2$ *of T satisfies certain structural properties, then an arbitrary unquantified formula* Ω *is a theorem of* $T_1 \cup T_2$ *iff the leaves of any proof tree for* Ω *are theorems of* T_1.

Key vords and phrases: conditional revrite rules, inference rules, proof search, reduction, case splitting, strong completeness, confluency, finite termination, decision algorithms.

1. INTRODUCTION

Term rewriting systems (TRSs for short) and the Knuth-Bendix completion algorithm [KB, HuOp] provide an effective search space-free decision mechanism for some equational theories. During the last decade a considerable progress has been achieved in generalizing the scope of rewriting techniques applicability. Among the main trends are: the complete rewrite rule-based strategies for the first-order predicate calculus [Hs], the "inductionless" inductive proof methods via the completion algorithm [HuHu, JoKo], various generalizations of rewriting systems, such as equational and conditional TRSs [Jo, ZR, Ka].

A kind of the problem we address in this paper is as follows. Suppose, we have two theories (sets of formulae) T_1 and T_2 and the related sets of inference rules R_1 and R_2. When can we prove theorems of $T_1 \cup T_2$ first applying the rules of R_1 to reduce the proof problem to T_2, and then using the rules of R_2 to complete the proof in T_2 alone? Axioms of T_2 may possess some undesirable properties, so isolating them from the axioms of T_1 can simplify proofs, otherwise too complex or even impossible. The separation may be useful if we want to use some specific proof methods for T_2 inapplicable in the general case. Among well known results of that sort are: Gentzen's midsequent theorem, the reducibility of multiple inductions into the unique one, permutability of inferences in proofs, completeness of different resolution strategies.

We consider this problem from the viewpoint of term rewriting and related techniques such as confluency (Church- Rosser property), finite termination, etc. As a result, for a class of theories we obtain a strongly complete proof search (decision) system. Strong completeness means that we are able to apply inference rules in an arbitrary order without the risk to miss a proof if it exists at all. The paper is organized as follows. In Section 2 we give a brief account of the main notions of term rewriting and in Section 3 we define a class of quantifier-free logical theories together with the related inference (proof search) rules. An axiom of a theory under consideration is a conditional equivalence of an atomic and an unquantified formulae, the condition is a formula of an arbitrary decidable subtheory, and the whole axiom is universally quantified. Such a structure gives possibility to treat axioms operationally, and in Section 4 we transform the axioms into conditional rewrite rules and introduce the related notions. In Section 5 we state the main result and connect it with a problem of inductive definitions and proofs by induction. Section 6 is devoted to the proof of the main theorem.

2. PRELIMINARIES

The aim of this section is twofold. Assuming the minimal familiarity with term rewriting theory (see [KB, HuOp]), we briefly remind the basic definitions. Also, we introduce denotational conventions used throughout the paper.

2.1. Term rewriting systems. Let $W(\Sigma, V)$ denote the set of terms built of variables from a set V and function symbols of a many-sorted signature Σ. *A reduction relation* $\xrightarrow{R}$ is a binary relation on $W(\Sigma, V)$, such that $t \xrightarrow{R} s$ implies $F(t) \xrightarrow{R} F(s)$ for any term $F(t)$, and $t \xrightarrow{R} s$ implies $\sigma(t) \xrightarrow{R} \sigma(s)$ for any substitution σ of terms for variables.

By $\xrightarrow{R}^{+}$, $\xrightarrow{R}^{*}$ and $\xrightarrow{R}^{-1}$ we denote the transitive closure, the reflexive-transitive closure and the inverse of the $\xrightarrow{R}$ relation respectively. We say that a term t is in the *R-normal form* iff there does not exist a term s such that $t \xrightarrow{R} s$. A term t^{*} is called the *R-normal form of a term* t iff $t \xrightarrow{R}^{*} t^{*}$ and the term t^{*} is in the R-normal form.

The simplest way to generate reduction relations is by means of term rewriting systems. *A rewrite rule* is an oriented pair $l \longrightarrow r$ of terms such that $l, r \in W(\Sigma, V)$, l is different from a variable and every variable occuring in r occurs also in l. *A term rewriting system* (TRS for short) is a set of rewrite rules $R = \{ l_i \longrightarrow r_i \}_{i \in I}$. The *reduction relation* $\xrightarrow{R}$ *generated by* a TRS R is defined as follows. Let a term t contain an occurence of a subterm s such that for some rule $l \longrightarrow r \in R$ and a substitution σ the term $\sigma(l)$ coincides with s. Then $t \xrightarrow{R} t'$ where t' is obtained from t by the replacement of the occurence of s with $\sigma(r)$. Similarly, the reduction relations generated by an equtional and conditional TRSs may be defined, see [Jo, ZR, Ka].

A reduction relation $\xrightarrow{R}$ is said to be

1) *noetherian*, iff there are no infinite chains of the form $t_0 \xrightarrow{R} t_1 \xrightarrow{R} t_2 \xrightarrow{R} \ldots$;

2) *confluent*, iff whenever $t \xrightarrow{R}^{*} t_1$ and $t \xrightarrow{R}^{*} t_2$ there exists a term s satisfying $t_1 \xrightarrow{R}^{*} s$ and $t_2 \xrightarrow{R}^{*} s$;

3) *canonical*, iff it is both noetherian and confluent.

An equational theory (ET for short) is a set of identities $T = \{ t_i = s_i \}_{i \in I}$, every identity being universally quantified. An ET T generates the binary relation $=_T$ on $W(\Sigma, V)$ defined as follows: $t =_T s$ iff the identity $t = s$ is a valid consequence of the theory T. A reduction relation $\xrightarrow{R}$ and an ET T are called *equivalent* iff $=_T$ coincides with $(\xrightarrow{R}^{*} \cup \xrightarrow{R}^{-1})^{*}$. When an ET T possesses an equivalent canonical reduction relation $\xrightarrow{R}$ then $=_T$ relation is decidable by simple normal form reductions: $t =_T s$ iff R-normal forms

t^* and s^* of t and s are syntactically equal.

2.2. Denotational conventions. The following notation and conventions will be used throughout the text.

Let S_0 and S ($S_0 \subseteq S$) be two sets of sorts, Σ_0 be an S_0-sorted similarity type (or signature) and Σ be an S-sorted similarity type such that the difference $\Sigma \setminus \Sigma_0$ *does not contain* function symbols with the S_0-sorted codomain. T_0 will denote an arbitrary quantifier-free decidable theory (say Presburger arithmetic PAR), further referred to as the *basic theory*. By B and C we will denote T_0-consistent quantifier-free formulae of the Σ_0 similarity type, further called *contexts*, whereas ϕ and ψ will stand for atomic formulae of the similarity type Σ with leading predicate symbols from $\Sigma \setminus \Sigma_0$. Quantifier-free formulae of Σ will be denoted by Φ, Γ. By $\Gamma[\psi]$ we will express the fact that a formula Γ contains a distinguished occurence of an atomic formula ψ, and by $\Gamma[\Psi/\psi]$ we will denote the result of the replacement of this distinguished occurence of ψ in Γ by Ψ.

Let $Var(\Phi)$ denote the set of variables in Φ, σ denote a substitution of terms for variables, $\sigma(\Phi)$ denote the result of substituting to a formula, &, $\vee$, $\supset$, $\neg$, and $\Leftrightarrow$ denote the usual boolean connectives. $\Sigma_{PAR} = \langle 0, Succ, +, -, \leq \rangle$ is the signature of *Presburger arithmetic PAR*, $Stm(PAR)$ is the *standard model* $\langle \omega, \Sigma_{PAR} \rangle$ of PAR, ω being the set of natural numbers. Let $T_0^=$ denote the basic theory T_0 enriched with the (uninterpreted) symbols of $\Sigma \setminus \Sigma_0$, augmented with equality axioms of the form $x = y \,\&\, \phi(x) \supset \phi(y/x)$ for each atomic formula ϕ. By $STM(PAR^= \cup T)$ we will denote the class of *standard models of* $PAR^= \cup T$, with the carriers of the arithmetic sort isomorphic to $Stm(PAR)$.

3. E-THEORIES AND INFERENCE RULES

The aim of this section is to define a class of quantifier-free logical theories with axioms easily transformable into inference (proof search) rules as well as into conditional rewrite rules.

Definition. An *$E(T_0)$-theory* T consists of axioms of the form

$$B \supset (\phi \Leftrightarrow \Phi), \qquad (E\text{-}axiom)$$

such that $Var(B) \cup Var(\Phi) \subseteq Var(\phi)$. ■

A large number of interesting and nontrivial data types and problems in artificial intelligence, program specification and verification (e.g. elementary matrices in linear algebra, array sorting, etc) can be axiomatized via $E(T_0)$-theories, as it is shown in [Vo2].

To formulate a notion of an inference rule generated by an E-axiom we need an auxiliary definition of equality modulo the basic theory.

Definition. Let α, β be terms or formulae of the similarity type Σ. The condition $Eq(\alpha, \beta)$ is defined inductively as follows:

- if $\alpha \equiv f(\alpha_1 \ldots, \alpha_n)$ and $\beta \equiv f(\beta_1 \ldots, \beta_n)$ for $f \in \Sigma \setminus \Sigma_0$, then $Eq(\alpha, \beta) \equiv_{df} \&_{i=1}^{n} Eq(\alpha_i, \beta_i)$;
- if α, β are terms of the similarity type Σ_0, then $Eq(\alpha, \beta) \equiv_{df} \alpha = \beta$;
- if α, β are formulae of the similarity type Σ_0, then $Eq(\alpha, \beta) \equiv_{df} \alpha \Leftrightarrow \beta$;
- otherwise $Eq(\alpha, \beta)$ is defined as the **false** condition.■

Definition. *A reduction inference rule* generated by an E-axiom $B \supset (\phi \Leftrightarrow \Phi)$ is a figure of the form

$$\frac{C \supset \Gamma [\sigma(\Phi) / \psi]}{C \supset \Gamma [\psi]},$$

such that C is T_0-consistent and

$$T_0 \models C \supset (\sigma(B) \,\&\, Eq(\psi, \sigma(\phi))).$$

Let $Red(T)$ denote the set of all possible (for appropriate C, $\Gamma[\psi]$, σ) reduction inference rules generated by E-axioms of T. ■

Definition. *A case splitting inference rule* generated by an E-axiom $B \supset (\phi \Leftrightarrow \Phi)$ is a figure of the form

$$\frac{C_1 \supset \Gamma[\sigma(\Phi)/\psi] \qquad\qquad C_2 \supset \Gamma[\psi]}{C \supset \Gamma[\psi]},$$

such that for the formula $Cond \equiv_{df} \sigma(B) \,\&\, Eq(\psi, \sigma(\phi))$ both formulae $C_1 \equiv_{df} C \,\&\, Cond$ and $C_2 \equiv_{df} C \,\&\, \neg Cond$ are T_0-consistent.

Let $CSR(T)$ denote the set of all possible (for appropriate C, σ, $\Gamma[\psi]$) case splitting rules generated by E-axioms of T. ■

The notion of provability is now straightforward.

Definition. Let $\mathfrak{R}$ be a set of reduction and case splitting inference rules. We say that an unquantified formula Ω is an *$\mathfrak{R}$-axiom* iff Ω does not occur below the line in the rules of $\mathfrak{R}$ (neither $\frac{\Omega'}{\Omega}$, nor $\frac{\Omega' \quad \Omega''}{\Omega}$ belongs to $\mathfrak{R}$).

An $\mathfrak{R}$-proof tree of a formula $C \supset \Gamma$ is a tree satisfying the following three properties:

1) the nodes of the tree are labelled with unquantified formulae, the root of the tree is labelled with $C \supset \Gamma$ and the leaves are labelled with $\mathfrak{R}$-axioms;

2) each node of the tree has at most two successors;

3) whenever a node labelled with Ω has exactly one successor Ω', the rule $\frac{\Omega'}{\Omega}$ belongs to $\mathfrak{R}$; whenever a node labelled with Ω has two successors Ω' and Ω'', the rule $\frac{\Omega' \quad \Omega''}{\Omega}$ belongs to $\mathfrak{R}$.

We say that a formula Ω is *$\mathfrak{R}$-provable from a set of formulae $\aleph$* iff there exists an $\mathfrak{R}$-proof tree of Ω with the leaves labelled by the formulae of $\aleph$. ■

4. E-THEORIES AND CONDITIONAL REWRITE RULES

We introduce now a class of conditional rewrite rule systems closely related to $E(T_0)$-theories as well as to the inference rules defined above. We also define the notions concerning rewriting techniques, such as *redex*, *convolution*, *conditional reducibility*, *normal form*.

Definition. An expression ρ of the form

$$B :: \phi \;---->\; \Phi$$

such that $Var(B) \cup Var(\Phi) \subseteq Var(\phi)$ is called *a conditional rewrite rule*; the formulae B, ϕ, Φ are called *the condition*, *the left-hand side* and *the right-hand side* of the rule. A set of rewrite rules (usually finite) is called *a conditional rewriting system*. Subsequently the letter R will denote an arbitrary rewrite system, and R_T will stand for the system obtained by the obvious transformation of the axioms of an $E(T)$-theory T. ■

Definition. Let ψ be an occurence of an atomic formula in a formula Γ. We say that ψ is the *$R(\rho, \sigma)$-redex* in Γ, iff

1) the rule $\rho \equiv B :: \phi \;---->\; \Phi$ belongs to R,

2) the formula $Cond(\psi, \rho, \sigma) \equiv_{df} \sigma(B)\ \&\ Eq(\psi, \sigma(\phi))$ is T_0-consistent.

If ψ is an $R(\rho, \sigma)$-redex in a formula Γ, then the formula Γ' obtained by the replacement of ψ with $\sigma(\Phi)$ is called the *$R(\psi, \rho, \sigma)$-conversion of* Γ. ■

Definition. Suppose that Γ' is an $R(\psi, \rho, \sigma)$-conversion of Γ and the condition $T_0 \models C \supset Cond(\psi, \rho, \sigma)$ holds. We then say that the formula Γ *$R(C)$-reduces* to the formula Γ', and write $\Gamma \xrightarrow{R(C)} \Gamma'$. By $\xrightarrow{R(C)}{}^{+}$ and $\xrightarrow{R(C)}{}^{*}$ we denote the transitive and the reflexive-transitive closures of the $\xrightarrow{R(C)}$ relation. ■

It follows directly from the definitions that there exists an evident relationship between inference rules and conditional reducibility. Note that to reduce a formula we need a decision algorithm for the basic theory T_0. As it is proved in [Vol] $Th^{\forall}(PAR)$ cannot be axiomatized by a canonical (i.e. uniqly terminating, see below) context-free (i.e. containing no boolean connectives in the left-hand sides of the rules) term rewriting system, usual or conditional, finite or infinite. So, building-in decision algorithms essentially increases the expressive power of conditional term rewriting.

5. COMPLETENESS

In this section we examine sufficient conditions for the *strong completeness* of an inference rule system $\mathfrak{R}_T$ generated by an $E(T_0)$-theory T. Strong completeness (alias search space freedom) guarantees that a correct proof of every valid formula can be obtained whatever inference rule is applied at each step of a proof search process. We suppose in this paper that all possible $\mathfrak{R}_T$-proof trees are finite. The criteria of the finiteness will be studied elsewhere. Strong completeness and finiteness give possibility to use $\mathfrak{R}_T$-proof trees to decide unquantified formulae of T, and a result of the decision is independent of a proof tree chosen.

First of all we must formulate the decision problem precisely. We restrict ourselves to the case when the role of the basic theory T_0 plays Presburger arithmetic *PAR*. A straightforward generalization is left to the reader.

Definition. The *quantifier-free decision problem* for a $E(PAR)$-theory T consists in determining, whether an unquantified formula belongs to the set

$$Th^{\forall}(STM(PAR^{=}\cup T)) \equiv_{df} \{ \Omega \mid STM(PAR^{=}\cup T) \models \Omega,\ \Omega \text{ is unquantified} \}$$

One necessary condition for the strong completeness is obvious and well known in the framework of rewrite rule theory. This is *the confluence property* appropriately generalized for the class of conditional rewriting systems in consideration, cf. [Ka,ZRe].

Definition. A conditional rewriting system R is *confluent*, iff whenever $\Gamma \overset{R(C)}{\text{------}\!\!>}{}^{*} \Gamma_1$ and $\Gamma \overset{R(C)}{\text{------}\!\!>}{}^{*} \Gamma_2$, there exist such sets I, $\langle C_i \rangle_{i\in I}$, $\langle \Gamma_1^i \rangle_{i\in I}$ and $\langle \Gamma_2^i \rangle_{i\in I}$ that

a) for all $i \in I$ $Var(C_i) \subseteq Var(C)$;

b) $T_0 \vDash C \Leftrightarrow (\bigvee_{i \in I} C_i)$;

c) for all $i \in I$ $\quad \Gamma_j \xrightarrow{R(C_i)}{}^{*} \Gamma_j^i \quad (j = 1, 2)$;

d) for all $i \in I$ $\quad T_0^{=} \vDash C_i \Rightarrow (\Gamma_1^i \Leftrightarrow \Gamma_2^i)$.

We say that an $E(T_0)$-theory is *confluent* iff it generates a confluent conditional rewrite rule system. ■

Suppose, we have some method to complete (by adding axioms) non-confluent $E(T_0)$-theories, transforming them to equivalent confluent ones (this is, in fact, the well-known Knuth-Bendix completion method, not discussed here).

From now on let us fix an arbitrary partition of an $E(PAR)$-theory $T = T_c \cup T_v$, $T_c \cap T_v = \emptyset$. Consider the inference rule system

$$\mathfrak{R}(T_c, T_v) \equiv_{df} Red(T_c) \cup CSR(T_c) \cup Red(T_v).$$

Note the asymmetry between T_c and T_v - we prohibit case splitting via axioms of T_v. The main reason for such a restriction is the aspiration to make $\mathfrak{R}(T_c, T_v)$-proof trees finite. We will show in the example below that such prohibition is crucial.

The completeness theorem (see below) states that each leaf of every $\mathfrak{R}(T_c, T_v)$-proof tree of any unquantified theorem of $STM(PAR^{=} \cup T_v \cup T_c)$ is a theorem of $STM(PAR^{=} \cup T_v)$ (axioms of T_c are omitted). This is in fact the result on relative decidability. In particular, if $T_v = \emptyset$ then the conditions (1), (2) and (3) of the theorem become trivial, and we obtain the real decidability result, since $Th^{\forall}(STM(PAR^{=}))$ is decidable, e.g. using the algorithms described in [Sh1, Sh2].

The Structural Completeness Theorem. *Let $T = T_c \cup T_v$ $(T_c \cap T_v = \emptyset)$ be a partition of a confluent $E(PAR)$-theory T such that every $\mathfrak{R}(T_v, T_c)$-proof tree is finite and for each $\mathfrak{R}(T_v, T_c)$-axiom $C \Rightarrow \Gamma$ there exist such sets I, $\{ C_i \}_{i \in I}$ and $\{ \Gamma_i \}_{i \in I}$ that:*

(1) *for all $i \in I$ $Var(C_i) \subseteq Var(C)$, and $Stm(PAR) \vDash C \Leftrightarrow (\bigvee_{i \in I} C_i)$;*

(2) *for all $i \in I$ the formula $C_i \Rightarrow \Gamma_i$ is the $\mathfrak{R}(T_v \cup T_c, \emptyset)$-axiom, and the formula $C_i \Rightarrow \Gamma$ is $\mathfrak{R}(\emptyset, T_v \cup T_c)$-provable from $C_i \Rightarrow \Gamma_i$;*

(3) *for all* $i \in I$ $STM(PAR^{=} \cup T_{\upsilon}) \models C_i \supset (\Gamma_i \Leftrightarrow \Gamma)$.

If Ω *is an arbitrary unquantified formula of the similarity type* Σ *and* $\{\Omega_j\}$ $(j \in J)$ *are the leaves of an arbitrary* $\mathfrak{R}(T_{\upsilon}, T_c)$*-proof tree for* Ω, *then*

$$STM(PAR^{=} \cup T_{\upsilon} \cup T_c) \models \Omega \;\; iff \;\; STM(PAR^{=} \cup T_{\upsilon}) \models \Omega . \;\; \blacksquare$$

Remark. Note that the conclusion of the theorem is equivalent to the following one: *for every* $\mathfrak{R}(T_c, T_{\upsilon})$*-axiom* Ω $STM(PAR^{=} \cup T_{\upsilon} \cup T_c) \models \Omega$ *iff* $STM(PAR^{=} \cup T_{\upsilon}) \models \Omega$. It means that the truth of $\mathfrak{R}(T_c, T_{\upsilon})$-axioms is independent of T_c, and the cost of prohibition of case splitting with T_{υ} axioms is the presence of T_{υ} in both sides of the "*iff*".

Let us show that the permission or prohibition of case splitting is crucial for the finiteness of proof trees. Suppose, T consists of the unique axiom $x > 0 \supset (P(x) \Leftrightarrow P(x-1))$. Then there exists the infinite $\mathfrak{R}(T, \emptyset)$-proof tree of $z \geq 0 \supset P(z)$ built of the case splitting rules only. The immediate calculation shows that the leftmost branch of the tree contains the following *infinite* sequence of formulae: $z \geq 0 \supset P(z)$, $z \geq 0 \& z > 0 \supset P(z-1)$, $z \geq 0 \& z > 0 \& z-1 > 0 \supset P(z-1-1)$, etc.

At the same time there are no infinite $\mathfrak{R}(\emptyset, T)$-proof trees. If we suppose, on the contrary, that such a tree exists, then for some formula C we have $PAR \models C \supset x > i$ for every $i \in \omega$ by the definition of the reduction inference rule. Since C is the PAR-consistent formula (by the definition), and PAR is known to be *a complete theory*, the existential closure $\exists xC$ of C must be true in every model of PAR, particularly, in the standard model $Stm(PAR)$. Hence, there exists a variable interpretation ν validating C in the $Stm(PAR)$. But the same interpretation ν must validate *all* the formulae $x > i$ $(i \in \omega)$ in the $Stm(PAR)$. This means, however, that x is interpreted as a non-standard natural number in the $Stm(PAR)$. This is however a contradiction.

Note that axioms of the form $x > 0 \supset (\phi(x) \Leftrightarrow \Phi[\phi(x-1)])$ leading to infinite proof rules when allowed in unrestricted case splits, are nothing else but usual inductive definitions, indispensable in data type specifications. So, the argument above shows that inductive proofs are not easy, but the Structural Completeness Theorem allows to isolate inductive definitions and apply specific proof methods to purely inductive theories. We will report on this subject elsewhere, cf. [Vo2].

6. PROOF OF THE COMPLETENESS THEOREM

As it was noted in the remark to the completeness theorem (see previous section), it is sufficient to prove that for each $\mathfrak{R}(T_c, T_\upsilon)$-axiom Ω the following implication is true:

$$STM(PAR^{=} \cup T_\upsilon \cup T_c) \models \Omega \text{ implies } STM(PAR^{=} \cup T_\upsilon) \models \Omega$$

(the inverse is obvious). Let us assume, on the contrary, that the implication above is false. This means that for some $\mathfrak{R}(T_c, T_\upsilon)$-axiom Ω conditions (1) and (2) below hold:

(1) there exists a model $\mathfrak{M}_0 \in STM(PAR^{=} \cup T_\upsilon)$ such that Ω is not valid in $\mathfrak{M}_0$, i.e. for some ground instance Ω^c of Ω we have $\mathfrak{M}_0 \models \neg \Omega^c$;

(2) for all $\mathfrak{M} \in STM(PAR^{=} \cup T_\upsilon \cup T_c)$ we have $\mathfrak{M} \models \Omega$.

(Call a formula ground iff it contains no variables, an instance of a formula is a result of applying a substitution to the formula, a ground instance is an instance being a ground formula).

To obtain the desired contradiction we will describe how to transform an arbitrary standard model $\mathfrak{M}_0$ of $PAR^{=} \cup T_\upsilon$ into the standard model $\mathfrak{M}_T(\mathfrak{M}_0)$ of $PAR^{=} \cup T_\upsilon \cup T_c$ preserving validity and invalidity of all $\mathfrak{R}(T_c, T_\upsilon)$-axioms.

We can assume that the signature of $\mathfrak{M}_0$ is $\Sigma' = \Sigma \cup \Sigma_c$, where Σ_c consists of constant symbols only, and every element of $\mathfrak{M}_0$ is denoted by (i.e. is a value of) some ground term of Σ. Moreover, we can suppose that for any sort s different from the sort of natural numbers and for any pair t, r of distinct ground terms of the sort s the values of t and s are the distinct elements of $\mathfrak{M}_0$. This can be achieved by dublicating the elements of the carrier of $\mathfrak{M}_0$ denoted by different terms, preserving the values of predicates.

To descibe the needed transformation we need an auxiliary

Definition. Let ψ^c be an $R_T(\rho, \sigma^c)$-redex in a ground formula Γ^c

such that $\mathfrak{M}_0 \models Cond(\psi^c, \rho, \sigma^c)$, then Γ^c $R_T(\mathfrak{M}_0)$*-reduces* to $\Gamma^{c'}$, in symbols $\Gamma^c \xrightarrow{R_T(\mathfrak{M}_0)} \Gamma^{c'}$, the last formula is called the $R_T(\psi^c, \rho, \sigma^c)$*-conversion* of Γ^c. A ground formula Ψ is called an $R_T(\mathfrak{M}_0)$*-normal form* of a ground formula Γ iff Γ $R_T(\mathfrak{M}_0)$-reduces (may be in multiple steps) to Ψ and Ψ is $R_T(\mathfrak{M}_0)$-irreducible. ■

Now we are ready to describe the transformation of $\mathfrak{M}_0$ into $\mathfrak{M}_T(\mathfrak{M}_0)$ with the desired properties as follows:

1) the signature and the carrier of $\mathfrak{M}_T(\mathfrak{M}_0)$ coincide with those of $\mathfrak{M}_0$;

2) a ground formula Γ^c is true in $\mathfrak{M}_T(\mathfrak{M}_0)$ iff its $R_T(\mathfrak{M}_0)$-normal form is true in the model $\mathfrak{M}_0$.

Note that for every ground formula all its $R_T(\mathfrak{M}_0)$-reduction chains must be finite, since we have assumed that all proof trees are finite.

To show that $\mathfrak{M}_T(\mathfrak{M}_0)$ possesses all necessary properties we must prove the following two lemmas.

Lemma 1. $\mathfrak{M}_T(\mathfrak{M}_0) \models PAR^= \cup T_v \cup T_c$.

Proof. *PAR* remains valid in $\mathfrak{M}_T(\mathfrak{M}_0)$ since arithmetic formulae are $R_T(\mathfrak{M}_0)$-irreducible and, by definition, their validity (invalidity) is preserved by the transformation described.

All ground instances of equality axioms of $PAR^=$ remain valid in $\mathfrak{M}_T(\mathfrak{M}_0)$ (and hence $\mathfrak{M}_T(\mathfrak{M}_0) \models PAR^=$), because of the definition of $R_T(\mathfrak{M}_0)$-reduction: if x and y are instantiated by the same constant in $x = y \supset (\psi \supset \psi[x/y])$ then both ψ and $\psi[x/y]$ are $R_T(\mathfrak{M}_0)$-reducible (resp. irreducible) by the same sequences of reductions.

Let us show that $\mathfrak{M}_T(\mathfrak{M}_0) \models T_v \cup T_c$. Take an arbitrary axiom $B \supset (\phi \Leftrightarrow \Phi)$. It is necessary to prove that $\mathfrak{M}_T(\mathfrak{M}_0)$ validates every ground instance $B^c \supset (\phi^c \Leftrightarrow \Phi^c)$ of this axiom. Suppose B^c is true (otherwise the instance is vacuously true). If Φ^c is true then ϕ^c is also true, since $\phi^c \xrightarrow{R_T(\mathfrak{M}_0)} \Phi^c$. And if ϕ^c is true then $\phi^c \xrightarrow{R_T(\mathfrak{M}_0)} \Phi_1^c \xrightarrow{R_T(\mathfrak{M}_0)}{}^* \Psi_1^c$, the last formula is true in $\mathfrak{M}_T(\mathfrak{M}_0)$ and Φ_1^c does not necessarily coincide with Φ^c. Nevertheless, we must prove that in this case Φ^c is also true in $\mathfrak{M}_T(\mathfrak{M}_0)$. Using the fact that the theory under consideration is confluent, we can conclude that Φ^c is $R_T(\mathfrak{M}_0)$-reducible (may be in multiple steps) to some formula Ψ^c which is equivalent to Ψ_1^c (in $\mathfrak{M}_0$). So, Ψ^c is true in $\mathfrak{M}_0$ and hence Φ^c is true in $\mathfrak{M}_T(\mathfrak{M}_0)$. ■

Lemma 2. $\mathfrak{M}_T(\mathfrak{M}_0) \models \neg\Omega^c$.

Proof. Let us assume that Ω^c is of the form $C^c \supset \Gamma^c$. As it was discussed above, the formula C^c remains true in $\mathfrak{M}_T(\mathfrak{M}_0)$, since it is in the $R_T(\mathfrak{M}_0)$-normal form (and C^c is necessarily true in $\mathfrak{M}_0$, because Ω^c is false in $\mathfrak{M}_0$). The first condition of the theorem guarantees that there exists $i_0 \in I$ such that $\mathfrak{M}_0 \models C^c \Leftrightarrow C^c_{i_0}$. It follows directly from the definition of reducibility that Γ^c $R_T(\mathfrak{M}_0)$-reduces to $\Gamma^c_{i_0}$, and, moreover, $\Gamma^c_{i_0}$ is in the $R_T(\mathfrak{M}_0)$-normal form, because $\Omega^c_{i_0} \equiv C^c_{i_0} \supset \Gamma^c_{i_0}$ is the $\mathfrak{R}(T_c \cup T_v, \emptyset)$-axiom. Since Γ^c is equivalent to $\Gamma^c_{i_0}$ modulo $Stm(PAR^= \cup T_v)$ (see the third condition of the theorem) and $\mathfrak{M}_0 \in Stm(PAR^= \cup T_v)$, Γ^c remains false in $\mathfrak{M}_T(\mathfrak{M}_0)$. Hence Ω^c remains false in $\mathfrak{M}_T(\mathfrak{M}_0)$. This contradiction completes the proof of the lemma and the theorem. ■

7. REFERENCES

[Hs] Hsiang J. Refutational Theorem proving using term rewriting systems. - Artificial Intelligence, 1985, vol. 25, no.2, pp. 255-300.

[HuHu] Huet G., Hullot J.M. Proofs by Induction in Equational Theories with Constructors. - Proc. 21st Symp. on Foundations of Computer Science, 1980, pp. 96-107.

[HuOp] Huet G., Oppen D.C. Equations and Rewrite Rules: A Survey. - In: Formal Language Theory: Perspectives and Open Problems. - New-York, Academic Press, 1980, pp. 349-406.

[Jo] Jouannaud J.-P. Confluent and Coherent Equational Term Rewriting Systems: Applications to Proofs in Abstract Data Types.- Lecture Notes in Computer Science, 1983, vol. 159, pp. 256-283.

[JoKo] Jouannaud J.P., Kounalis E. Automatic Proofs by Induction in Equational Theories Without Constructors. - Proc. Symp. "Logic in Computer Science", 1986, pp. 358-366.

[Ka] Kaplan S. Conditional Rewrite Rules. - Theoretical Computer Science, 1984, vol. 33, no. 2-3, pp. 175-193.

[KB] Knuth D.E., Bendix P.B. Simple Word Problems in Universal Algebras. - In: Computational Problems in Universal Algebras. - Pergamon Press, 1970, pp. 263-297.

[Sh1] Shostak R.E. On the SUP-INF Method for Proving Presburger Formulas. - Journal of the ACM, 1977, vol. 24, no. 4, pp. 529-543.

[Sh2] Shostak R.E. A Practical Decision Procedure for Arithmetic with Function Symbols. - Journal of the ACM, 1979, vol.26, no. 2, pp. 351-360.

[Vo1] Vorobyov S.G. On the Arithmetic Inexpressiveness of Term Rewriting Systems. - Proc. 3rd Symp. on Logic in Computer Science, 1988.

[Vo2] Vorobyov S.G. Conditional Rewrite Rule Systems with Built-inArithmetic and Induction. - Lecture Notes in Computer Science, 1989, vol. 355.

[ZRe] Zhang H., Remy J.-L. Contextual Rewriting. - Lecture Notes in Computer Science, 1985, vol. 202, pp. 46-62.

A PROOF-SEARCH METHOD FOR THE FIRST ORDER LOGIC

A.A. Voronkov
Institute of Mathematics
630090 Novosibirsk 90
USSR

Abstract. A new proof search method for the first order logic based on Gentzen-like calculus is presented. Neither conjunctive nor disjunctive normal forms are used. Soundness and completeness of some strategies are proved. The method essentially uses the general ideas of Maslov's inverse method [8].

1 Introduction

Most of known proof search methods for the first order logic use either resolution method or its modifications [4,7,10] or modifications of Gentzen-like cut-free calculus [8,2,5] (see also [1] for higher order logic). At present resolution provers are most popular. In our opinion it is closely connected with the following shortages of existing methods based on Gentzen-like systems: they either make use of conjuctive normal form of formulae [8,5] or efficient complete strategies that can clear up search space are unknown for these methods.

In this article we introduce a proof search method for the first order logic based on Gentzen-like calculus. In section 2 two formal systems S_1 and S_2 are presented to prove completeness of our method. In section 3 the computational version S_{2A} of S_2 specialized to prove formula A is introduced. In section 4 we give an example of proof in S_{2A} of the formula proposed by Andrews (see [1]). In section 5 subsumption strategy and conjunct strategy are considered.

2. Systems S_1 and S_2

In this section we will formulate two proof systems needed to prove completeness of our method. The first is a modification of systems developed by Schutte [9] from Gentzen's LK [6], the second is a modification of the first systems which is closer to the computational version defined in the next section.

In what follows we will consider only formulae constructed from literals (i.e. atomic formulae and its negations) using connectives &,v and the existencial quantifier E. It is easy to show that any formula of the first order predicate calculus can be effectively transformed into provable equivalent formula of our calculus using scolemizing and translation to negate normal form.

We treat disjunction and conjunction as associative and commutative. Thus we do not distinguish formulae (AvB)vC and (CvB)v A.

<u>Sequent</u> is a finite set of formulae. Formulae in sequents will be separated by commas. In systems S_1 and S_2 derivable objects are sequents.

<u>System</u> S_1

<u>Axioms</u>: $P(\bar{t}),\ \sim P(\bar{t})$

where P is predicate symbol, t is sequence of terms.

<u>Inference rules</u>:

$$(v)\ \frac{A,\ \Gamma}{AvB,\ \Gamma} \qquad (\&)\ \frac{A,\ \Gamma_1 \quad B,\ \Gamma_2}{A\&B,\ \Gamma_1,\Gamma_2} \qquad (E)\ \frac{A(\bar{t}),\ \Gamma}{(E\bar{x})A(\bar{x}),\ \Gamma}$$

Provable sequents are defined as usual. We say that the formula A is <u>provable</u> iff the sequent {A} is provable.

<u>Theorem 1</u>. If a formula A is provable in the first order predicate calculus then it is provable in S_1.

We call a formula A <u>special</u> iff it has no subformulas of the form (Ex)(BvC) and all bound variables in A are different. It is easy to see that any formula can be effectively transformed into equivalent special one using transformations (Ex)(BvC)--->(Ex)B v (Ex)C and renaming of bound variables.

In system S_2 sequents consist only of special formulae.

<u>System</u> S_2

<u>Axioms</u>:

$$(E\bar{x})P(\bar{t}_1)\ v\ A,\quad (E\bar{y})\sim P(\bar{t}_2)\ v\ B$$

where P is a predicate symbol, A, B are arbitrary formulae, possibly empty (formula AvB with an "empty" A means the formula B), $(E\bar{x})$ is a quantifier prefix, possibly empty (formula $E\bar{x}A$ with an "empty" $(E\bar{x})$ means the formula A), provided there is a substitution θ of the form $[\bar{x},\bar{y}<-\bar{r}]$ such that $\bar{t}_1*\theta=\bar{t}_2*\theta$.

<u>Inference rules</u>:

$$\frac{A(\bar{t}),\ \Gamma_1 \qquad B(\bar{t}),\ \Gamma_2}{(E\bar{x})(A(\bar{x})\&B(\bar{x}))\ v\ C,\Gamma_1,\Gamma_2}$$

where A, B, C are formulae, C may be empty, (Ex) is a quantifier prefix, possibly empty.

<u>Theorem 2</u>. Let Γ be a sequent, consisting only of special formulas, and Γ is provable as the first order predicate calculus. Then there exists a sequent $\Gamma'\subseteq\Gamma$ which is provable in S_2.

<u>Corollary</u>. A special formula A is provable in the first order predicate calculus iff it is provable in S_2.

<u>3 The calculus S_{2A}</u>

From now on we will tacitly presume that "subformula" means "subformula with fixed occurence". Let A be a special formula. Let us consider what inference rules of S_2 can be used in the proof of A. It is easy to see that such a proof can contain only such subformulae B of A that coinsides with A or B occurs in some subformula of the form B & C. Thus if we want to construct specialized version S_{2A} of the calculus S_2 intended for proving the formula A then from the very beginning we can consider only such subformulae of A.

We begin construction of this calculus from an example (all formal definitions will be given in the end of this section). Let A have the form

$$(Ex)\tilde{}F(x,f(x))\ v\ (Euvw)(F(u,v)\ \&\ F(v,w)\ \&\ \tilde{}G(u,w))\ v\ (Ey)G(a,y)$$

(father-grandfather formula). Let us first introduce shorthands for the subformulae of A of the form described above. There are five such subformulas:

$A_0 \equiv A$

$A_1(u,v,w) \equiv F(u,v) \& F(v,w)$

$A_2(u,v,w) \equiv \sim G(u,w)$

$A_3(u,v,w) \equiv F(u,v)$

$A_4(u,v,w) \equiv F(v,w)$

(Note: we could choose another formulae $A_0 - A_4$ if we take another ordering of bracketing in the formula F(u,v) & F(v,w) & ~G(u,w). However it does not matter for us).

All rules that can be applied to prove A take the form

$$\frac{A_3(t_1,t_2,t_3),\ \Gamma_1 \qquad A_4(t_1,t_2,t_3),\ \Gamma_2}{A_1(t_1,t_2,t_3),\ \Gamma_1,\ \Gamma_2} \quad \text{and}$$

$$\frac{A_1(t_1,t_2,t_3),\ \Gamma_1 \qquad A_2(t_1,t_2,t_3),\ \Gamma_2}{A_0,\ \Gamma_1,\ \Gamma_2} \quad \text{and}$$

The set of such rules will be called the <u>rule scheme</u> of $A(R_A)$. The suitable representation for R_A will be the following:

(R_1) $A_0 \leftarrow A_1(u,v,w),\ A_2(u,v,w)$

(R_2) $A_1(u,v,w) \leftarrow A_3(u,v,w),\ A_4(u,v,w)$

Besides these inference rules we need axioms. As can be seen from the form of axioms of S_2 candidates to form the axioms are formulae of the form P v A, where P is a literal. So we will use the set of <u>axiom schemes</u> for A (AS_A) which in our example are

(S_1) $A_0 \leftarrow \sim F(x,f(x))$
(S_2) $A_0 \leftarrow G(a,u)$
(S_3) $A_2(u,v,w) \leftarrow \sim G(u,v)$
(S_4) $A_3(u,v,w) \leftarrow F(u,v)$
(S_5) $A_4(u,v,w) \leftarrow F(v,w)$

To generate from S_1-S_5 axioms we have to make some substitutions for the variables u,v,w,x,y. Such substitutions can be found using unification algorithm. Thus we obtain the following axioms:

(1) A_0, A_3(x,f(x),w) [from S_1, S_4]

(2) A_0, A_4(u,x,f(x)) [from S_1, S_5]

(3) A_0, A_2(a,v,w) [from S_2, S_3]

Now using rules R_1-R_2 we can obtain from 1-3 the following proof of the goal A_0:

(4) A_0, A_1(u,f(u),f(f(u))) [R_2 from 1,2]

(5) A_0 [R_1 from 3,4]

We hope that the example above is a good illustration of our proof search algorithm. However, the exact definitions are needed. We give them below.

Let us define notions of verbal and usable subformulae.

1) B is a verbal subformula of any of the formulae B, B&C, BvC, (Ex)B;

2) if B is a verbal subformula of C and C is verbal subformula of D then B is a verbal subformula of D.

Let A be a special formula and B its subformula. Then B is called usable subformula of A iff B coincides with A or B occurs in A in some formula of the form B&C.

Not to keep the global structure of the formulae in derivations we give them names (in our example A_0-A_4). So we need a <u>naming function</u> which maps any usable subformula B of A into its name name (B).

The <u>rule scheme</u> R_A of a special formula A is the set of expressions constructed as follows. If there are subformulae B,C,D,E of A such that B takes the form $(E\bar{x})(C\&D)$ v E and $\bar{y}$ are all variables such that B is in the domain of some quantifier which bounds a variable y from $\bar{y}$, b=name(B), c=name(C), d=name(D), then the expression

$b(\bar{y})$<-$c(\bar{y},\bar{x}),d(\bar{y},\bar{x})$

belongs to R_A.

The set of <u>axiom schemes</u> AS_A of a special formula A is constructed as follows. Let B be a usable subformula of A of the form $Cv(E\bar{x})D$ where D is a literal. Let $\bar{y}$ be all variables such that B is in the domain of some quantifier which bounds a variable y from $\bar{y}$ and b=name(B). Then the expression

$b(\bar{x})$<-D

belongs to AS_A.

Now we are ready to introduce formal system $S2_A$ intended for proving a formula A.

System $S2_A$

Axioms

Let the expressions

$b_1(\bar{x})$<-$P(\bar{t}_1)$

$b_2(\bar{y})$<-~$P(\bar{t}_2)$

belong to AS_A and the sets of variables of these expressions be disjoint (we can rename variables in such expressions). Let a substitution θ be the most general unifier of $\bar{t}_1$, $\bar{t}_2$. Then the sequent

$[b_1(\bar{x}), b_2(\bar{y})]*\theta$

is the axiom of $S2_A$.

Inference rules

&-rule

Let the expression

$b(\bar{x})$<-$c(\bar{x},\bar{y}), d(\bar{x},\bar{y})$

belong to R_A, sets of variables of sequents

$c(\bar{t}_1,\bar{t}_2), \Gamma_1$

$d(\bar{t}_3,\bar{t}_4), \Gamma_2$

be disjoint and θ be the most general unifier of the sequences of terms $\bar{t}_1$, $\bar{t}_2$ and $\bar{t}_3$, $\bar{t}_4$. Then

$$\frac{c(\bar{t}_1,\bar{t}_2), \Gamma_1 \quad d(\bar{t}_3, \bar{t}_4), \Gamma_2}{[b(\bar{t}_1), \Gamma_1, \Gamma_2]*\theta}$$

is an inference rule of $S2_A$.

Factoring rule

Let θ be the most general unifier of the sequences of terms $\bar{t}_1$, $\bar{t}_2$. Then

$$\frac{b(\bar{t}_1),\ b(\bar{t}_2),\ \Gamma}{[b(\bar{t}_1),\ \Gamma]*\theta}$$

is an inference rule of $S2_A$.

<u>Theorem 3</u>. Let A be a special formula. Then the following conditions are equivalent:

1) A is provable in the first order predicate calculus

2) name(A) is provable in $S2_A$.

<u>4. An example of proof</u>

To illustrate the system $S2_A$ we give an example of a proof of the formula proposed by Andrews (see [1]). This formula has the form

(Ex)∀y(P(x)<->P(y))<->((Ex)Q(x)<->∀uP(y))-->

-->((Ex)∀y(Qx<->Qy)<->((Ex)P(x)<->∀yQ(y))

Our proof search system has a built-in formula preprocessor which makes the following transformations of the formula to be proved:

(i) translate into negate normal form;

(ii) maxiscoping of universal quantifiers (to decrease the number of various Scolem functions. E.g. the formula ∀xP(x) & ∀yQ(y) after scolemizing gives two various Scolem functions for x and for y whereas the formula ∀x(P(x) & Q(x)) - only one);

(iii) scolemizing;

(iv) miniscoping of existencial quantifiers.

After preprocessing the following R_A and AS_A had been obtained:

R_A: (R_1) $A_0 \leftarrow A_1, A_2$

(R_2) $A_0 \leftarrow Ag, A_{10}$

(R_3) $A_0 \leftarrow A_{17}, A_{18}$

(R_4) $A_1 \leftarrow A_3(x), A_4(x)$

(R_5) $A_2 \leftarrow A_5, A_6$

(R_g) $A_{10} \leftarrow A_{15}, A_{16}$

(R_{10}) $A_{17} \leftarrow A_{19}, A_{20}$

(R_{11}) $A_{17} \leftarrow A_{21}, A_{22}$

(R_{12}) $A_{17} \leftarrow A_{23}, A_{24}$

(R

(R_5) A_2<-A_5,A_6

(R_6) A_2<-A_7,A_8

(R_7) Ag<-A_{11},A_{12}

(R_8) A_{10}<-A_{13},A_{14}

(R13) A_{13}<-A_{18},A_{25},A_{26}

(R_{14}) A_{18}<-A_{27},A_{28}

(R_{15} A_{18}<-A_{29}(x),A_{30}(x)

AS_A: (S_1) A_3(x)<-~P(x)

(S_3) A_4(x)<-~P(g(x))

(S_5) A_5<-Q(x)

(S_7) A_7<-P(a)

(S_9) A_{11}<-~Q(b)

(S_{11}) A_{12}<-~P(x)

(S_{13}) A_{13}<-P(b)

(S_{15}) A_{15}<-P(x)

(S_{17}) A_{19}<-Q(d)

(S_{19}) A_{21}<-Q(x)

(S_{21}) A_{23}<-~P(e)

(S_{23}) A_{24}<-~Q(x)

(S_{25}) A_{25}<-P(x)

(S_{27}) A_{27}<-Q(d)

(S_{29}) A_{29}<-~Q(d)

(S_{31}) A_{30}(x)<-~Q(f(x))

(S_2) A_3(x)<-Pg(x))

(S_4) A_4(x)<-P(x)

(S_6) A_6<-~P(x)

(S_8) A_8<-~Q(a)

(S_{10}) A_{11}<-P(c)

(S_{12}) A_{12}<-Q(x)

(S_{14}) A_{14}<-~P(x)

(S_{16}) A_{16}<-~P(b)

(S_{18}) A_{20}<-~Q(x)

(S_{20}) A_{22}<-~Q(d)

(S_{22}) A_{23}<-Q(c)

(S_{24}) A_{24}<-P(x)

(S_{26}) A_{26}<-~Q(x)

(S_{28}) A_{28}<-~P(d)

(S_{30}) A_{29}(x)<-Q(f(x))

(S_{32}) A_{30}(x)<-Q(x)

Proof:

(1) A_3(x), A_{12} [S_2,S_{11}]

(2) A_3(b), A_{13} [S_1,S_{13}]

(3) A_3(x), A_{14} [S_2,S_{14}]

(4) A_3(x), A_{15} [S_1,S_{15}]

(5) A_3(x), A_{24} [S_1,S_{24}]

(6) A_3(x), A_{25} [S_1,S_{25}]

(7) A_4(x), A_{12} [S_4,S_{11}]

(8) A_4(x), A_{15} [S_3,S_{15}]

(9) A_4(b), A_{16} [S_4,S_{16}]

(10) A_4(x), A_{24} [S_3,S_{24}]

(11) A_4(x), A_{25} [S_4,S_{25}]

(12) A_5, A_{24} [S_5,S_{23}]

(13) A_5, A_{29}(x) [S_5,S_{29}]

(14) A_5, A_{30}(x) [S_5,S_{31}]

(15) A_6, A_{24} [S_6,S_{24}]

(16) A_6, A_{25} [S_6,S_{25}]

(17) A_7, A_{12} [S_7,S_{11}]

(18) A_7, A_{14} [S_7,S_{14}]

(19) A_8, A_{12} [S_8, S_{12}]

(20) A_8, A_{21} [S_8, S_{19}]

(21) A_{11}, A_{21} [S_9, S_{19}]

(22) A_{13}, A_{16} [S_{13}, S_{16}]

(23) A_{15}, A_{23} [S_{15}, S_{21}]

(24) A_{19}, A_{22} [S_{17}, S_{20}]

(25) A_{20}, A_{23} [S_{18}, S_{22}]

(26) A_{26}, A_{29} (x) [S_{26}, S_{30}]

(27) A_{26}, A_{30} (x) [S_{26}, S_{32}]

(28) A_3(b), A_{10} [R_8, 2,3]

(29) A_1, A_{12} [R_4, 1,7]

(30) A_1, A_{15} [R_4, 4,8]

(31) A_1, A_{24} [R_4, 5,10]

(32) A_1, A_{25} [R_4, 6,11]

(33) A_5, A_{18} [R_{15}, 13,14]

(34) A_2, A_{24} [R_5, 12,15]

(35) A_2, A_{12} [R_6, 17,19]

(36) A_{10}, A_{13}, A_{23} [Rg, 22,23]

(37) A_8, A_{17}, A_{19} [R_{11}, 20,24]

(38) A_{11}, A_{17}, A_{19} [R_{11}, 21,24]

(39) A_{18}, A_{26}, [R_{15}, 26,27]

(40) A_1, A_{10}, A_{16} [R_4, 9,28]

(41) A_0, A_{24} [R_1, 31,34]

(42) A_0, A_{12} [R_1, 29,35]

(43) A_8, A_{17}, A_{23} [R_{10}, 25,37]

(44) A_6, A_{18}, [R_{13}, 16,39]

(45) A_1, A_{18}, [R_{13}, 32,39]

(46) A_1, A_{10}, [R_9, 30,40]

(47) A_0, A_{17}, A_{20} [R_{12}, 25,41]

(48) $A_0, A_{10}, A_{13}, A_{17}$ [R_{12}, 36,41]

(49) A_0, A_9, A_{21} [R_7, 21,42]

(50) A_0, A_9, A_{17}, A_{19} [R_7, 38,42]

(51) A_0, A_8, A_{17} [R_{12}, 41,43]

(52) A_2, A_{18} [R_5, 33,44]

(53) A_0, A_{17}, A_{22} [R_{10}, 24,47]

(54) A_0, A_1, A_{21} [R_2, 46,49]

(55) A_0, Ag, A_{17} [R_{10}, 47,50]

(56) A_0, A_{18} [R_1, 45,52]

(57) A_0, A_2, A_8 [R_3, 51,52]

(58) A_0, A_1, A_{17} [R_{11}, 53,54]
(59) A_0, A_{10}, A_{13} [R_3, 48,56]
(60) A_0, Ag [R_3, 55,56]
(61) A_0, A_2, A_{14} [R_6, 18,57]
(62) A_0, A_1 [R_3, 56,58]
(63) A_0, A_2, A_{10} [R_8, 59,61]
(64) A_0, A_2 [R_2, 60,63]
(65) A_0 [R_1 62,64]

5. Proof search strategies

Our proof search method in the above form is not very suitable for practical applications. To make it practically usable we need some strategies to clear up search space. Here we will introduce two such strategies: The first is complete for all the first order predicate calculi and the second is complete only for propositional case. In what follows we shall tacitly presume that A is a fixed formula we are going to prove.

Subsumption strategy

Let C_1, C_2 be usable subformulae of A, c_1=name(C_1), c_2=name(C_2). We say that $c_1<c_2$ iff C_1 is a subformula of C_2. A sequent Γ_1 subsumes sequent Γ_2 ($\Gamma_1<\Gamma_2$) iff there is a substitution θ such that for any formula $b(t_1,\dots,t_m,\dots,t_n)\in\Gamma_1$ there is a formula $c(t_1,\dots,t_m)\in\Gamma_2\circ\theta$ such that $b<c$.

In what follows if there exists a proof P of the sequent Γ in S_{2A} such that any top sequent of P belongs to the set $\Gamma_1,\dots,\Gamma_n$ then we say that Γ is deducible from $\Gamma_1,\dots,\Gamma_n$.

Theorem 4. Let a sequent Γ be deducible from $\Gamma_1,\dots,\Gamma_n$ and $S_1<S_2$. Then there is a sequent S' such that S' is deducible from $S_2,\dots,S_n$ and S<S'.

Corollary. If the sequent name(A) is deducible from sequents $S_1,\dots,S_n$ and $S_1<S_2$ then it is deducible from $S_2,\dots,S_n$.

Thus we can delete from search space all sequents that can be subsumed by another sequents from the search space. It is the matter of our subsumption strategy. This strategy is very close to resolution subsumption strategy but in essence is stronger. For example the sequent b(a) can be subsumed not only by the sequent b(x) but also by any sequent of the form c(a) or c(x) such that b<c.

We also note that our subsumption strategy allows us to make factoring rule more efficient. Now we can deduce from the sequent

$b(t_1),\ c(t_2,t_3),\ \Gamma$

such that c<b the sequent

$[c(t_1,t_3),\ \Gamma]*\theta$

where θ is the most general unifier of the sequents of terms t_1,t_3. In particular we may replace the sequent

$b(t_1),\ c(t_1,t_2),\ \Gamma$

by the sequent

$c(t_1,t_2),\ \Gamma$

since any one of these sequents subsumes another.

Conjunct strategy

Let us begin with one assertion.

Theorem 5. Let the sequent name(A) be deducible from $\Gamma_1,\ldots,\Gamma_n$, Γ_1 contains formulae $b(t_1,\ldots,t_l,\ s_1,\ldots,s_m)$ and $c(t_1,\ldots,t_l,\ \Gamma_1,\ldots,\Gamma_k)$ and there exist d,e,f such that b<d, c<e and the expression $f(x_1,\ldots,x_l)\text{<-}d(x_1,\ldots,x_l,\ldots,y),\ e(x_1,\ldots,x_l,\ldots,z)$ belongs to R_A (if this condition holds then we call a pair of formulae $b(t_1,\ldots,t_l,\ s_1,\ldots,s_m)$ and $c(t_1,\ldots,t_l,\ \Gamma_1,\ldots,\Gamma_k)$ conjunct). Then name(A) is deducible from $S_2,\ldots,S_n$.

The use of conjuct strategy means that we do not include in the search space sequents containing conjunct.

Theorem 6. The union of conjunct strategy and subsumption strategy is complete.

Conjunct strategy is most efficient in the propositional case when we do not need verifying whether first l arguments of b,c coincide.

Conclusion

We omitted here many aspects of our method. For example we use an interesting representation of sequents based on skeletons of formulae and structure sharing, and efficient algorithms of unification and subsumption working on this representation. Using the techniques of proofs of Theorems 1-3 we can generalize Prolog Technology Theorem Prover [10] to obtain the procedure complete for arbitrary scolemized first order formulae. All of these developments will be published elsewhere.

Acknowledgements

I am grateful to S.G.Alexandrov, A.I. Degtjarev, A.I Kokorin, A.V.Mantzivoda, V.I. Martjanov, D.A. Miller and N. Shankar who all stimulated my work in automated theorem proving.

References

1. Andrews P.B.: Theorem proving via general matings. J.Assoc. Comp. Math. 2(1981), v.28, 193-214
2. Bibel W.: Automated theorem proving. Vieweg Verlag, (1982)
3. de Champeaux D.: Sub-problem finder and instance checker: two cooperating preprocessors for theorem-provers. 6th IJCAI, (1981), v.1, 191-196
4. Chang C., Lee R.C.: Symbolic logic and mechanical theorem proving. Academic Press, (1973)
5. Dragalin A.: A proof-search method. Proc. 8th Int. Congress on Logic, Methodology and Philosophy of Sci. Moscow, (1987)
6. Gentzen G.: Untersuchungen uber das logishe Schliessen. Math. Z. (1935), v.39, 176-210; 405-431
7. Loveland D.: Automated theorem proving: a logical basis. North Holland, (1978)
8. Maslov S.Y.: Inverse method of establishing deducibility for logical calculi. Trudy Math. Inst. (1968), v.98
9. Schutte K.: Beweistheorie. Springer Verlag, (1960)
10. Stickel M.: A Prolog technology theorem prover. 8th International Conf. on Automated Deduction, Lecture Notes in Comput. Sci. (1986), v.230
11. Voronkov A.A.: A proof search method. Vychislitelnye sistemy, Novosibirsk, (1985), v.107, 109-124 (in Russian)
12. Voronkov, A.A.: Degtjarev A.I.: Automated theorem proving I,II. Kibernetika, (1986), v.3, (1987) v.4 (In Russian)

Vol. 379: A. Kreczmar, G. Mirkowska (Eds.), Mathematical Foundations of Computer Science 1989. Proceedings, 1989. VIII, 605 pages. 1989.

Vol. 380: J. Csirik, J. Demetrovics, F. Gécseg (Eds.), Fundamentals of Computation Theory. Proceedings, 1989. XI, 493 pages. 1989.

Vol. 381: J. Dassow, J. Kelemen (Eds.), Machines, Languages, and Complexity. Proceedings, 1988. VI, 244 pages. 1989.

Vol. 382: F. Dehne, J.-R. Sack, N. Santoro (Eds.), Algorithms and Data Structures. WADS '89. Proceedings, 1989. IX, 592 pages. 1989.

Vol. 383: K. Furukawa, H. Tanaka, T. Fujisaki (Eds.), Logic Programming '88. Proceedings, 1988. VII, 251 pages. 1989 (Subseries LNAI).

Vol. 384: G. A. van Zee, J. G. G. van de Vorst (Eds.), Parallel Computing 1988. Proceedings, 1988. V, 135 pages. 1989.

Vol. 385: E. Börger, H. Kleine Büning, M. M. Richter (Eds.), CSL '88. Proceedings, 1988. VI, 399 pages. 1989.

Vol. 386: J.E. Pin (Ed.), Formal Properties of Finite Automata and Applications. Proceedings, 1988. VIII, 260 pages. 1989.

Vol. 387: C. Ghezzi, J. A. McDermid (Eds.), ESEC '89. 2nd European Software Engineering Conference. Proceedings, 1989. VI, 496 pages. 1989.

Vol. 388: G. Cohen, J. Wolfmann (Eds.), Coding Theory and Applications. Proceedings, 1988. IX, 329 pages. 1989.

Vol. 389: D.H. Pitt, D.E. Rydeheard, P. Dybjer, A.M. Pitts, A. Poigné (Eds.), Category Theory and Computer Science. Proceedings, 1989. VI, 365 pages. 1989.

Vol. 390: J.P. Martins, E.M. Morgado (Eds.), EPIA 89. Proceedings, 1989. XII, 400 pages. 1989 (Subseries LNAI).

Vol. 391: J.-D. Boissonnat, J.-P. Laumond (Eds.), Geometry and Robotics. Proceedings, 1988. VI, 413 pages. 1989.

Vol. 392: J.-C. Bermond, M. Raynal (Eds.), Distributed Algorithms. Proceedings, 1989. VI, 315 pages. 1989.

Vol. 393: H. Ehrig, H. Herrlich, H.-J. Kreowski, G. Preuß (Eds.), Categorical Methods in Computer Science. VI, 350 pages. 1989.

Vol. 394: M. Wirsing, J.A. Bergstra (Eds.), Algebraic Methods: Theory, Tools and Applications. VI, 558 pages. 1989.

Vol. 395: M. Schmidt-Schauß, Computational Aspects of an Order-Sorted Logic with Term Declarations. VIII, 171 pages. 1989. (Subseries LNAI).

Vol. 396: T.A. Berson, T. Beth (Eds.), Local Area Network Security. Proceedings, 1989. IX, 152 pages. 1989.

Vol. 397: K.P. Jantke (Ed.), Analogical and Inductive Inference. Proceedings, 1989. IX, 338 pages. 1989. (Subseries LNAI).

Vol. 398: B. Banieqbal, H. Barringer, A. Pnueli (Eds.), Temporal Logic in Specification. Proceedings, 1987. VI, 448 pages. 1989.

Vol. 399: V. Cantoni, R. Creutzburg, S. Levialdi, G. Wolf (Eds.), Recent Issues in Pattern Analysis and Recognition. VII, 400 pages. 1989.

Vol. 400: R. Klein, Concrete and Abstract Voronoi Diagrams. IV, 167 pages. 1989.

Vol. 401: H. Djidjev (Ed.), Optimal Algorithms. Proceedings, 1989. VI, 308 pages. 1989.

Vol. 402: T.P. Bagchi, V.K. Chaudhri, Interactive Relational Database Design. XI, 186 pages. 1989.

Vol. 403: S. Goldwasser (Ed.), Advances in Cryptology – CRYPTO '88. Proceedings, 1988. XI, 591 pages. 1990.

Vol. 404: J. Beer, Concepts, Design, and Performance Analysis of a Parallel Prolog Machine. VI, 128 pages. 1989.

Vol. 405: C.E. Veni Madhavan (Ed.), Foundations of Software Technology and Theoretical Computer Science. Proceedings, 1989. VIII, 339 pages. 1989.

Vol. 407: J. Sifakis (Ed.), Automatic Verification Methods for Finite State Systems. Proceedings, 1989. VII, 382 pages. 1990.

Vol. 408: M. Leeser, G. Brown (Eds.) Hardware Specification, Verification and Synthesis: Mathematical Aspects. Proceedings, 1989. VI, 402 pages. 1990.

Vol. 409: A. Buchmann, O. Günther, T.R. Smith, Y.-F. Wang (Eds.), Design and Implementation of Large Spatial Databases. Proceedings, 1989. IX, 364 pages. 1990.

Vol. 410: F. Pichler, R. Moreno-Diaz (Eds.), Computer Aided Syste Theory – EUROCAST '89. Proceedings, 1989. VII, 427 pages. 1990.

Vol. 411: M. Nagl (Ed.), Graph-Theoretic Concepts in Computer Scien Proceedings, 1989. VII, 374 pages. 1990.

Vol. 412: L. B. Almeida, C. J. Wellekens (Eds.), Neural Networks. Proce ings, 1990. IX, 276 pages. 1990.

Vol. 413: R. Lenz, Group Theoretical Methods in Image Processing. V 139 pages. 1990.

Vol. 414: A. Kreczmar, A. Salwicki (Eds.), LOGLAN '88 – Report on the P gramming Language. X, 133 pages. 1990.

Vol. 415: C. Choffrut, T. Lengauer (Eds.), STACS 90. Proceedings, 19 VI, 312 pages. 1990.

Vol. 416: F. Bancilhon, C. Thanos, D. Tsichritzis (Eds.), Advances in Da base Technology – EDBT '90. Proceedings, 1990. IX, 452 pages. 1990

Vol. 417: P. Martin-Löf, G. Mints (Eds.), COLOG-88. Proceedings, 19 VI, 338 pages. 1990.